ONE WEEK LOAN

Today, at one and the same time, scholarly publishing is drawn in two directions. On the one hand, this is a time of the most exciting theoretical, political and artistic projects that respond to and seek to move beyond global administered society. On the other hand, the publishing industries are vying for total control of the ever-lucrative arena of scholarly publication, creating a situation in which the means of distribution of books grounded in research and in radical interrogation of the present are increasingly restricted. In this context, MayFlyBooks has been established as an independent publishing house, publishing political, theoretical and aesthetic works on the question of organization. MayFlyBooks publications are published under Creative Commons license free online and in paperback. MayFlyBooks is a not-for-profit operation that publishes books that matter, not because they reinforce or reassure any existing market.

1. Herbert Marcuse, *Negations: Essays in Critical Theory*
2. Dag Aasland, *Ethics and Economy: After Levinas*
3. Gerald Raunig and Gene Ray (eds), *Art and Contemporary Critical Practice: Reinventing Institutional Critique*
4. Steffen Böhm and Siddhartha Dabhi (eds), *Upsetting the Offset: The Political Economy of Carbon Markets*

UPSETTING THE OFFSET

Upsetting the Offset:
The Political Economy of Carbon Markets

Steffen Böhm and Siddhartha Dabhi (eds)

www.mayflybooks.org

First published by MayFlyBooks in paperback in London and free online at
www.mayflybooks.org in 2009.

Printed in Great Britain by the MPG Books Group, Bodmin and King's Lynn

Contents

ALTERNATIVES

AFTERWORDS

Contributors

Cristián Alarcón is a PhD candidate at the department of Urban and Rural Development, Swedish University of Agricultural Sciences (SLU). My research project focuses on the structures of political economy, political ecology and environmental communication of forest sectors in Chile and Sweden. I also have a research position within the project GloPat – Global Patterns of Production and Consumption: Current Problems and Future Possibilities. I am affiliated to the CEFO Research Forum at the center for sustainable development at SLU and Uppsala University Sweden. Research interests: politics of forests and forest sectors; environmental communication; political economy; political ecology; salmon aquaculture and exploitation of marine resources; climate change and its relationships with power structures, politics and science; discourses on sustainable development. cristianalarconferrari@yahoo.com

Sally Andrew has worked as an activist, educator and writer in South Africa. She has a Masters in Adult Education (UCT) and a certificate in Environmental Education (Rhodes). She is the author of The Fire Dogs of Climate Change – An inspirational call to action (Findhorn Press, 2009). sally@mail.ngo.za

Ida Auken is a member of the Danish Parliament, Spokesperson for Environmental Affairs. Member of the Standing Comittees on Environment and Planning, Climate and Energy and the Standing Committee on Culture. Cand.theol. and author of several publications on the relationship between religion and politics.

Walden Bello is a member of the House of Representatives of the Philippines representing the political party Akbayan. Associated with the Climate Justice Now! Network, he is a senior analyst of Focus on the Global South and president of the Freedom from Debt Coalition. He is the author or co-author of numerous articles and 15 books on environmental, economic, and political issues. His latest book is *Food Wars* (New York: Verso, 2009). In 2007, he received the Right Livelihood Award – also known as the Alternative Nobel Prize – for his work on corporate-driven globalization. www.waldenbello.org

Steffen Böhm is Reader in Management at the University of Essex. He is a member of the Interdisciplinary Centre for Environment and Society based at Essex. He holds a PhD from the University of Warwick. His research focuses on the critique of the political economy of organization and management. He is co-founder of the open-access journal *ephemera: theory & politics in organization* (www.ephemeraweb.org), and co-founder and co-editor of the new open publishing press MayFlyBooks (www.mayflybooks.org). He has also published *Repositioning Organization Theory* (Palgrave) and *Against Automobility* (Blackwell). steffen@essex.ac.uk

Patrick Bond is senior professor at the University of KwaZulu-Natal School of Development Studies in Durban, South Africa, where since 2004 he has directed the Centre for Civil Society. His work presently covers aspects of economic crisis, environment (energy, water and climate change), social mobilization, public policy and geopolitics. His recent books include: *Climate Change, Carbon Trading and Civil Society; Looting Africa; Talk Left, Walk Right;* and *Elite Transition.* He was a founding member of the Durban Group for Climate Justice and is active in Climate Justice Now!'s South Africa branch. In service to the new South African government from 1994-2002, Patrick authored/edited more than a dozen policy papers, including the *Reconstruction and Development Programme.* Patrick earned his doctorate in economic geography under the supervision of David Harvey at Johns Hopkins. pbond@mail.ngo.za

Joanna Cabello is a researcher at the Carbon Trade Watch project of the Transnational Institute in Amsterdam, The Netherlands. She has an MA in Politics of Alternative Development from the Institute of Social Studies and a Bachelor Degree in Social Communications from *Universidad de Lima*, Peru. Previously she has worked with different projects concerning the Peruvian rainforest and at the UN office for Peace and Disarmament in Latin America. joanna@tni.org

Ricardo Carrere is the International Coordinator of the World Rainforest Movement, an international network of citizens' groups of North and South involved in efforts to defend the world's rainforests and to secure the lands and livelihoods of forest peoples. Since the late 80s, Ricardo has concentrated his energies on research and campaigning at national and international levels to both protect forests and its peoples and to oppose the spread of large-scale tree monocultures. Ricardo is a Uruguayan forester and the author of numerous publications on forests and tree plantations as well as co-author – with Larry Lohmann – of 'Pulping the South: Industrial Tree Plantations and the World Paper Economy'. rcarrere@wrm.org.uy

Melissa Checker is an Assistant Professor of Urban Studies at CUNY, Queens College. Her research focuses on environmental justice activism and issues of urban sustainability in the United States. More recently she has added environmental gentrification in New York City and the global struggle for climate justice to her areas of study. Her publications include *Polluted Promises: Environmental Racism and the Search for Justice in a Southern Town* (NYU Press, 2005), and the edited volume, *Local Actions: Cultural Activism, Power and Public Life* (with Maggie Fishman, Columbia University Press, 2004). In addition, she has published a number of academic and journalistic articles. mchecker@qc.cuny.edu

Ricardo Coelho is a PhD student in Environmental Economics, specializing in carbon trading, in Universidade do Porto, Portugal. Personal website: http://sites.google.com/site/ricardosequeiroscoelho; dr.kandimba@gmail.com

Philippe Cullet is a Reader in Law at the School of Oriental and African Studies – University of London (SOAS) where he teaches law related to the environment, natural resources and intellectual property. He is also a Founding Programme Director of the International Environmental Law Research Centre (IELRC) and the Editor-in-Chief of the Journal of Law, Environment and Development (LEAD-journal.org). He studied law at the University of Geneva and King's College London (LLM). He received an MA in development studies from SOAS and his doctoral degree in international environmental law from Stanford Law School. He is the author of *Differential Treatment in International Environmental Law* (Ashgate, 2003), *Intellectual Property and Sustainable Development* (Butterworths, 2005) and *Water, Law and Development in the context of Water Sector Reforms in India* (Oxford University Press, forthcoming 2009) and the editor of *The Sardar Sarovar Dam Project: Selected Documents* (Ashgate, 2007). pc38@soas.ac.uk

Fábio Renato da Silva is an MSc student in Management at the Federal University of Rio Grande do Sul, Porto Alegre, Brazil, where he also received his Bachelor in Social Communication and his Bachelor in Public Relations. He also holds a Bachelor degree in Management Systems and Health Services at the State University of Rio Grande do Sul.

Siddhartha Dabhi is a researcher from India, currently based at the University of Essex, UK, working on critiques of the political economy of carbon markets. He did his post-graduate studies in Economics at the University of Essex (UK) and his undergraduate studies in Economics at St. Xavier's College – Ahmedabad (India). sdabhi@essex.ac.uk

Simon Dale has spent several years living low impact and permaculture lives. He has particular interest and experience in low impact building, woodlands, permaculture design and creative learning.

Soumya Dutta is a social activist and researcher. Apart from jointly founding Science Communicators' Forum in Kolkata, he took up the responsibility of national organizing secretary of Bharat Jan Gyan Vigyan Jatha in 1992 and helped build-up a nation-wide people's science program, along with publishing a large number of science communication material. He was in the core team – as national convener and later, co-convener – that started and nurtured the National Children's Science Congress, till the year 2002. He also worked in Vigyan Prasar as project coordinator for information systems division for a year, and as in charge of Indira Gandhi Paryavaran Kendra in Delhi from 1997-2002. He has contributed to several books/training manuals, produced several print and electronic media material, edited several teachers training manuals on innovative science projects, developed/innovated experiments for hands-on training and organized a large number of training camps & workshops for science communication. He has conducted several teacher training workshops on creativity in science, on analyzing environmental issues etc – both for teachers and senior students, in nine States of India. soumyadutta_delhi@yrediffmail.com

John Fenwick is a Reader in Organization and Public Management at Newcastle Business School, Northumbria University, Newcastle upon Tyne NE1 8ST, UK. He teaches and researches in public sector management and has a particular interest in the relationship between local government and local community. He also has active interests in the practical governance of education at school level and is involved in the broad co-operative and mutual movement. john.fenwick@northumbria.ac.uk

Rafael Kruter Flores is a PhD student in Management at the Federal University of Rio Grande do Sul, Porto Alegre, Brazil, where he also received his MSc and Bachelor, both in Management. rkflores@ea.ufrgs.br

Gender CC – Women for Climate Justice is a global network of women and gender activists and experts from all world regions working for gender and climate justice. WRM is part of Gender CC core group and acts as its Latin American focal point. http://www.gendercc.net/

Soumitra Ghosh is social activist and researcher, Soumitra works among the forest communities of India. A co-editor of *Mausam*, the recently e-launched Indian journal on climate change and related issues, Soumitra is associated with NESPON, an environmental group based in sub-Himalayan West Bengal, India, National Forum of Forest People and Forest Workers (NFFPFW), India and also the Durban Coalition for Climate Justice. soumitrag@gmail.com

Jane Gibbon is a Lecturer in Accounting at Newcastle University Business School, Newcastle upon Tyne NE1 7RU, UK. Her research focuses on social accounting practice. She is working with community based organizations in both the third and public sector. She is currently completing her PhD supervised by Professor Jan Bebbington at the Centre for Social and Environmental Accounting Research (CSEAR) at St Andrews University. jane.gibbon@newcastle.ac.uk

Tamra L. Gilbertson is a Coordinator of the Environmental Justice Project at the Transnational Institute, Amsterdam, The Netherlands. She is a co-founder of Carbon Trade Watch and the Durban Group for Climate Justice. She has been active in the project since 2003 and has worked on socially focused environmental issues since 1992. She holds degrees in Marine Biology and Zoology from Humboldt State University, USA. She received scholarships from The Teamsters Union, 1995-1998 and the Samuel Rubins Young Fellowship Award in 2004. She has spent much of her time over the past six years documenting the affects of carbon offset project on community struggles. tamra@tni.org

Mike Hannis is currently completing a PhD thesis on the relationship between sustainability and freedom, and working as a sessional lecturer at Birkbeck College, University of London. Over the last decade he has acted as a consultant in a number of low impact planning cases, after winning a lengthy battle to gain planning permission for his own off-grid community in Somerset. mhannis@ukonline.co.uk

Chris Land is a lapsed biketivist who now lives shackled to a desk at the University of Essex where he writes about activism instead of actually doing anything. He wonders whether his cynical mockery of the paradox captured in Gerard Winstanley's words – 'words and writings were all nothing, and must die, for action is the life of all, and if thou dost not act thou dost nothing' – is enough to reconcile him to his keyboard-bound existence. cland@essex.ac.uk

Chris Lang is an activist and researcher. He has worked for many years with the World Rainforest Movement, mainly focussing on the expansion of the pulp and paper industry in the global South. In addition to regular articles in the WRM Bulletin, he has written several books and reports, including *'Plantations, poverty and power: Europe's role in the expansion of the pulp industry in the South', 'Banks, Pulp & People: A Primer on Upcoming International Pulp Projects', '"A funny place to store carbon": UWA-FACE Foundation's tree planting project in Mount Elgon national park, Uganda'* and *'Genetically modified trees: The ultimate threat to forests'*. He is currently working on a website (www.redd-monitor.org) looking at the rapid developments in the world of 'avoided deforestation'. reddmonitor@googlemail.com

Larry Lohmann works with The Corner House, a UK-based research and advocacy organization. His books include *Pulping the South: Industrial Tree Plantations in the Global Paper Economy* (with Ricardo Carrere) (Zed, 1996) and the edited volume *Carbon Trading: A Critical Conversation on Climate Change, Privatization and Power* (Dag Hammarskjold Foundation, 2006). He is a founding member of the Durban Group for Climate Justice and his articles have appeared in journals such as Accounting, Organizations and Society; Asian Survey; Carbon & Climate Law Review; Bulletin of Concerned Asian Scholars; Race & Class; International Journal of Environment and Pollution; Development; New Scientist; Development and Change and Science as Culture. larrylohmann@gn.apc.org

Vito De Lucia is an independent researcher with a background in Law from the University of Rome 'La Sapienza'. He is a founding member of the research network Eco Pax Mundi, and research associate of the Centro Internazionale per La Cultura e i Diritti dell'Uomo. His research focuses primarily on ecological and climate policy, theories of (climate) justice, post-global social and legal theory and critical approaches to international law. vitodelucia@gmailcom

Nishant Mate teaches in a college in Nagpur, (Mahastrasta), India. He is also a social activist associated with National Forum of Forest People and Forest Workers (NFFPFW). He investigated the impacts of several CDM projects in India. nishant24@gmail.com

Larch Maxey has been practising and researching sustainability since 1986. l.maxey@swan.ac.uk

Sunita Narain, has been with the Centre for Science and Environment (CSE) from 1982. She is currently the director of the Centre and the director of the Society for Environmental Communications and publisher of the fortnightly magazine, *Down To Earth*. She is a writer and advocate, conducting her research with forensic rigour and passion, so that knowledge can lead to change. sunita@cseindia.org

Raquel Núñez is a member of the WRM Secretariat where she has been working for eight years as WRM Bulletin coordinator and writer. She has been part of the Gender CC group that participated in the Bonn Climate Change Talks, June 2009.

Isaac 'Asume' Osuoka is a Vanier Scholar in the doctoral program of the Faculty of Environmental Studies at York University, Toronto. He has been director of Social Action, a Nigerian project for education and solidarity for communities and activists working for environmental justice and democracy. He is a joint coordinator of the Gulf of Guinea Citizens Network (GGCN). asumeo@yahoo.co.uk

Matthew Paterson is Professor of Political Science at the University of Ottawa. His research focuses on the political economy of global environmental change. In addition to a book developing a general theoretical approach out of these interests, he has developed them in relation to global climate change and the politics of the automobile. His publications include *Global Warming and Global Politics* (Routledge 1996), *Understanding Global Environmental Politics: Domination, Accumulation, Resistance*, (Palgrave 2000), and *Automobile Politics: Ecology and Cultural Political Economy* (Cambridge University Press 2007). He is currently co-writing a book with Peter Newell (University of East Anglia, UK) provisionally entitled *Climate Capitalism*, as well as a series of articles on the political economy of climate change governance, especially its 'market-led' character. matthew.paterson@uottawa.ca

Ann Marie Sidhu is a Lecturer in Accounting at Newcastle University Business School, Newcastle upon Tyne NE1 7RU, UK. She teaches and researches auditing and ethics in accounting. She is currently pursuing a PhD on the impact of accounting for carbon emissions at Newcastle University. ann.sidhu@newcastle.ac.uk

Kevin Smith is a London-based writer/researcher with Carbon Trade Watch. He is the author of 'The Carbon Neutral Myth: Offset Indulgences for your Climate Sins' and the co-author of 'Hoodwinked in the Hothouse: The G8, climate change and free-market environmentalism'. He has published articles on the subject for publications including the Guardian, the Economist, Environmental Finance, Resurgence, Red Pepper and the Tatler. In his spare time he is an enthusiastic participant in the UK Camp for Climate Action. kevin.smith@gmx.net

Sian Sullivan is a Lecturer in Environment and Development at Birkbeck College, University of London. She is working on a book manuscript entitled *An Ecosystem At Your Service? Culture, Nature and Service Provision in Global Environmental Governance.* s.sullivan@bbk.ac.uk

Pedro Volkmann is MSc student in Management at the Federal University of Rio Grande do Sul, Porto Alegre, Brazil. He received his Bachelor in Social

Communication at the Catholic University of Rio Grande do Sul, Porto Alegre, Brazil.

World Rainforest Movement is an international network committed to the defence of world's rainforests and their forest-related communities and indigenous peoples. The WRM has a campaign against the industrial tree plantations that are being promoted under the disguise of forests. www.wrm.org.uy

Hadida Yasmin is a scientist and researcher, Hadida is associated with North Bengal University, India and Nespon. She has been a part of the NESPON and NFFPFW team that has been researching Indian CDM projects for last 3 years.

Zoe Young is a London-based researcher, writer and low-budget film-maker. She published a major critical study of the Global Environment Facility (http://www.newgreenorder.info) and promotes mycelial media networks that expose international public finance (http://ifiwatch.tv). She also dances. zoe@esemplastic.net.

Acknowledgments

First and foremost we would like to thank all of our authors for their hard work and cooperation without which we would have not been able to compile this fairly large book within less than one year. Many people have gone out of their way to help this project along; we would particularly like to mention Larry Lohmann, Kevin Smith, Emma Dowling, Simon Fairlie, Oscar Reyes, Chris Lang, Sian Sullivan, Zoe Young, Ricardo Carrere and Ricardo Coelho. We would also like to thank Armin Beverungen, the co-editor of MayFlyBooks, for all his support and hard work to get this book ready in time for the Copenhagen climate summit in December 2009. We would also like to thank Soumitra Ghosh, Radical Anthropology, Zed Books, Trusha Reddy, Jenny Pickerill and Findhorn Press for allowing us to reprint some material. We are very grateful to the British Academy and Essex Business School (University of Essex) for providing funding for this project. Last but not least we would like to thank all the people and communities who have provided information, data and evidence, exposing the scam of carbon markets. This book is dedicated to all people who are struggling for real climate justice!

Steffen Böhm and Siddhartha Dabhi
Colchester, UK, 13 November 2009

Foreword

The Business of Carbon is Different

Sunita Narain

Climate change is clearly the biggest challenge our world has ever faced. It is now clear that if the world does not find ways of reducing its emissions of greenhouse gases drastically, it is faced with deadly catastrophe. What is also clear is that this task is not easy, because the emissions are linked to growth. The challenge, therefore, is to reinvent economic growth so that it does not cost us the earth.

This is why the world needs the business of carbon. It is a business, which works to reduce the carbon intensity of economies across the world. It is also a business, where it makes sense to 'avoid' pollution, not to first pollute and then clean up. In other words, the business is about investing in the opportunities, which re-engineer growth so that it is more efficient and less dependent on fossil fuels.

But this business of carbon, we must remember, is not about business as usual. It is about ensuring that the world takes effective steps to reduce emissions. This will mean that the business must not fall into the trap of cheap or corrupt emissions reduction deals. It must not become (as is the current practice) a simple market mechanism – an agreement between private parties looking to make a fast buck. I believe that the design of the Clean Development Mechanism (CDM) is flawed because it promotes really creative carbon accounting projects. It is also designed to keep prices low and so it forces the South to discount its advantage in reducing emissions. It is designed to invest in the small change and not in high end technological options which would drastically reduce pollution. In this way, it does little to seriously or effectively combat climate change. This is unacceptable.

If the corporate world has to be part of the solution, as I believe it must, then it must partner in the idea of the new tomorrow, which is both equitable and sustainable. Let us be clear, climate change teaches us that there are limits to growth as we know it. It also teaches us that the world is one; if the rich world pumped excessive quantities of carbon dioxide into the atmosphere yesterday, the emerging rich world will do so today. It also tells that we must build technological and economic futures which will share the benefits of growth with all. Without fairness and equity, cooperation will not be possible. The bottom line is that if climate change is the market's biggest failure, then the market (as it works today) cannot be the solution.

Foreword

Offsets Under Kyoto: A Dirty Deal for the South

Kevin Smith

In Western Panama, the Naso and Ngobe peoples are fighting against the construction of four hydroelectric dams being built on the land of Indigenous Peoples, saying that they will destroy their homelands. In Okhla, India, members of the local community have been turning up in large numbers to protest against the construction of a waste incinerator in a residential area. Across Indonesia, small farmers are being driven from their land to expand palm oil plantations, with the palm oil finding its way to the refinery in Riau owned by PT Murini Samsam.

All of these projects are receiving, or are in the process of being approved for carbon financing through the Clean Development Mechanism (CDM). This controversial and increasingly discredited finance mechanism, which enables countries and companies in the Global North to buy offset credits from projects in the Global South, has become an emblem of the wider climate injustice being exacerbated by the Kyoto Process.

There is a widespread crisis of confidence in the CDM. All but the most dogmatically market oriented NGOs are no longer willing to entertain it as being any part of the solution. In 2008, the US Government Accountability Office, the 'audit, evaluation and investigative arm of Congress,' released a detailed report that questions the credibility of the scheme. A statement from the International Forum of Indigenous Peoples on Climate Change to the UN climate talks in Bali testified that CDM projects were being carried out 'without the free prior and informed consent of Indigenous Peoples.'

Upsetting the Offset is yet another voice in the mounting chorus of criticism of the CDM, which is supposed to serve three purposes under the Kyoto Protocol: assist in the achievement of sustainable development, contribute to attaining the environmental goals of the broader climate change treaty, and assist Northern countries in complying with their emissions reduction commitments.

The first two objectives have been abysmal failures. The third has been a resounding success, but paradoxically so. The CDM has largely been rewarding big industrial polluters in the global South that contribute nothing towards sustainable development. Meanwhile numerous studies have cast profound doubt on the ability of the CDM to bring climate benefits. A 2008 study from Stanford's Energy and Sustainability Program suggested that up to two thirds of CDM projects didn't bring about any emissions cuts.

The CDM has provided a means for Northern governments and companies to 'outsource' their responsibility for taking necessary steps towards a low-carbon economy. This aspect of the CDM's 'success' highlights the climate injustice underpinning the system. The winners are energy intensive companies, whose profit margins have benefited enormously in the short term through the lucrative trade in the credits themselves. Because of fundamental flaws in the design of the CDM, industry has been able to buy cheap carbon credits to meet their emissions commitments and avoid the cost of shifting to low carbon technologies. Add these savings to potential windfalls from new trading options in derivatives and other exotic financial services and it's no surprise there is such a 'gold rush' for this lucrative market.

Conversely, Southern countries have lost out enormously. Many projects, such as the waste incinerator in India, have been imposed on communities without their prior, informed consent. CDM financing has entrenched dirty development by acting as a financial subsidy for big industrial polluters such as chemical factories, coal fired power stations and pulp and paper mills. The CDM has been promoted at the expense of an existing adaptation fund and the truly clean technology transfer that is so urgently needed.

In 1997, the CDM was initially presented in a 'lump it or leave it' manner in lieu of the more substantial funds that had been proposed by Southern countries. The CDM and the other flexible mechanisms were included as a condition to get the US to commit to binding emissions targets, which they then refused to ratify anyway.

Aside from the emissions reductions targets undertaken by the Global North, the Convention in 1997 also acknowledged the principle of 'ecological debt' in putting the historic responsibility for climate change on Northern countries and committing them as a result to providing funds for mitigation, adaptation and technology transfer.

One of the biggest frustrations for the G-77 countries in the negotiations has been the failure of the Northern countries in making progress on these commitments. More than a decade later as we approach Copenhagen, almost nothing has been seen in the way of hard cash.

More substantial sums have been pledged in Climate Investment Funds, but these are to be administered through the World Bank, an institution with an appalling human rights and environmental track record that increased its lending to fossil fuel projects by 94 percent over the course of 2008. In addition the vast majority of this money will be administered in the form of loans rather than as outright grants, and many commentators foresee that the US-dominated Bank will again be used as a tool of political agency in the relationship between Northern and Southern countries.

Instead of developing climate funds with representative governance that have substantial, obligatory and automatic finance and that are accessible to the most vulnerable, the many years of wrangling within the negotiations have resulted in a market based mechanism that is only benefiting the North and a handful of Southern industrial elites.

While Northern governments do what they can to avoid making commitments to the necessary adaptation and technology transfer funds, the World Bank and industry lobby groups such as the International Emissions Trading Association (IETA) will be out in force in Copenhagen extolling the virtues of the CDM and looking for opportunities to expand and deregulate the market. In Poznan, a World Bank representative took the floor in a plenary and said that 'in order to maintain the successes of the CDM, we need to expand it and make it more flexible.' The demand for offsets credit is increasingly dramatically in order to meet national commitments and for countries and institutions that are taking part in various existing and proposed carbon markets, and so the expansion and deregulation of the CDM process will be necessary in order to meet that demand.

The proposed expansions include technologies that are controversial, such as nuclear power, and unproven, such as Carbon Capture and Storage, while IETA has presented a set of recommendations to the CDM Executive Board with the intention of speeding up the approval process and relaxing rules around 'additionality' and 'absolute assurance'. And even industry insiders are frightened at the impact that the enormous volume of potential credits under the various REDD schemes being proposed could have on the carbon market.

Maintaining the 'successes' of the CDM through these proposals will inevitably mean even more Southern communities facing unwelcome projects that threaten their livelihoods, and it will mean more opportunities for project developers and carbon traders to profit from bogus operations.

With the climate talks mired in political stalemate and hijacked by corporate interests, the most positive sign in the post-Copenhagen landscape is what could be the most articulate expression yet of climate justice in social movements and organizations rejecting the false solution of carbon trading. *Upsetting the Offset* is informed by, and informing this, struggle to expose the corrupt, market-obsessed and ineffective reality of carbon trading and offsetting. The invisible hand of the free market is not going to neatly sweep up the mess that it has created in the first place, the transition to a low carbon planet is only going to happen through the scale of political will and intervention that we saw characterize the financial crisis.

Across the world, communities are engaged in struggles to halt the expansion of the fossil fuel frontier – from indigenous communities in Peru blockading oil companies, to the alliance of First Nation activists and environmentalists fighting against the devastation of tar-sands in Canada, to mass-mobilizations in Europe shutting down coal-fired power stations, to peasant organizations fighting for the right to maintain existing low-carbon lifestyles. Any hope for a sustainable and just future lies in the success of these struggles rather than in the boardrooms and balance sheets of carbon brokers and bureaucrats.

Foreword

Carbon Markets: A Fatal Illusion

Walden Bello

Carbon trading, also known as the 'cap-and-trade' system, was introduced in the Kyoto negotiations as a mechanism for regulating carbon emissions mainly to please the United States. Despite the fact that the US did not sign up to Kyoto, carbon trading as a mechanism to manage greenhouse gas emissions was able to get its foot in the door and eventually dominated the mitigation agenda, marginalizing more direct and drastic measures such as carbon taxes.

Carbon trading is now the preferred scheme, with the prominent climate economist Nicholas Stern envisaging a global cap and trade system in place by 2020. This reliance on the market to deal with the most threatening problem of our time is incongruous, given the massive market failures of the last decade, in international finance, in dealing with poverty, in promoting development. The salience of carbon markets owes more to corporate lobbying than to any proof of superiority over state-imposed regulations. The most ambitious effort at carbon trading so far, the European Emissions Trading Scheme (ETS), has been a patent failure. A very damaging verdict on this scheme has been delivered by the New Labor academic Anthony Giddens, an analyst that one cannot accuse of being anti-market:

> A lot of money has changed hands within the ETS, but the scheme has been ineffective for the purposes for which it was set up. Early on in its history, the carbon price reached as much as 31 euros per tonne. Later it dropped so dramatically that it was worth .001 of that sum. It lost its value completely as it became clear that there was a large surplus of allowances because of the slack built into the national allocation plans. In addition, some power generating companies made windfall profits by passing onto consumers the price of carbon credits, even though they were allocated free of charge.[1]

With heavy corporate lobbying to make the 'cap and trade' scheme currently before the US Senate as friendly to corporations as possible, it is difficult to imagine a different fate for it as that which befell the ETS. Both are likely to go down as classic cases of regulatory capture by powerful interests that were meant to be regulated.

The articles in this volume provide a formidable and comprehensive critique of carbon trading. They underline the fact that the ideological fetishism of the market is one of the greatest obstacles to coming up with a viable global strategy to deal with global warming.

Carbon trading is one of those mechanisms, just like biofuels, carbon sequestration, clean coal, and nuclear power, which climate activists have called false solutions to climate change. It is an approach associated with a perspective that posits that there can be a relatively painless transition to a post-carbon economy, one that will not significantly affect the bottom line of all of those interests that have benefited from the fossil fuel civilization. This perspective assumes that market-fixes and techno-fixes on the energy side will allow production and consumption to proceed with the minimum of disruption.

These are comforting illusions, especially to people in the North, many of whom believe that their lifestyle is not, to quote George W. Bush, 'up for negotiation'. But they are dangerous illusions, and we can ill afford to hang on to them at a time that the future of the planet hangs in the balance.

Notes

1 Giddens, A. (2009) *The Politics of Climate Change*. Cambridge: Polity Press, p. 199.

I

INTRODUCTION

In Part I of this book carbon markets are introduced, focusing specifically on the logic of the Clean Development Mechanism (CDM), which is one of the most prominent carbon markets administered and controlled by the United Nations. The first introductory chapter by Steffen Böhm and Siddhartha Dabhi gives a broad overview of the most recent climate change science and the political steps taken so far towards its mitigation. The main aim of this chapter is to form a premise for why we might want to 'Upset the Offset' and engage in a critique of carbon markets. The second introductory chapter by Larry Lohmann talks about the formation of carbon markets and how they commodify the atmosphere, creating perverse incentives for money making and exploitation.

1

Upsetting the Offset: An Introduction

Steffen Böhm and Siddhartha Dabhi

Why 'Upset the Offset'?

It seems to be widely accepted nowadays that global climate change is one of the biggest and most urgent problems the world is currently facing. Yet, despite this urgency, which most scientists tell us is needed to address and deal with climate change, all we seem to have been able to do since the groundbreaking Kyoto summit in 1997 is to try to set up a few carbon markets – for example, the European Emissions Trading System (EU ETS) – which have been working at a sub-optimal level, to say the very least. Why sub-optimal? Because there is now growing evidence that these markets haven't achieved what they were set out to do: reduce global greenhouse gas (GHG) emissions. Many critics would go further and argue that these carbon markets have been completely counter-productive, as they have helped to legitimize the actual growth of carbon emissions. And if there are some countries that are on the road to meet their Kyoto targets for reducing emissions, then only because there is a lot of creative accounting involved.

One of these creative accounting systems, it seems, is the Clean Development Mechanism (CDM), which allows countries and companies to offset their GHG emissions, in order to meet their Kyoto obligations and targets. The CDM's logic is that, if a country or company is not able to meet their emissions reduction targets efficiently at home – that is, it would cost them too much to change their own production processes – then they are able to buy special permits from countries of the 'Global South', i.e. those underprivileged countries that are sometimes referred to as 'developing countries' or even the 'third world'. The idea of the CDM is for the rich countries of the 'Global North' to pay poorer states to implement clean or cleaner technologies that would reduce the overall carbon emissions of the world. As the name 'clean development mechanism' suggests, the logic is for these countries to develop along 'clean' or 'cleaner' lines with the help of the rich, polluting North, which would provide finances and technologies. Hence, the CDM not only tries to be a vehicle to reduce carbon emissions, but also contribute to an agenda of 'sustainable development'.

Now, a host of critiques of the CDM has emerged over recent years. That is, a lot of data and evidence has been compiled by analysts and activists from around the world, showing that the CDM has failed to both reduce overall

GHG emissions and promote sustainable development goals in the South. For example, in India, which is one of the countries with most CDM projects registered, the journal *Mausam* has emerged, documenting, what they call, the 'CDM scam'.[1] In The Netherlands, the Transnational Institute has been running a project called Carbon Trade Watch, which has compiled a lot of evidence of CDM failures from around the world.[2] And already in 2006, Larry Lohmann, one of the most renowned experts and critics of carbon markets, published a book, outlining the case against carbon markets.[3] The present volume builds on this excellent work, while updating and extending it. Given that the first Kyoto compliance period didn't start until 2008, a lot of new data has emerged over the past two or three years, providing fresh evidence for the failures of the CDM and carbon markets in general.

This book thus aims to provide the most up-to-date cases and critiques of carbon markets, with a particular focus on the Clean Development Mechanism (CDM), as it is this tool that has the most profound implications for the global order and climate change mitigation agenda. Given that most rich countries of the North will not come anywhere near meeting their Kyoto obligations – or, as in the case of the USA, as the world's biggest polluter, they haven't signed up to the Kyoto Protocol in the first place – the CDM is likely to play an even more significant role in any post-Kyoto agreement. It is therefore of utmost importance to compile these cases and critiques and make them available to a wide variety of people around the world, for everyone to see what is being done – often in our names – in terms of meeting the immense climate change challenges.

For us, the introduction of carbon markets has been a big delaying tactic, introduced by the world's biggest polluters in the North, in order for them to continue with their 'way of life', to recall the famous maxim of the George W. Bush dynasty, 'The American way of life is not up for negotiation'. Needless to say, of course, that it is exactly this 'way of life' that has brought us into this desperate situation in the first place. But we can go further. As the cases in this book show, carbon markets, and particularly the CDM and other carbon offsetting instruments, have also presented new money making opportunities for the elites in both North and South, benefiting those countries and corporations that can be labelled the most unsustainable. All this has meant that we have effectively lost a decade – if not more – in the fight against climate change. By introducing the complex – and many would argue unworkable – web of carbon markets, we have lost valuable time to switch to a fossil-fuel-free economy and society. Carbon markets simply don't address the underlying and root causes of climate change, which is an over-consumption of finite fossil fuels. We are addicted to oil, gas, coal and a whole range of other fossil fuels, which, when burned for heating, electricity generation or other usages, release greenhouse gases. It is now time to make up for the lost decade, starting to deal with our underlying reliance on fossil fuels.

What are Carbon Markets?

The United Nations Framework Convention on Climate Change (UNFCCC), through the Kyoto Protocol, prepared the ground for establishing a range of carbon markets. The whole idea behind creating carbon markets is to achieve maximum possible emissions reductions at the lowest possible cost.[4] This follows, as Lohmann shows in detail,[5] the nowadays mainstream economic theory that human behaviour is determined by the relation between rational ends and scarce resources. So, GHG emissions – that is, pollution – are made into an economically scarce resource, which is supposed to maximize the efficiency in which this pollution is reduced on a global basis.

In order to do this, you need to create a market for GHG emissions reductions, which are called 'Certified Emissions Reductions' (CERs) in the case of the CDM. These CERs can then be traded between countries, companies or even individuals. The idea is that if somebody in the polluting North cannot reduce their GHG emissions efficiently – that is, it would cost them too much – then they could simply buy some CERs from a country or company in the South which would have a comparative cost advantage in implementing GHG emissions cuts. For example, one ton of emissions reduction in China is thought to be equivalent to one ton of emissions reduction in the USA. So, basically the assumption is that the atmosphere doesn't care where cuts are made, as long as the cuts take place. But what is hoped is that carbon markets deliver efficient emission cuts, because cutting emissions in China might be cheaper (because of the low cost of labour in that country, for example) than say in the USA. Hence, by turning GHG emissions into a commodity, 'efficiency gains' can be made in terms of how these emissions are cut at a global level.

The Kyoto Protocol aims to reduce six greenhouse gases (GHG), namely carbon dioxide, methane, nitrous oxide, sulphur hexafluoride, hydrofluorocarbons (HFCs), and perfluorocarbons (PFCs). In order to reduce all of these GHG, they have to be turned into one commodity that can be traded globally. This is achieved by creating equivalents of greenhouse gases with respect to carbon dioxide. For instance, 1 ton of HFC-23 is equivalent to 11,700 tons of carbon dioxide. Hence, all greenhouse gases can now be calculated in terms of the equivalent power to pollute the earth in comparison to carbon dioxide, or 'carbon' in short, and all of these gases – at least this is the idea – can be cut in an 'efficient' way across the globe.

The Kyoto Protocol introduced and legitimized the set up of three market mechanisms for emissions reduction:

– Emissions Trading (ET)
– Joint Implementation (JI)
– Clean Development Mechanism (CDM)

The first two mechanisms are currently operational only in the developed countries (Annex I countries), and the CDM allows the developing countries (non-Annex I) and developed countries to jointly reduce emissions.[6] Now, there is also something called 'voluntary carbon offset market', which is more or less similar to the CDM and also aimed at reducing carbon emissions. However, this voluntary market is not governed by the Kyoto Protocol. As the name suggests, it is voluntarily set up and run, and it is not formally controlled by governments or international bodies such as the UNFCCC. Let us now introduce these carbon markets in a bit more detail.

Emissions Trading and the European Emissions Trading System (EU-ETS)

Emissions trading works on the principle of 'cap and trade'. For example, if a government has obligations – set by the Kyoto Protocol – to reduce a country's GHG emissions, then emissions trading provides a tool for introducing a 'cap' on all GHG emissions. So, the government would then give out allowances to certain industries and big polluters, effectively saying: 'you are allowed to emit x number of tons of GHG, but not more'. Hence this 'cap' acts as an upper limit to which a company or entire industry is allowed to pollute. The idea is that if this cap is sufficiently low, there would be an incentive and need for this company or industry to introduce changes to its production processes (e.g. adopt green technologies). However, if these changes could not be made efficiently, that is, it would comparatively cost a lot of money to invest in green technologies, for example, then this company or industry would be allowed to buy extra allowances in order to meet its obligations. Equally, if a company finds it very easy to change its production processes, that is, it can reduce its GHG emissions far beyond the allowances it has been given, then it would be able to sell these allowances, also called 'carbon credits', to other polluters. Hence, the 'trade' of carbon credits and pollution allowances is introduced with the overall aim of efficient cuts of GHG emissions in an industry or country and ultimately at a global level.

The underlying assumption of this system is to achieve maximum possible emissions reduction at the lowest possible cost by: first, quantifying emissions caused by industrial activities; second, setting a cap on all GHG emissions; and, third, incentivizing companies and entire industries to make decisions on how to meet their caps in the cheapest possible way. They could do this by bringing in 'green technologies' or by buying credits/allowances from other polluters or, in fact, from offsetting schemes (see below). The 'cap and trade' system assumes that cuts made anywhere are globally equivalent (which is technically speaking and from a climate science point of view correct, but in a wider socio-economic sense does not hold true). Hence, this carbon market could allow a company or industry to continue in its polluting business-as-usual ways, as long as it keeps buying allowance or carbon credits from other sources.

The EU-ETS, which is based on the 'cap and trade' system, is today the world's largest multi-national emissions trading system. Currently, 15 EU countries are part of the ETS, covering more than 10,000 industrial installations. Their aim is to collectively reduce their emissions by 8% below the 1990 levels during the first Kyoto compliance period between 2008 and 2012.

Joint Implementation (JI)

JI was established under Article 6 of the Kyoto Protocol. Under JI an Annex I (developed) country, which is not able to reduce its emissions efficiently, i.e. at a sufficiently low cost, can invest in an emissions reduction project in some other Annex I country where the GHG reductions can be implemented comparatively cheaply. Most JI projects are being implemented in 'economies in transition', such as Russia and Ukraine, which fall under Annex B of the Kyoto Protocol.[7] In many ways JI is very much similar to CDM, with the only difference that projects take place strictly among developed countries that are legally bound under the Kyoto Protocol.

Clean Development Mechanism (CDM) and Voluntary Carbon Offset Markets

The CDM was introduced under pressure from the US delegation during the final week of the Kyoto negotiations. The idea of a CDM was based on a Brazilian proposal for a 'clean development fund', which would collect penalties from nations that fail to meet their targets for emissions reduction, and this fund would then be used to help developing nations to fund the introduction of technologies necessary for climate change mitigation. But some Annex I parties, especially the USA, were not happy with the idea of penalizing countries which are not able to meet their targets. Hence, the USA proposed the CDM, where countries that are not able to meet their emissions reduction targets 'cost effectively', could invest in 'green technologies' in the South and thus help developing countries to mitigate climate change.[8]

The idea behind the creation of the CDM was to create a 'win-win' situation both for the North and South, where, on one hand, developed countries could meet their emissions targets relatively cheaply by investing in the developing world (where, labour and other costs are usually lower), and, at the same time, the South would benefit from the transfer of 'clean technologies'. In the subsequent negotiations two very important conditions were introduced without which no CDM project would be given the go-ahead by the UNFCCC: First, all CDM projects need to be 'additional', i.e., they should not be 'business-as-usual'. In other words, no CDM project should have been possible without the investment coming in from the North. Second, all projects need to show how they contribute to sustainable development.[9]

Voluntary carbon offsetting is to some extent very similar to the CDM, the main difference being that it has developed independently from the Kyoto process and is therefore not controlled by governmental or inter-governmental institutions, such as the UNFCCC. The voluntary markets do not normally aim to meet any binding targets, and anyone (corporations, NGOs, individuals) can participate in them – although there are some voluntary offset markets that are legally binding and have self-imposed emissions reduction targets, like the Chicago Climate Exchange.[10]

The basic idea behind voluntary offset markets is to enable individuals or companies to offset their carbon emissions by investing in green projects outside the formal Kyoto-based mechanisms, such as the EU-ETS and CDM.

So, if you are, for example, a business person flying a lot and you would like to do something about your carbon emissions (without changing your travelling habits), then you can go onto one of the many carbon offsetting websites, where you pay a few pounds to offset your, say, one-way trip from London Heathrow to New York's JFK airport (which comes to 0.7 tons of carbon dioxide, and you can offset this by paying between £5 to £14 depending on the type of offsetting project[11]). This money would then be spent by the carbon offsetting company or broker on green projects around the world. These projects can range from the usual tree planting in Scotland or Brazil to reducing landfill gas emissions in the developing world.[12]

Voluntary offset markets, which have been labelled the 'wild west' of carbon trading,[13] have attracted a good deal of controversy precisely because all sorts of 'rogue traders' have started to enter carbon markets, trying to make a quick buck. As a result, many dubious claims about offsetting projects and their emissions reduction and contributions to sustainable development have been made. In response to these 'cowboys' and the completely unregulated nature of voluntary carbon offsets, a range of different standards have emerged, trying to introduce some level of regulation and quality control. One of the best known ones is the so-called 'Gold Standard',[14] which is supported by a large number of NGOs. Yet, many critics argue that even these standards will not 'address the more fundamental problem of paying others to clean up after us'.[15]

Carbon offsetting, in the guise of the CDM or the voluntary market, is a mushrooming business. Currently there are 1815 registered CDM projects, producing about 315,582,965 CERs annually, and there are more than 4200 CDM projects in the pipeline.[16] Equally, the voluntary carbon offsetting market doubled in size in 2008 alone and was then worth about $700 million.[17] This is projected to become a multi-billion industry over the next few years, already attracting large market entrants, such as the bank JP Morgan Chase, which recently bought the offset provider Climate Care.[18]

Precisely because of this exponential growth, it becomes very important to analyze what is actually going on in these carbon offsetting markets. Are these markets really leading to cuts in GHG emissions? Who is benefiting from the money raised? Are claims of sustainable development met by the realities on the ground? It is these and other questions that need to be asked, if the creators and proprietors of these carbon markets want to be seen as credible. This is exactly what this book sets out to do, by collecting the most up-to-date cases and critical engagements with carbon offsetting markets.

Critiques of Carbon Markets

There is nothing new about critiques of carbon markets, as they must be seen in connection to a wider process of commodification and expansion of capitalist markets, which has marked the neo-liberal project for the past two to three decades.[19] At the most basic level, the ideology of carbon markets, claiming that they are the most efficient way of reducing greenhouse gas (GHG) emissions, has been attacked. Critics have shown that, by turning GHG into a commodity, you might create an incentive to horde and speculate with this commodity, and you hence might encourage more pollution, because with pollution one can now earn money.[20] This is very much in line with recent critiques of the sub-prime mortgage market that has led to the implosion of the global financial system in 2008 and 2009. There, too, a commodity – or rather a string of commodities – was created literally out of nothing, leading to a huge speculative bubble. When a trader bought a credit derivative, s/he could not know, by default, the real, underlying mortgage deal(s). S/he was only interested in making a quick buck by creating complex financial arrangements that could be sold in the financial markets. In the end nobody seemed to be able to understand the underlying risks involved, leading to the implosion of the sub-prime bubble – and the rest is history, as they say (see Lang in this volume).

Critics argue that this is precisely the future that will await us, if the current expansion of carbon markets continues (see, for example, the next chapter by Lohmann in this volume).[21] The only problem is that this time we are speculating with the future of the planet, as the climate cannot be bailed out even by a concerted effort of all governments put together. There comes a stage when climate change will be irreversible; hence drastic action is needed now. Instead, what we are seeing is carbon market speculation. Take, for example, HFC-23, one of the most potent GHG, which is equivalent to 11,700 tons of CO2. Precisely because of this extraordinarily high CO2 equivalent, and the associated high earning potential associated with it, it is now feared that new HFC-23 production facilities are being set up in places like China only to profit from the sale of CERs.[22] That is, rather than 'efficiently' reducing the production of this highly potent GHG, the newly created carbon markets have introduced a perverse incentive to produce and emit even more GHG.

Another fundamental critique of carbon markets is that they have failed to bring about structural changes in the countries and industries of the rich North for three main reasons:

First, the introduction of 'cap and trade' systems, such as the EU-ETS, have suffered from lack of political will and heavy industry lobbying, which resulted in the over-allocation of emissions allowances. When the EU and its participating countries/ governments first introduced the EU-ETS, quite generous allowances or caps were distributed to the big polluters, which provided no incentive whatsoever for them to reduce their GHG emissions. On the contrary, as has been widely reported, big polluters have enjoyed windfall profits by selling these carbon credits in the marketplace.[23]

This has meant, second, that the creation of an economic scarcity of carbon hasn't actually taken place, resulting in a drastic fall of the price of carbon allowances. For example, the price of the EU-ETS allowances was around €30 until April 2006, but it fell to €10 in May 2006, further declining to €0.10 in September 2007. It seems obvious, as many critics argue that such prices and price volatilities don't provide any incentive to reduce emissions by bringing in the much needed structural changes. All in all, many critics have raised serious doubts over the effectiveness and seriousness of the EU-ETS, arguing that it is very questionable that any actual emissions reduction have taken place at all.[24] Instead, what seems to have happened is that the newly created markets have, besides generating windfall profits for the big polluters, incentivized companies and countries of the North to continue with 'business-as-usual', which, as history shows, hasn't been particularly sustainable.

Third, with the inclusion of the Kyoto trading mechanisms, JI and CDM, Northern companies and countries can also offset their emissions by buying cheap overseas credits, again contributing to the failure of not bringing about real structural changes in the industry and the economy at large.[25]

Yet, the underlying problem of carbon offsetting markets is an ethical and political one. As with the CDM and voluntary offset markets the burden of emissions reduction is shifted from the North to the South, the North is enabled to continue in its polluting 'way of life', while developing countries are supposed to become 'clean'. Here we have to bear in mind that the overwhelming majority of people who live in developing countries (that is, the majority of the world population) emit only a fraction of GHG compared to those people living in the rich North (see Dutta in this volume). Do developed countries not have a moral duty to realize that 'their way of life' has got us into the current climate change mess in the first place, and does this not mean that it is the rich who have to do something about climate change mitigation first? This introduces the dimension of history into this debate, which doesn't feature very often, unfortunately. Is it, possible, one could ask, for example, to offset the North's carbon emissions over a period of the past say 200 years?[26] And if so, perhaps any discussion of carbon offsetting should start with such a premise (see Bond in this volume).

But most contributions to this book deal with the here and now, documenting the often disastrous effects of carbon offsetting projects on communities and eco-systems in the 'Global South'. These range from monoculture tree plantations to polluted water sources, from air pollution to the loss of employment opportunities. What the cases and critiques of carbon offsetting schemes collected in this book show is that climate change mitigation and the reduction of greenhouse gases (GHG) is not simply a number crunching game – this is not just about counting and the accounting of carbon emissions. This always involves questions of accountability as well, which, in most cases seems to be missing.

The CDM is explicitly geared towards sustainable development. One way to account for this is by involving the local community, which would be, in most cases, the most knowledgeable about local issues of sustainability and development. Hence, the CDM registration process requires a report on community involvement, which, unfortunately doesn't seem to be taken very seriously (see, for example, Dabhi in this volume). Once a CDM has been given the go-ahead, often without the knowledge of a broad spectrum of the local population, there doesn't seem to be a transparent system in place for a continuous involvement of those communities who are affected by the CDM projects. If people want to raise any issues, they are confronted with a web of technical jargon, such as CERs, PDDs, DOEs, DNAs, COP/MOPs, SBIs, and SBSTAs, which is extremely off-putting, if not outright elitist. This web effectively suppresses the general public from participating in climate discussions and decision making.[27]

In addition, although the website of the UNFCCC, which is the governing body of the carbon markets and overall responsible for the functioning of the CDM, includes a string of documentation for each project, it seems to be extremely difficult to obtain real figures from the companies involved in these CDM projects. In fact, the UNFCCC does not even want to take the responsibility for monitoring and has left this to the governments in the developing countries, which have, by and large, embraced the CDM as a business opportunity that they do not want to see hampered by complicated local democratic processes. As a result, several researchers contributing to this book have reported that there seems to be a veil of secrecy and opaqueness that surrounds many CDM projects, which is hard to understand, given that this is an official UN process, dealing with one of humankind's most urgent problems, namely climate change.

But perhaps this lack of transparency in the CDM process is quite understandable, given the manifold human rights violations and negative social, economic, political as well as environmental implications associated with many CDM projects around the world. The pattern that seems to be emerging from the cases collected in this book is that the CDM has been a welcome boost for local business elites in developing countries, enabling them to increase their profits from their existing production processes. This often involves extremely big and powerful companies that have had an extremely questionable social, economic and environmental track record in the past. Hence, the CDM in its current form is, unfortunately, nothing more than a money making opportunity for polluting industries, rather than a tool for climate change mitigation. Kevin Smith quotes the managing director of SRF, which is a big chemical and technical textiles manufacturing company in India, who told the Economic Times that

> strong income from carbon trading strengthened us financially, and now we are expanding into areas related to our core strength of chemical and technical textiles business.[28]

It is statements like these that confirm the real impact of the CDM. Rather than contributing to the clean and sustainable development of developing countries, they have become a money making tool that is used primarily to expand existing and polluting productions. Needless to say that the CDM profits that are used for the expansion of existing businesses will be turned into GHG emissions – bearing in mind that the whole point of this process is for Northern companies and countries to offset their increased carbon emissions. So, one quickly realizes that the CDM doesn't actually work; by default it cannot achieve its main goal of reducing global GHG emissions, because of a design fault. Rather than incentivizing the reduction of GHG emissions, carbon markets seem to do the opposite. Hence, carbon markets seem to have enabled and legitimized the continuation of an agenda of business-as-usual with two main resulting outcomes: no progress with global climate change mitigation – some might even talk about regress; and the solidification of unequal and unjust relations between North and South, leading to the continued suffering of poor communities in developing countries.

Running Out of Time: What Climate Change Scientists Tell Us

Leaving the grave consequence of ill thought out and implemented CDM projects for communities in the South to one side for the moment, why is it so important to make urgent progress with reducing global greenhouse gas (GHG) emissions anyway? The answer is: because this is what the overwhelming majority of climate change scientists tell us we need to do!

In 1869 the Swedish scientist Svante Arrhenius was the first to put forward the theory that accumulation of carbon dioxide in the atmosphere by human activities would lead to warming.[29] One and a half centuries later this is not a theory anymore. In 2007 The Fourth Assessment Report (AR4) of the Intergovernmental Panel on Climate Change (IPCC), which is the world's prime scientific body that deals with climate change issues, states:

> Warming of the climate system is unequivocal, as is now evident from observations of increases in global average air and ocean temperatures, widespread melting of snow and ice and rising global average sea level.[30]

The increase in global temperatures is leading to melting of glaciers at the poles, rise in sea water levels in turn threatening human and animal life and biodiversity around coastal regions. It is leading to increased floods, cyclones, heat waves, droughts, ocean acidification, change in seasonal patterns, and changes in life cycle events like blooming and insect emergence, and the plant and animal ranges are shifting towards the poles which eventually is affecting the food chain and specially the species that are on top of the food chain, including human beings. Global warming will also lead to faster spread of diseases and species extinction as well as the loss of biodiversity.

Global warming, which leads to climatic changes of various types, many of which are still unforeseeable, is caused by both climatic drivers (climatic variations like Pacific Decadal Oscillation, El Nino-Southern Oscillation and North Atlantic Oscillation) and non-climate drivers of change, such as industrialization, urbanization, agriculture and land use as well as pollution that influences the climate both directly as well as indirectly.[31] But according to the IPCC, human activities (industrialization, urbanization, etc.) are primarily responsible for global warming and climate change. The IPCC clearly states that

> global GHG emissions due to human activities have grown since pre-industrial times, with an increase of 70% between 1970 and 2004... Global atmospheric concentrations of CO_2, CH_4 and N_2O have increased markedly as a result of human activities since 1750 and now far exceed pre-industrial values determined from ice cores spanning many thousands of years. Global increases in CO_2 concentrations are due primarily to fossil fuel use, with land-use change providing another significant but smaller contribution. It is very likely that the

observed increase in CH_4 concentration is predominantly due to agriculture and fossil fuel use. The increase in N2O concentration is primarily due to agriculture.[32]

According to estimates by NASA, fossil fuel burning roughly contributes to 5.5 gigatons of carbon[33] per year (giga = 1 billion); other activities like land use, such as deforestation and agriculture, contribute roughly 1.6 gigatons of carbon per year. Since 1957 human activities have approximately contributed to 7.1 gigatons of carbon per year, of which approximately 3.2 gigatons of carbon remains in the atmosphere, resulting in an increase of atmospheric carbon dioxide. In addition, the oceans absorb roughly 2 gigatons of carbon. Scientists are not yet sure about where the remaining 1.9 gigatons of carbon go. But evidence suggests that it might be getting absorbed by the land surface. However, there is no consensus on this yet.[34]

So, what happens with the 3.2 gigatons of carbon in the atmosphere? Amongst all the greenhouse gases, carbon dioxide has the longest (atmospheric) life.[35] The carbon dioxide produced today sticks around in the atmosphere for almost a century, but 20% of this will still exist for almost another 800 years. It seems clear that whatever we are doing to the atmosphere today will continue to be a problem for our children, grandchildren and many more generations to come.[36] Some even argue that the climate change we cause today will be 'largely irreversible for 1000 years after emissions stop'.[37]

Another way of looking at the global carbon cycle is to see it as a 'sink' where all the waste is disposed, or as a 'pool' where carbon keeps flowing in, constantly. But then there is a point when this 'pool' is full and no more carbon can be absorbed. This is the point when things start to go wrong. For example, the pre-industrial levels of carbon dioxide were 280 ppm which has now gone up to approximately 387 ppm.[38] Many scientists now argue that, in order for us to have any hope of avoiding the most dangerous effects of run-away climate change, we need to get below 350 ppm. It is this figure which has hence inspired a political campaign in the run-up to the climate change summit in Copenhagen in December 2009.[39]

The Problem is Our Addiction to Fossil Fuels

Given the scientific facts about the longevity of carbon in the atmosphere and the grave situation we are facing in terms of the carbon pool of the earth being full, the only solution is to identify and deal with the root causes of 'human induced' global warming. Now, the IPCC, which is the world's primary scientific body dealing with climate change issues, says it clearly: the root cause is our over-usage of fossil fuels.

The problem with fossil fuels is that their use is like a 'one-way street'. Once fossil fuels are brought out of the ground and are used, they start circulating in the active carbon pools, such as the air, oceans, vegetation and soil. The second problem with fossil fuels is that it is comparatively easy to drill them out and burn them, but it takes millions of years for the earth to create them.[40] Hence, the climate crisis we are facing is not really a crisis created by 'natural' climatic factors; it is a human induced fossil fuel crisis.

According to the World Energy Council, currently there are about 847.5 billion tons of coal left in the ground, which would last for about 150 years, considering the current rate of production. Crude oil accounts for 36.4% of the world's primary energy and the 'Estimated Ultimate Recovery' (EUR) at the end of 2005 was 387 billion tons. So far 47% total reserves of conventional oil discovered have been consumed. Within the next 10 to 20 years half of the EUR will have been recovered, and afterwards a decline in the production of conventional oil is inevitable. Global reserves of natural gas are 177,000 billion cubic metres.[41] Others would be more radical and argue that we have, in fact, already gone beyond 'peak oil', meaning that we are already witnessing a decline in global oil production. Given that it is oil we are mostly dependent on, it is grassroots movements such as the Transition Town Network that urges us to increase the resilience of communities by becoming less dependent on fossil fuels.[42]

Now, if all of our fossil fuel reserves are burnt, it would produce roughly 2800 gigatons of carbon dioxide, which would heat up the earth by about 3°C,[43] bearing in mind that this is a very conservative estimate. Given that the IPCC urges us to limit the increase of the global average of surface temperatures to 2°C, then we can afford to use only 22% of the current reserve of fossil fuels between now and 2050.[44] And if we add non-conventional types of fossil fuels, such as tar sands, oil shales, bitumens and methane hydrates, then the amount of fossil fuels we could afford to use would be even less.[45] All this means that we better start becoming less addicted to fossil fuels soon.

The Failure of Carbon Markets

All the data and evidence presented in this book, as well as many other publications, suggest that carbon markets will not help us to reduce our addiction to fossil fuels. In fact, they seem to have the opposite effect, providing an incentive for business-as-usual and the continued growth of fossil fuel usage.

The key to understand this apparent failure of carbon markets is in analyzing the connections between climate change science, carbon emissions and the political economy of development. The first thing for us to realize about these connections is that climate change can be understood as earth having a fever.[46] That is, climate change is not the main problem; it is a symptom of a much deeper problem, which is to do with our lifestyle and our addiction to fossil fuels. As we have argued, carbon markets don't address these root problems; instead, they try to offset the problem to either poorer communities or future generations. Is this what we are doing when we run a fever ourselves? When our body is run down and our natural immune system isn't coping anymore, then we need to rest and let our body recover from the strain we have put it under. We certainly don't want to continue to go to work or live a hectic lifestyle; and we certainly can't ask somebody else or our children to rest for us. We have to do it ourselves.

Given the dominant ideology of the current politico-economic system, our earth doesn't seem to be allowed to get a rest. As most economic policies seem to be, more than ever, plagued into a logic of continued and relentless economic growth, what we are effectively doing is to keep our earthship working hard, while it is already running a high fever. Carbon markets, with all of their rhetoric of market efficiencies and cost and benefit analyses, are seemingly the continuation of a system that has got the earth-body into the situation it is in at the moment. With all the hustling about the technical and economic ways of introducing carbon markets we are given the impression that something is really being done about climate change. Unfortunately, this doesn't seem to be more than an illusion. In fact it is worse; carbon markets are the emperor's new 'green clothes'. Hence, they are not just about 'buying time' in order to delay the introduction of the necessary structural changes needed to 'our way of life'. Instead, they are the active attempt to provide a new system of legitimation and accumulation that enables the status quo of capitalism to continue during an era when humankind is extremely concerned about climate change and other grave environmental as well as social degradations.

It should therefore become obvious that larger things are at stake here. Carbon markets and climate change mitigation are not merely about the reduction of greenhouse gas emissions. Instead, we have to talk about wider issues of social, economic, environmental and climatic justice. The key problem is that, through the commodification of carbon, climate change is turned into a numbers game, inviting all sorts of creative accounting techniques that don't actually correspond with the reality on the ground. As business school academics, we know that accounting can be used to make all sorts of claims about reality. The problem is that the reality often looks different.

So, if anything, this book calls on us to start facing reality. Climate change is already affecting millions of people's lives around the world in very real terms. What can we do to help them? What can we do to reduce our fundamental addiction to fossil fuels? What techniques do we have at our disposal to bring about radical changes to the way we run our economies and societies, given that without fossil fuel most 'developed' nations would simply collapse? It is questions like these that we need to start to address, realizing that climate change is an immense social, economic and political issue. Carbon markets are merely a 'market fix', and, as Larry Lohmann points out, 'fixes do no fix'.[47] We need to start dealing with the underlying symptoms of climate change.

A Ray of Hope and a Call to Action

What is often forgotten in the hustling and extremely technical discourse of the science and political economy of climate change is that there are many alternatives to a fossil fuel centred economy that are already functioning in many communities around the world. That is, we don't need to reinvent the wheel here. Humanity has, by at large, lived fairly sustainably alongside the world's ecosystems for many thousands of years. Perhaps there is something we can learn from our ancestors and the way they have engaged with nature. Perhaps we need to dare looking to the past, in order to learn something for our own future.

Of course, the danger is that this looking to the past is simply labelled – by the carbon elites – as a regressive step, accusing everybody who actively decides to live without a car, for example, to be some kind of ecological primitivist who wants us all to go back into the woods (see Land in this volume). We must not allow this stereotypical labelling to take place. What the contributions to the 'Alternatives' section of this book show is that even in fossil fuel dependent economies of the North, low impact communities are sustainable alternatives to sprawling urban conglomerations. That is, real alternatives to the dominant 'way of life' that has got us into the current climatic mess are possible, and they are existing in many parts of the world, if we dare to look for them.

In fact, reading Andrew's chapter at the end of the book, we can realize that there is an excess of creativity and resilience that people and communities show. The sustainable campaigns, initiatives and practices Andrew lists are not part of some kind of sci-fi novel; they are really happening and are already working for many communities around the world. If we follow their lead, we can deal with the structural changes needed to mitigate climate change and deal with the socio-economic and other environmental factors that are closely connected with it. There is hope, and we have hope that the world can come together to do something constructive about climate change. And we hope that this book will contribute to such a project of hope.

But hope is not enough. There might be already many alternatives to the fossil fuels centred 'way of life', practiced by millions of people around the world. But these alternative and more sustainable 'ways of life' have to be extended, and brought into the mainstream, so to say, in order for the earth to notice any effect. This can only happen through action and struggle. That is, it won't happen all by itself. We need to actively make a different world possible, otherwise there will be plenty of fossil fuelled elites who will shape the world in their mirror image. We must not allow this to happen, and hence we hope this book is, more than anything, a call to action.

Notes

1   We have included several cases from India in this book, most of which have been previously published in *Mausam*.

2   www.carbontradewatch.org.

3   Lohmann, L. (2006) *Carbon Trading: A critical conversation on climate change, privatisation and power.* Uppsala: Dag Hammarskjold Foundation, http://www.thecornerhouse.org.uk/pdf/document/carbonDDlow.pdf.

4   http://unfccc.int/kyoto_protocol/mechanisms/emissions_trading/items/2731.php.

5   Lohmann, L. (2006) *Carbon Trading: A critical conversation on climate change, privatisation and power*, pp. 45ff.

6   Annex I countries are the developed countries (North) and non-Annex I countries are the developing countries (South).

7   http://ji.unfccc.int; see also http://en.wikipedia.org/wiki/Joint_Implementation.

8   Lohmann, L. (2006) *Carbon Trading: a critical conversation on climate change, privatisation and power.*

9   http://cdm.unfccc.int/about/index.html.

10 http://www.publications.parliament.uk/pa/cm200607/cmselect/cmenvaud/ 331/331.pdf.

11 http://www.carbonneutral.com/cncalculators/flightcalculator.asp; calculations made on 11 Nov. 09.

12 Smith, K. (2007) *The Carbon Neutral Myth: Offset Indulgences of your climate sins.* Amsterdam: Carbon Trade Watch, http://www.carbontradewatch.org/pubs/ carbon_neutral_myth.pdf.

13 Fahrenthold, D.A. and S. Mufson (2007) 'Cost of saving the climate meets real-world hurdles', *Washington Post*, 16 August, http://www.washingtonpost.com/wp-dyn/content/article/ 2007/08/15/AR2007081502432.html.

14 http://www.cdmgoldstandard.org/.

15 Smith, K. (2008) 'Offset standard is off target', *Red Pepper*, http://www.redpepper.org. uk/Offset-standard-is-off-target.

16 http://cdm.unfccc.int/Statistics/index.html.

17 http://apps3.eere.energy.gov/greenpower/markets/certificates.shtml?page=2& companyid=746.

18 http://www.jpmorganclimatecare.com/.

19 See Lohmann, L. (2006) *Carbon Trading: A critical conversation on climate change, privatisation and power.*

20 Lohmann, L. (2005) 'Making and Marketing Carbon Dumps: Commodification, Calculation and Counterfactuals in Climate Change Mitigation', *Science as Culture*, 14(3): 203-235, http://www.thecornerhouse.org.uk/pdf/document/carbdump.pdf; Lohmann, L. (2006) *Carbon Trading: A critical conversation on climate change, privatisation and power*; see also Lohmann's Chapter 2 in this volume.

21 Lohmann, L. (2009) 'When Markets are a poison: Learning about climate policy from the financial crisis', *The Corner House*, http://www.thecornerhouse.org.uk/pdf/ briefing/40poisonmarkets.pdf.

22 McCully, P. (2008) 'Kyoto's Great Carbon Offset Swindle', http://www.renewable energyworld.com/rea/news/print/article/2008/06/kyotos-great-carbon-offset-swindle-52713.

23 WWF (2008) 'EU ETS Phase II – The potential and scale of windfall profits in the power sector', http://assets.panda.org/downloads/point_carbon_wwf_windfall_profits_mar08 _final_report_1.pdf.

24 Lohmann (2008) 'Hold the applause: A critical look at recent EU claims'.

25 Please see Table 4 in Lohmann, L. (2008) 'Hold the applause: A critical look at recent EU claims', *The Corner House*, p.6, http://www.thecornerhouse.org.uk/pdf/document /HoldtheApplause.pdf.

26 See Smith, K. (2007) *The Carbon Neutral Myth: Offset Indulgences of your climate sins*, p.24.

27 Lohmann, L. (2008) 'Carbon Trading, Climate Justice and the Production of Ignorance: Ten Examples', *Development*, 51: 359-365, http://www.thecornerhouse.org.uk/pdf/ document/IgnoranceFinal.pdf.

28 Smith, K. (2007) 'Carbon Trading: The limits of free-market logic', *China Dialogue*, 20 September, http:// www.tni.org/detail_page.phtml?act_id=17350.

29 Cowie, J. (2007) *Climate Change: Biological and Human Aspects.* New York: Cambridge University Press.

30 IPCC (2007) *Fourth Assessment Report*, http://www.ipcc.ch/pdf/assessment-report/ar4/ syr/ar4_syr.pdf.

31 Rosenzweig, C., G. Casassa, D.J. Karoly, A. Imeson, C. Liu, A. Menzel, S. Rawlins, T.L. Root, B. Seguin, P. Tryjanowski (2007) *Climate Change 2007: Impacts, Adaptation and Vulnerability.*

Contribution of Working Group II to the Fourth Assessment Report of the Intergovernmental Panel on Climate Change, p.84.

32 IPCC (2007) *Fourth Assessment Report,* p. 36 and 37.

33 Please note there is difference between the terms carbon and carbon dioxide. The molecular weight of carbon dioxide is 3.667 times that of carbon.

34 http://earthobservatory.nasa.gov/Features/CarbonCycle/.

35 Methane (CH4) takes about a decade to leave the atmosphere. It gets converted to carbon dioxide and Nitrous Oxide takes about a century to leave the atmosphere.

36 http://www.ucsusa.org/global_warming/science_and_impacts/science/CO2-and-global-warming-faq.html.

37 http://www.guardian.co.uk/commentisfree/2009/sep/01/global-warming-emissions-fossil-fuels.

38 Lohmann, L. (2006) *Carbon Trading: a critical conversation on climate change, privatisation and power.*

39 http://www.350.org/.

40 Lohmann, L. (2006) *Carbon Trading: a critical conversation on climate change, privatisation and power.*

41 These calculations are based on 2005 figures; http://www.worldenergy.org/publications/survey_of_energy_resources_2007/crude_oil_and_natural_gas_liquids/638.asp.

42 For more information, see http://www.transitiontowns.org.

43 http://www.climateinstitute.org.au/images/stories/CI056_EACC_Report_v1.pdf.

44 Meinshausen, M., N. Meinshausen, et al. (2009) 'Greenhouse-gas emission targets for limiting global warming to 2C', *Nature,* 458(7242): 1158-1162. http://www.nature.com/nature/journal/v458/n7242/abs/nature08017.html; Also see http://www.carbonequity.info/PDFs/QandA-Meinshausen.pdf for a better understanding of the paper by Meinshausen, et al. Why is the global warming target for 2050 set to 2°C? 'For avoiding dangerous climate change, the ultimate objective of the United Nations Framework Convention on Climate Change, limiting warming to below 2°C is the most prominently discussed target in science and policy circles alike. The countries supporting a 2°C or lower temperature target comprise together a total of 110 countries and represent approximately 20% of the World's population in 2005. 2°C is not a safe level though, and significant impacts, like major long-term sea level rise are likely to occur even below 2°C warming', http://www.carbonequity.info/PDFs/QandA-Meinshausen.pdf.

45 Monbiot, G. (2009) 'We're pumping out CO2 to the point of no return. It's time to alter course', *The Gaurdian,* 1 September, http://www.guardian.co.uk/commentisfree/2009/sep/01/global-warming-emissions-fossil-fuels.

46 Earlier fever was considered to be a disease, but later it was discovered that it is not a disease but a body's indication that something is wrong. We are here trying to make a connection between the concept of fever and global warming, suggesting that global warming is a sign that something is wrong with the body that we call 'earth'.

47 Lohmann, L. (2006) *Carbon Trading: a critical conversation on climate change, privatisation and power.*

2

Neoliberalism and the Calculable World: The Rise of Carbon Trading[*]

Larry Lohmann

Introduction

Neoliberalism can be a vague, even incoherent concept when it becomes entangled in the false dichotomies between market and state that are habitually thrown up by its adherents. It is often said, for example, that neoliberalism promotes free markets and reins in the state; yet, as Karl Polanyi pointed out long ago, laissez faire itself is an interventionist state project ('laissez faire was planned; planning was not').[1] It is said, too, that neoliberalism looks to economic growth rather than the state to solve many social problems; yet the quantifiable entity called 'the economy' was created in the 20th century largely by reorganizing and redistributing knowledge and embedding new practices of description and calculation in governmental practice, and can at no point be sharply marked off from official coercion, state corruption and 'non-economic' institutions.[2] Similarly, the neoliberal attempt to simulate efficient market outcomes by deploying cost-benefit analysis in policymaking depends on calculation and regulation undertaken by the state.[3]

Nowhere is the state/market dichotomy more misleading than in the analysis of one of the last, most ambitious manifestations of neoliberalism – the carbon markets that began to emerge in the 1990s as the main international policy response to climate change. While carbon markets are typically defended using neoliberal rhetoric ('What is the best way to tackle climate change? If we have a global carbon price, the market sorts it out';[4] 'Carbon trading is seen as a market-based alternative to either direct taxation or a "command and control approach"'[5]), the commodity in which the biggest carbon markets trade owes its very existence to government fiat and regulation. In tracing the causes of the havoc carbon markets are in the process of creating and abetting, it is useful to look beyond the misleading market/state, choice/coercion, efficiency/inefficiency dualisms commonly used to justify them. This chapter focuses instead on the power dynamics implicated in abstraction, commensuration and commodification as the features of the neoliberal approach to climate change that will most repay study. In so doing, it hopes to provide an introduction to one of neoliberalism's potentially greatest class projects: the attempt to privatize the climate itself.

What is Carbon Trading?

First proposed in the 1960s, pollution trading was developed by US economists and derivatives traders in the 1970s and 1980s and underwent a series of failed policy experiments in that country before becoming the centrepiece of the US Acid Rain Programme in the 1990s at a time of deregulatory fervour. In 1997, the Bill Clinton 2 regime successfully pressed for the Kyoto Protocol to become a set of carbon trading instruments (Al Gore, who carried the US ultimatum to Kyoto, later became a carbon market actor himself). In the 2000s Europe picked up the initiative to become the host of what is today the world's largest carbon market, the EU Emissions Trading Scheme (EU ETS) – although under Barack Obama the US may soon take over that position. Carbon markets now trade over US$100 billion yearly, and are projected to rival the financial derivatives market, currently the world's largest, within a decade. Pioneered by figures such as Richard Sandor of the Chicago Board of Trade and Ken Newcombe, who relinquished leadership of the World Bank's carbon funds to become a carbon trader at firms such as Goldman Sachs, carbon markets have recently become a magnet for hedge funds, banks, energy traders and other speculators.

Carbon trading treats the safeguarding of climatic stability, or the earth's capacity to regulate its climate, as a measurable commodity. After being granted or auctioned off to private firms or other polluters, the commodity can then be allocated 'cost-effectively' via market mechanisms. Obviously, the commoditized capacity in question was never produced for sale. Rather than being consumed, it is continually reused. Although difficult to define or even locate, the capacity forms part of the background 'infrastructure' for human survival. Framing it as a commodity, moreover, involves complex contradictions and blowbacks.[6] Current efforts to assemble carbon markets are likely, when carried beyond a certain point, to engender systemic crises. The earth's climate-regulating capacity is thus a quintessential Polanyian 'fictitious commodity'. Accordingly, illuminating comparisons and contrasts can be drawn with Polanyi's original 'fictitious commodities' of land, labour and money, as well as with other candidates for 'fictitious commodity' status that have been proposed since, including knowledge, health, genes and uncertainty.

The attempt to build a climate commodity proceeds in several steps (see BOX). First, the goal of maintaining the earth's capacity to regulate its climate is conceptualized in terms of numerical greenhouse gas emissions reduction targets. Governments determine – although currently more on explicitly political than on climatological grounds – how much of the world's physical, chemical and biological ability to regulate its own climate should be enclosed, 'propertized', privatized and made scarce. They then give it out (or, sometimes, sell it) to large polluters, before 'letting the market decide' on its final distribution.[7]

Making climate benefits and disbenefits into quantifiable 'things' opens them up to the possibility of exchange. For example, once climate benefit is identified with emissions reductions, an emission cut in one place becomes climatically 'equivalent' to, and thus exchangeable with, a cut of the same magnitude elsewhere. An emissions cut owing to one technology becomes climatically equivalent to an emissions cut that relies on another. An emissions cut that is part of a package that brings about one set of social effects becomes climatically equivalent to a cut associated with another set of social effects. Where emissions permit banking is allowed, an emission cut at one time becomes climatically equivalent to a cut achieved at another. Once all these identities are established, it becomes possible for a market to select for the emissions reductions (and, ipso facto, the climate benefits) that can be achieved most cheaply.

Carbon market construction in brief

Step 1: The goal of overcoming fossil fuel dependence by entrenching a new historical pathway is changed into the goal of placing progressive numerical limits on emissions (cap).

Step 2: A large pool of 'equivalent' emissions reductions is created through regulatory means by abstracting from place, technology, history and gas, making a liquid market and various 'efficiencies' possible (cap and trade).

Step 3: Further tradeable emissions reductions 'equivalents' are invented through special compensatory projects, usually in regions not covered by any cap, for additional corporate cost savings, and added to the commodity pool for enhanced liquidity and further 'efficiencies' (offsets).

Step 4: Project bundling, securitization, financial regulation, rating agencies, 'programmatic CDM' etc. add new layers of obscurity and complexity.

At first glance, these equivalences may seem uncontroversial. Market proponents tend to repeat, with the air of someone airing a tautology, that (for example) 'a carbon dioxide molecule released in Samarkand has the same climatic effect as one released in Sandusky'. A moment's reflection will show, however, that, in producing such equivalences, carbon traders are already drifting away from the climate problem. That problem consists mainly of the challenge of initiating a new historical pathway that leads away from dependence on fossil fuels, which are by far the major contributor to human-caused climate change. Once taken out of the ground and burned, coal, oil and gas add to the carbon burden cycling between the atmosphere and the oceans, soil, rock and vegetation. This transfer is, for human purposes, irreversible: once mined and burned, fossil carbon cannot be locked away safely underground again in the form of new deposits of coal, oil or gas, or in the form of carbonate rock, for millions of years. The transfer is also unsustainable: there is simply not enough 'space' in above-ground biological and geological systems to park safely the huge mass of carbon that is coming out of the ground without carbon dioxide building up catastrophically in the air and the seas. As biologist Tim Flannery puts it, 'There is so much carbon buried in the world's coal seams [alone] that, should it find its way back to the surface, it would make the planet hostile to life as we know it'.[8] Most un-mined coal, oil and gas, in other words, is going to have to stay in the ground. Accordingly, industrialized societies, currently 'locked in' to fossil fuels, need instead to 'lock in' non-fossil energy, transport, agricultural and consumption regimes within at most a few decades. Because this shift is structural, the first steps need to be undertaken immediately to minimize future dangers and costs.[9]

It follows that short-term actions can be assessed for their climatic effectiveness only by determining the part they play in a longer-term shift away from reliance on fossil fuels. For example, the choice of technology used in making a short term billion-ton emissions cut will make a large difference to long-term climatic outcomes. If the technology is one that reinforces overall societal addiction to fossil fuels, it will be more climatically damaging than one which contributes toward a pathway that keeps most remaining fossil fuels in the ground. Similarly, a billion-ton reduction in one place may have social effects which have a different impact on long-term fossil fuel use (and thus on future reductions) than a supposedly 'identical' billion-ton reduction in another place. Workable climate solutions, in short, are embedded in future history.

A commodity approach, by contrast, abstracts from where, how, when and by whom the cuts are made, disembedding climate solutions from history and technology and re-embedding them in neoclassical economic theory, trade treaties, property law, risk management and so forth. For example, carbon trading gives emissions-reduction technologies that are likely to result in unquantifiable but important 'spillovers'[10] leading to radically-lessened long-term dependence on fossil fuels equal weight with technologies lacking such effects, as long as both achieve the same numerical emissions reduction over the short term in a particular locality. While carbon trading encourages ingenuity in inventing measurable 'equivalences' between emissions of different types in different places, it does not select for innovations that can initiate or sustain a historical trajectory away from fossil fuels (the effectiveness of which is less easy to measure). Indeed, once the carbon commodity has been defined, merely to weigh different long-range social and technological trajectories or evaluate and 'back-cast' from distant goals is to threaten the efficiency imperative.

A commodity approach also functions to detach the global warming problem from climatological uncertainties and indeterminacies. This is because the sum of fungible greenhouse gas pollution rights that governments create and distribute for purposes of trade are implied to approach, in principle if not in practice, an economically optimal, 'climatically safe' level of overall greenhouse gas pollution. As work by the Harvard economist Martin Weitzman and others suggests, this move engenders a degraded conception of the climate problem: the commensuration process inherent in multi-equation, computerized Integrated Assessment Models that aggregate economic growth with simple climate dynamics heightens systemic hazards by 'presenting a cost-benefit estimate for what is inherently a fat-tailed situation with potentially unlimited downside exposure as if it is accurate and objective'.[11]

Disembedding: A Second Stage

The disembedding/re-embedding process inherent in carbon trading then ramifies and proliferates through a succession of further acts of commensuration and abstraction. After the state creates a divisible, tradeable commodity whose 'efficient' allocation in the form of pollution rights can become a coherent, 'apolitical' programme for action ('cap and trade'), its status as asset, grant, or financial instrument is engineered to fit various accounting standards.[12] Grants of pollution rights are made to industrialized countries (under the Kyoto Protocol) or private firms or other polluters (under the EU ETS), according to their existing pollution levels. Due to industrial lobbying efforts and measurement difficulties, these grants are often more generous than the polluters need to cover their existing level of emissions. Corporations receiving EU ETS grants are then allowed to pass on to their customers the nominal market cost of the asset they have received for free. (Auctioning may become more common in the future, but so far has not been widespread.) In this way, the bulk of the earth's carbon-cycling capacity is in effect made into property and distributed to the industrialized North, and in particular to the heaviest corporate polluters.

A second class of measurable, thing-like climate-benefit units called 'offsets' is then developed to be pooled together with 'reductions' for further 'efficiency' gains. These offsets are manufactured by special projects requiring special expertise, most located in the global South, that are claimed to result in less greenhouse gases accumulating in the atmosphere than would be the case in the absence of carbon finance, such as tree plantations (which are supposed to absorb carbon dioxide emissions) or fuel switches, wind farms and hydroelectric dams (which are argued to reduce or displace fossil energy). Schemes for generating still more saleable greenhouse gas pollution licenses – including projects involving agrofuels, biochar, nuclear energy, forest conservation and the capture, liquefaction and storage of carbon dioxide from coal-fired power plants – are also under consideration. Such 'project-based' credits, no matter what their origin, are designed to be fungible with the emissions allowances created and distributed by governments in the industrialized North. Indeed, in an act of commensuration-by-fiat, the Kyoto Protocol stipulated in Articles 3 and 12 that these offset credits are emissions reductions, thus legislating into existence a new, abstract, non-situated, omnibus category of reductions/offsets. It thus helped open a niche for a new corps of specialists and consultants – analogous to the 'quants' who helped develop advanced financial derivatives – to seek profits working out the needed commensuration procedures. Such 'carbon quants' produce calculations claiming, for example, that reducing carbon emissions from a power plant in Britain is 'the same as' building a wind farm in India or Brazil because the wind farm displaces fossil fuel use.

Since the carbon dioxide resulting from fossil fuel combustion is only one of many greenhouse gases, it is possible to create still more equivalences, making possible yet further supposed 'efficiencies' in attaining any particular cap. In the 1990s, the Intergovernmental Panel on Climate Change (IPCC) devised a new abstraction called 'global warming potential' that commensurates an entire basket of climate-forcing gases according to how they compare to carbon dioxide in their climate impact. That ultimately enabled corporations to arrange to make spectacular savings in meeting emissions targets under the EU

Emissions Trading Scheme. Instead of cutting its own carbon dioxide emissions, for example, the German-based generating firm RWE could plan on investing in United Nations-certified 'offset' projects destroying small amounts of nitrous oxide (a greenhouse gas stipulated to be 298 times more powerful than carbon dioxide over a 100-year time horizon) at factories in Egypt and South Korea and even smaller amounts of HFC-23 (a climate-forcing gas with a 'global warming potential' set at 14,800 times that of carbon dioxide over a 100-year horizon) at chemical plants in China.[13] It could also explore the possibility of buying carbon credits from projects that would capture and burn methane (yet another greenhouse gas stipulated to be more harmful than carbon dioxide, especially over the short term) from landfills and coal mines in China and Russia. Commensurating all these gases was hard work, since they vary in their effects along many different axes and time scales. In one reflection of the un-clarities and disputes involved, in 2007 the IPCC increased the 100-year factor for HFC-23 by over 23 per cent, enabling at a keystroke the production of millions of tons more carbon credits.

Using offsets to achieve increased liquidity and 'efficiency' distances carbon markets from the global warming problem not only because it ignores the importance of achieving a transition away from fossil fuels, but also because it tends to suppress, in a class- and culturally-biased way, concrete practices likely to play a significant part in those solutions. Carbon offset accounting necessarily frames the political question of what would have happened without carbon projects as matter of expert prediction in a deterministic system, while at the same time framing (usually wealthy) project proponents non-deterministically, as free decision-makers whose initiatives are capable of changing 'business as usual'. Activists in Minas Gerais, Brazil, called attention to this contradiction early on when they contested an attempt by a local charcoal and pig iron company, Plantar (see Chapter 9 in this volume), to get carbon credits for the environmentally-destructive eucalyptus plantations it had established on seized land: 'The argument that producing pig iron from charcoal is less bad than producing it from coal is a sinister strategy ... What we really need are investments in clean energies that at the same time contribute to the cultural, social and economic wellbeing of local populations'.[14] After insisting that 'the claim that without carbon credits Plantar ...would have switched to coal as an energy source is absurd,' the activists went on to characterize the accounting procedure as a 'threat': 'It is comparable to loggers demanding money, otherwise they will cut down trees'.[15]

Typically, offset income supports conventional developments that harm local low-carbon livelihoods and sources of agricultural knowledge while at the same time doing little if anything for local transitions to a non-fossil society. In the mountainous river valleys of Uttaranchal, India, for example, scores of dam projects in line to be part-financed through selling carbon credits to Northern industry are damaging local low-carbon irrigation systems. In China, 763 hydroelectric dams have applied or are planning to apply to the United Nations to be allowed to sell more than 300 million tons of carbon dioxide pollution rights to Northern industry through the Kyoto Protocol's Clean Development

Mechanism, yet they do not replace fossil-fuelled generation, but merely supplement it, and were arguably going to be built anyway.[16] In November 2008, the US Government Accountability Office warned that such carbon projects can allow industries in the North 'to increase their emissions without a corresponding reduction in a developing country'.[17]

Nigeria's oil-extraction zone offers another good example of carbon markets' tendency to encourage private corporations and technical experts to expend ingenuity on inventing novel, geographically far-flung market 'equivalents' for emissions reductions rather than finding ways to implement a structural shift away from fossil fuels. For 50 years, energy companies operating in the Niger Delta have burned off the great bulk of the methane they find in underground oil reservoirs. Although methane is a valuable fuel, it is cheaper for corporations such as Shell and Chevron simply to flare it on site than to use it in power plants or reinject it underground. As a result, local people are subjected to continuous noise, light and heat, acid rain, retarded crop yields, corroded roofs, and respiratory and skin diseases.[18] Although flaring is prohibited by law in Nigeria (in 2005 the Nigerian Federal High Court confirmed that gas flaring was illegal and a gross violation of human rights), oil companies have so far contented themselves with paying penalties for non-compliance. In this context, one focus of local and international environmental activism is simply to insist on the rule of law. The Clean Development Mechanism, however, takes breaches of the law in Nigeria as the 'baseline' for carbon accounting. The Italian oil corporation Eni-Agip, for example, plans to buy some 1.5 million tons per year of cheap carbon dioxide equivalent pollution rights from a project at an oil-gas installation at Kwale that was registered with the UN in November 2006.[19] Eni-Agip and its validator, the Norwegian consultant DNV, claim that the project will be reducing emissions by putting gas which would otherwise be flared to productive use (although it is difficult to verify whether the gas in question will come from oil wells or dedicated gas extraction operations also present in the region, whose production is not flared). The core of the calculation is that

> whilst the Nigerian Federal High Court recently judged that gas flaring is illegal, it is difficult to envisage a situation where wholesale changes in practice in venting or flaring, or cessation of oil production in order to eliminate flaring will be forthcoming in the near term.[20]

Accordingly, the project creates a new incentive for the Nigerian authorities to replace legal sanctions with prices and the rule of law with markets for environmental services. It would be difficult to imagine a purer expression of neoliberal doctrines. Isaac Osuoka, the joint coordinator of the Gulf of Guinea Citizens Network, believes that 'carbon trading reflects one of the worst forms of neoliberal fanaticism and attempts at re-legitimating corporate rule experienced in the past decades'.[21]

Current proposals to allow industrialized countries and their corporations to compensate for continued fossil fuel use by pressing millions of hectares of land in the global South into service as biotic carbon stores or dumps further highlight carbon offsets' tendency toward regressive redistribution. In one proposed scheme, REDD ('Reducing Emissions from Deforestation and Degradation'), billions of dollars would be invested in acquiring and preserving carbon in the world's native forests, which would then be traded for permission to continue greenhouse gas pollution elsewhere. Land grabs have already begun in central Africa, Indonesia and Papua New Guinea in order to feed the expected need for forested land of the US's proposed carbon trading system under the Waxman-Markey Act. State forestry departments, conservation organizations, local authorities, indigenous communities or logging or plantation companies would serve as onsite security staff for this global carbon warehouse. REDD advocates include ex-World Bank chief economist Nicholas Stern, who sees it, ton for ton, as one of the cheapest ways of keeping carbon dioxide molecules out of the atmosphere; Wall Street firms such as Merrill Lynch (now owned by Bank of America), which see high potential in trading such new 'carbon assets'; the Food and Agriculture Organization, which welcomes it as an opportunity to expand its political role; and, often in the forefront, carbon consultants, forest scientists, technicians and master planners with careers in forest conservation, who are working on the ground in countries such as Indonesia to secure local authorities' consent to the schemes. The large sums of money potentially on offer have split indigenous peoples' movements, some of whom see REDD as an opportunity for advancement, others of whom see it as an enclosure movement; and environmentalists, who divide between large, Washington-based proponents such as Conservation International and The Nature Conservancy and less well-funded opponents who see REDD as disempowering forest peoples in favour of acquisitive corporations and state agencies.[22] Although its role and political nature are often misunderstood by traders and activists alike, commensuration is again central to this struggle: for trading to be possible, emissions arising from the combustion of fossil carbon must be made quantitatively comparable with tree carbon. This becomes an endless task due to the different roles played by fossil and biotic carbon in the climate system, as well as uncertainties and unpredictabilities in forest carbon absorption, which are being exacerbated by global warming itself.[23]

Finance and Securitization

A final step in the carbon markets' abstraction from the climate problem comes with securitization. Financial market actors have always been prominent in the carbon trade and today dominate the buyers' side of the credit market. Among the financial institutions that have set up desks to speculate in carbon permits are Deutsche Bank, Morgan Stanley, Barclays Capital, Rabobank, BNP Paribas Fortis, Sumitomo, Kommunalkredit, and Cantor Fitzgerald. JP Morgan Chase has snapped up the carbon offset firm Climate Care, while Credit Suisse has acquired a stake in the troubled carbon consultancy and accumulator EcoSecurities and Goldman Sachs has announced plans to buy Constellation Energy's carbon trading business. By 2008 there were about 80 carbon investment funds set up to finance offset projects or buy carbon credits, most oriented more toward speculation than toward helping companies comply with regulated carbon caps. Trading companies are also active, including Vitol, a major energy-market speculator, and while ENRON, an early enthusiast for the Kyoto Protocol carbon market, is no longer in business, some of the firm's ex-staff have moved into the carbon sector. Before the financial crash, even certain industrial companies, such as Arcelor Mittal (the world's largest steelmaker), opened departments specifically to seek profits in the carbon trade, just as companies such as General Electric opened finance divisions in the 1990s.[24] As with financial derivatives, a host of specialized new institutions have also been set up that deal in the commodity, with names like Sindicatum Carbon Capital, NatSource Asset Management, New Carbon Finance, Carbon Capital Markets, Trading Emissions plc, South Pole Carbon Asset Management, Noble Carbon, and so forth.

One of the tasks of such firms is to bundle together various types of small offset projects for buyers. With increased investment, securitization is likely to follow. Already in November 2008, Credit Suisse announced a securitized carbon deal that would bundle together carbon credits from 25 offset projects at various stages of UN approval, sourced from three countries and five project developers. The bank then split these assets into three tranches, allegedly representing different risk levels, before marketing them to investors. In this way, products which already had only the most tenuous relation to the climate problem they were designed to tackle, and had been further disconnected from underlying values through a cascade of contested commensuration processes, were transformed through yet further disaggregation and reassembly. Evaluation of such securities, whether by credit rating agencies or regulators, is certain to be even more challenging, and even less amenable to modeling, than was the evaluation of the mortgage-backed securities that played such an important part in the onset of the financial crisis. If carbon permit products are 'toxic' to climate change mitigation policy, they may prove to be no less so to financial stability, given the projected trillion-dollar scale of the market. The dangers of what Friends of the Earth analyst Michelle Chan calls 'subprime carbon' are obvious.[25]

Insofar as it is aimed merely at improving carbon market practice rather than at fossil fuel use, and relies on a theory-practice dualism, regulation tends to become yet another moment in the neoliberal disembedding/re-embedding process, adding further layers of attempted calculation to an unstable structure and further concealing the problematic nature of the underlying abstractions. A case in point is the continuing attempt of the Clean Development Mechanism's Executive Board and government regulators in various countries to tackle the riddle of 'additionality' in offset markets (that is, how to prove that a project goes beyond business as usual), to which, as carbon trader Mark Trexler noted years ago, there is no correct answer.[26] Constantly manufacturing and reaffirming the notion that offset projects' shortcomings are due either to imperfect methodology or incorrect implementation, ten years of regulatory effort have only further skewed the political economy of the offset markets further in favour of corporations locked into fossil fuel use, since it is only they who have the resources necessary for navigating the regulatory mazes that the additionality debate has made ever more intricate. Ironically, of course, this is an effect which, logically speaking, should itself enter into calculations of carbon saved and lost – one more example of the 'moving horizon' characteristic of the market-environmentalist project of 'internalizing externalities'. The recent establishment of a private carbon rating agency, as well as proposals for 'programmatic' and 'sectoral' carbon credits, which would help sidestep impossible 'additionality' requirements, reflect a continuing commitment to 'better calculation' in the face of irresolvable tensions between the needs for high-volume, predictable carbon credit output and for market credibility.

Conclusion

Like the neoclassical shibboleths (the efficient markets hypothesis, rational expectations and the like) that have so picturesquely come to grief during the financial crisis, the carbon credit prices flashing on electronic screens in trading rooms on Wall Street or in the City of London reflect a complex political movement to reorganize and redistribute knowledge and power. Spelling out another notable chapter in the political history of commensuration[27], they form a part of one of neoliberalism's last and greatest class projects: the attempt to appropriate the climate itself. Carbon trading thus takes its place alongside other movements of recent decades that have invented new possibilities of accumulation through the creation of fresh objects of calculation and the intensified commodification of some of the more hidden aspects of the infrastructure of human existence. Examples include attempts to expand credit by mathematizing and privatizing an unprecedented variety of uncertainties through derivatives markets[28], to privatize creativity through global intellectual property rights, and to transform health, health care and even biological species into measurable, tradeable commodities.[29]

All these efforts to appropriate involve abstraction and commensuration as part of wider processes involving deregulation, banking and land law, treaty negotiation, structural adjustment, police work, mapping, resource seizures, export subsidies and so on. This abstraction and commensuration can never be completed any more than politics or the evolution of a language can be completed. As Mitchell observes, internalizing all externalities would make exchange impossible.[30] Ideals of calculability, continually being developed and undermined in the course of attempts to carpenter together new structures of property and trade, are part of conflicted processes that can generate both profits and crisis. The largely unchecked pursuit of liquidity in risk markets, furthered by the achievements of quants, led in the end to a financial stampede for the exits and a drying up of liquidity, and may eventually do the same in the carbon markets. An unrestrained quest to 'internalize' the benefits of innovation leads in the end to the sapping of innovative forces and resources.[31] Cost-benefit analysis's attempt to isolate an uncontroversial basis for social choice in the calculation of individual preferences itself generates heightened controversy. Headlong attempts to implement 'market solutions' for global warming end up exacerbating the climate crisis as well as social dislocations of diverse kinds and wide geographical reach.

The troubled trajectory of such initiatives hints at the continuing relevance of earlier traditions of crisis analysis: Polanyi's[32] observation that the complete commodification of land would result in the 'demolition of society'; Marx's descriptions of the 'contradictions' of capitalism; Keynes's warning about finance's 'fetish of liquidity' that 'there is no such thing as liquidity of investment for the community as a whole'.[33] Yet, as this chapter's sketch of carbon trading has suggested, analytical space must also be made for newer concepts such as Michel Callon's 'overflows',[34] Timothy Mitchell's treatment of the theory/practice divide as a mode of modern power, and science scholars' emphasis on nonhuman agents, whether the recalcitrant rainforest trees now being pressed into service as carbon stores or the 'black swans' and 'monsters' of nonlinearity now routinely referred to by both financial analysts and climatologists.[35] Study of the arcane particularities of manifestations of neoliberalism such as carbon trading can both inform and transform analyses of contemporary politics generally. As Lydgate famously observed in *Middlemarch*, there must be a 'systole and diastole in all inquiry' aimed at 'continually expanding and shrinking between the whole human horizon and the horizon of an object-glass'.

The unfolding disaster of carbon trading prefigures the disintegration of the picture of a thoroughly calculable world to which neoliberalism clings more stubbornly than any state socialist project of the past. The important question is how this disintegration is to be effected politically. What sort of alliances can be fashioned among, for example, grassroots resisters of offset projects in the South, environmental justice movements battling fossil fuel extraction and pollution, and a Northern public frustrated at the largesse being lavished by their governments and the United Nations on the creation of yet another dysfunctional speculative market? The answers are not yet clear, but here as elsewhere the fall of neoliberalism will be something to be achieved through patient movement-building and a long series of political struggles, not something automatically given by the mechanics of yet another crisis.

Notes

*   Lohmann, L. (2010, forthcoming) 'Neoliberalism and the Calculable World: The Rise of Carbon Trading', in K. Birch, V. Mykhnenko, and K. Trebeck (eds) *The Rise and Fall of Neoliberalism: The Collapse of an Economic Order?* London: Zed Books. © Zed Books 2010. Reproduced here by permission of Zed Books, London & New York, www.zedbooks.co.uk.

1   Polanyi, K. (2001 [1944]) *The Great Transformation*. Boston: Beacon.

2   Mitchell, T. (2002) *Rule of Experts: Egypt, Technopolitics, Modernity*. Berkeley: University of California Press.

3   Lohmann, L. (2009) 'Towards a Different Debate in Environmental Accounting: The Cases of Carbon and Cost Benefit', *Accounting, Organizations and Society*, 34(3-4): 499-534.

4   Scott, M. (2008) 'Market Meltdown? Carbon Trading is just Warming up', *Independent on Sunday Business*, 27 July.

5   Milner, M. (2007) 'Global Carbon Trading Market Triples to $30b', *The Hindu*, 5 May.

6   Lohmann, L. (forthcoming) 'Uncertainty Markets and Carbon Markets: Variations on Polanyian Themes'.

7   Lohmann, L. (2005) 'Marketing and Making Carbon Dumps: Commodification, Calculation and Counterfactuals in Climate Change Mitigation', *Science as Culture,* 14 (3): 203-235; Lohmann, L. (2006) *Carbon Trading: A Critical Coversation on Climate Change, Privatization and Power.* Uppsala: Dag Hammarkjold Foundation.

8   Flannery, T. (2005) 'Monstrous Carbuncle', *London Review of Books,* 27(1), 6 January.

9   Unruh, G. C. (2000) 'Understanding Carbon Lock-In', *Energy Policy,* 28: 817–30.

10   Frischmann, B. and M. Lemley (2006) 'Spillovers', *Columbia Law Review,* 107.

11   Weitzman, M. (2008) 'On Modeling and Interpreting the Economics of Catastrophic Climate Change', REStat, http://www.economics.harvard.edu/faculty/weitzman/files/REStatFINAL.pdf.

12   MacKenzie, D. (2009) 'Making Things the Same: Gases, Emission Rights and the Politics of Carbon Markets', *Accounting, Organizations and Society,* 34(3-4): 440-455.

13   Lancaster, R. (2007) 'Mitigating Circumstances', *Trading Carbon,* December; Forster, P., V. Ramaswamy et al. (2007) 'Changes in Atmospheric Constituents and in Radiative Forcing', in *Intergovernmental Panel on Climate Change, Working Group 1, IPCC Fourth Assessment Report,* Cambridge: Cambridge University Press.

14   FASE et al. (2003) 'Open Letter to Executives and Investors in the Prototype Carbon Fund', *Espirito Santo,* Brazil, 23 May.

15   Suptitz, A. P. L. et al. (2004) 'Open Letter to the Clean Development Mechanism Executive Board', *Minas Gerais,* Brazil, 7 June.

16   McDonald, J., C. Hanley and P. McGroarty (2009) 'China Dams Reveal Flaws in Climate-Change Weapon', *Associated Press,* 29 January.

17   United States General Accounting Office (GAO) (2008) 'Lessons Learned from the European Union's Emissions Trading Scheme and the Kyoto Protocol's Clean Development Mechanism', *GAO,* Washington, DC, November.

18   Osuji, L.C. and G.O. Avwiri (2005) 'Flared Gases and Other Pollutants Associated with Air Quality in Industrial Areas of Nigeria: An Overview', *Chemistry and Biodiversity,* 2.

19   United Nations Environment Programme (UNEP) Risoe Centre (2009) 'CDM Pipeline', *UNEP,* Risoe, http://www.cdmpipeline.org.

20   Det Norske Veritas (DNV) (2004) 'Clean Development Mechanism Project Design Document Form for Recovery of Associated Gas that Would Otherwise be Flared at Kwale Oil-Gas Processing Plant, Nigeria', http://www.dnv.com/focus/climate_change/upload/final%20pddnigeria%20ver.21%20%2023_12_2005.pdf.

21   Osuoka, I. (2009) 'Paying the Polluter? The Relegation of Local Community Concerns in "Carbon Credit" Proposals of Oil Corporations in Nigeria', ms., April.

22   Griffiths, T. (2008) 'Seeing "REDD"? Forests, Climate Change Mitigation and the Rights of Indigenous Peoples and Local Communities', *Update for Poznan (UNFCCC COP 14).* Forest Peoples Programme, Moreton-on-Marsh, 3 December.

23   Philips O. L. et al. (2009) 'Drought Sensitivity of the Amazon Rainforest', *Science,* 323, 6 March: 1344-1347; Lindroth, A. et al. (2009) 'Storms Can Cause Europe-Wide Reduction in Forest Carbon Sink', *Global Change Biology,* 15: 346-355.

24   Cleantech Investor (2008) 'ArcelorMittal Clean Technology Venture Capital and Carbon Fund', *Cleantech Magazine,* September.

25   Chan, M. (2009) 'Subprime Carbon? Rethinking the World's Largest New Derivatives Market', *Friends of the Earth,* Washington, March.

26   Trexler, M. (2006) 'A Statistically Driven Approach to Offset-Based GHG Additionality Determinations: What Can We Learn?', *Sustainable Development, Law and Policy,* 6(2).

27   Espeland, W. and M.L. Stevens (1998) 'Commensuration as a Social Process', *Annual Review of Sociology,* 24.

28 Lohmann, L. (forthcoming) 'Uncertainty Markets and Carbon Markets: Variations on Polanyian Themes'.

29 Frischmann, B. and M. Lemley (2006) 'Spillovers'.

30 Mitchell, T. (2002) *Rule of Experts: Egypt, Technopolitics, Modernity.*

31 Frischmann, B. and M. Lemley (2006) 'Spillovers'.

32 Polanyi, K. (2001 [1944]) *The Great Transformation.*

33 Keynes, J. M. (2008 [1936]) 'Speculation, Cyclicality and the Euthanasia of the Rentier', in I. Erturk, and J. Froud et al. (eds) *Financialization At Work: Key Texts and Commentary.* London: Routledge.

34 Callon, M. (1998) 'An Essay on Framing and Overflowing: Economic Externalities Revisited by Sociology', in M. Callon (ed) *The Laws of the Markets.* Oxford: Blackwell: 244–269.

35 Mitchell, T. (2002) *Rule of Experts: Egypt, Technopolitics, Modernity.*

II

CASES

Part II of the book comprises a range of case studies from Thailand to Chile, from Uruguay to India, presenting rich details of the often negative effects of CDM and voluntary offset projects on local communities in the 'Global South'. Part II begins with papers by Melissa Checker, Tamra Gilbertson, Cristián Alarcón and Isaac 'Asume' Osuoka, showing how 'developed' and 'developing' countries and their respective governments, corporations and local communities are interlocked in a complex web of carbon market relations, which, rather than promoting sustainable development, help to increase inequalities between North and South. The next set of chapters – written by Ricardo Carrere, Raquel Nuñez, Rafael Kurter Flores and colleagues, and Steffen Böhm – are aimed at breaking our illusion of considering industrial tree plantations to be real forests that would help us fight climate change. The last set of cases – written by Soumitra Ghosh, Hadida Yasmin, Siddhartha Dabhi, Nishant Mate and Soumya Dutta – come from India, which is one of the largest hosts of CDM and voluntary offset projects.

3

Double Jeopardy: Pursuing the Path of Carbon Offsets and Human Rights Abuses[*]

Melissa Checker

Introduction

On the East Coast of Scotland, one of Europe's largest oil refineries flares excess gas into the sky, sending sulfur dioxide, nitrogen dioxide and other particles into the nearby town of Grangemouth.[1] Six thousand miles away in eastern Brazil, the villagers of Sao Jose do Buriti struggle as their water sources dry up and the plants they have subsisted on for generations disappear. Several years ago, NGOs, the Transnational Institute Environmental Justice Project and Carbon Trade Watch sent representatives to these two disparate places to help their inhabitants create a documentary film. Each community created a video diary that detailed their daily efforts to cope with the industries surrounding them and then shared it with the other. Residents found that their struggles had much in common – pollution in Grangemouth and water shortages in Sao Jose do Buriti were intimately linked through carbon offsets.[2]

In the late 1990s, British Petroleum (BP), then owners of the Grangemouth refinery, launched a major effort to 'green' their image, in part by offsetting their carbon emissions through investments in projects that reduce the production of greenhouse gases. Around the same time, a foundry near Sao Jose do Buriti publicized its plan to switch from using charcoal to carbon-intensive coal, due to a dwindling supply of charcoal-producing eucalyptus trees. Enter the World Bank, which gathered funding from various sources including BP, and initiated a project to expand the foundry's eucalyptus forest and generate carbon offsets. In addition to providing a renewable raw material, the trees would absorb carbon in the atmosphere. Each ton of carbon absorbed would then offset, or neutralize, a ton of carbon dioxide produced in Scotland. Only, this formula left out one very important factor – human beings. Not only did the plan allow the oil refinery to continue emitting noxious chemicals in Grangemouth, but the enormous roots of the carbon-absorbing, charcoal-producing trees in Brazil also drained local water resources, essential to the lives and livelihoods of nearby villagers.

Drawing on this and similar cases, this essay argues that rather than balancing out carbon emissions, carbon offset projects create equal measures of human injustice, for the communities that host them and for the communities surrounding the facilities that buy them. I also argue that offsets set off a chain reaction, creating multiple harms to human beings, in both the near and long term. I base my argument on three case studies. Each one focuses on a project meant to generate carbon offset credits and then tracks those credits to the industry whose emissions they were supposed to have counterbalanced. In so doing, I demonstrate that the path from offset credit producer to offset credit consumer is strewn with violations to both human rights and the environment.

Since their inception in the late 1990s, proponents of carbon offset projects have promoted them as a win-win scenario. Carbon-producing industries can make up for their emissions by investing in activities that reduce greenhouse gases, and at the same time, those greenhouse gas-reducing activities stimulate sustainable development.[3] However, over the past few years, a growing body of academic and popular literature contends that the premises behind carbon offsets are inherently flawed.[4] For example, on a scientific level, emissions equivalences are difficult to measure. Economically, when the value of carbon credits falls (as it has over the past year), incentives to generate them are drastically reduced. On a broader level, many critics of offsets focus on the problem of additionality, or the degree to which offset monies fund new emissions reductions versus projects that would have been done anyway.

Only recently, have researchers begun to gather evidence delineating the *social* costs of offset projects, including the ways in which they sometimes generate development that is more harmful than it is sustainable for local communities. This chapter combines existing and new research to build on such human-centered critiques and take them a step further. More specifically, in each of my three case studies, I first compile published data to describe how a carbon offset project has negatively affected the local community hosting it. Then, I gather data from web sites, newspapers, magazine articles and governmental and non-governmental reports to follow the offsets to the industry that benefited from them. This juxtaposing of offset producers and consumers places the compounded costs of carbon offsets schemes in stark relief. In the end, I find that, from a human rights perspective, carbon offsetting creates a double-jeopardy in terms of both space and time. On a short-term, geographic level offset projects exacerbate poor environmental, economic, political and/or social conditions for local communities in both the Global South and the Global North. On a longer term level, offset projects diminish human rights in several more indirect ways: First, in boosting the fiscal success of the corporations participating in them, they facilitate any unsustainable practices a corporation might sponsor. Second, an emphasis on greenhouse gas emissions has ignored the fact that offset projects sometimes produce highly toxic, non-greenhouse gas pollutants. Third, the profitability and positive publicity surrounding offset programs provides a disincentive for governments and corporations to develop practices that would significantly reduce (rather than just neutralize) carbon emissions. Fourth, because offsets are designed to balance out greenhouse gas emissions they ultimately exacerbate global warming, which in turn threatens the lives of economically and socially vulnerable communities.[5] On a more optimistic note, however, I conclude this chapter by arguing that once clarified, connections between communities surrounding offset producers and consumers open new opportunities for transnational alliance-building and opposition to carbon trading emerge.

Carbon Offsets Come of Age

In 2005, the Kyoto Protocol went into force, obligating its signers to cut their greenhouse gas emissions to a combined 5% of 1990 levels. To facilitate those goals, Kyoto negotiators established offsets, or mechanisms by which industries in developed nations could cut their emissions indirectly, by investing in programs that reduce, avoid, or sequester CO_2 or other greenhouse gases in some other place. For Kyoto signers, offsetting is largely conducted through the Clean Development Mechanism (CDM), a process supervised by various arms of the United Nations Framework Convention on Climate Change (UNFCCC). CDM projects usually take place in developing countries and can include implementing alternative energy sources, 'green' technology designed to reduce CO_2 emissions, the trapping or destroying of greenhouse gases or, to a limited extent, programs that trap carbon through forest growth. Once certified, such programs earn Certified Emissions Reduction (CER) credits. Credits may be awarded to a specific project investor and applied towards reducing that investor's carbon footprint, or traded on a carbon market. Today, experts predict that CDM-generated offsets will deliver more than half of the European Union's planned carbon reductions to 2020.[6]

Those not obligated by Kyoto can partake in the emerging, self-regulated voluntary carbon market. Thanks in part to the extraordinary success of the 2006 film, *An Inconvenient Truth*, the voluntary carbon market reached $705 million in 2008.[7] Approximately 5% of this self-regulated market comes from individuals wishing to assuage their guilt over carbon intensive activities such as airplane travel. The rest is comprised of businesses entities. Some of those include eco-conscious entertainers such as Leonardo DiCaprio, the band Coldplay and makers of the film Syriana, all of which famously offset airline travel and other carbon emissions generated by their professional projects or pursuits.

But, the bulk of offsets are bought by businesses wishing to bolster and 'green' their image.[8] At the end of 2009, for example, Dell Computers announced (or claimed) that it had achieved 100 percent carbon-neutrality, mostly by purchasing offsets. The U.S. National Football League used offsets to help 'neutralize' the carbon generated by the last two Superbowls. And in January 2009, Motorola unveiled what it called the world's first carbon neutral phone; besides making the phone of recycled plastic, the company vowed to offset the carbon produced during the phone's manufacture, distribution and operation.[9]

Certainly, national, individual and/or corporate concern for the planet is laudable, and corporations that go green might as well generate some good publicity for it along the way. The problem is that using offsets to reduce greenhouse gas emissions is like trying to lose weight by paying someone else to go on a diet and then claiming the weight loss as your own. In other words, when EU countries count up their emissions reductions, they include those that happened in foreign places (which is the whole point of offsetting). If they only counted local emissions reductions, the numbers would obviously be quite different. For instance, the European Commission highlights the fact that EU emission levels fell approximately 3.6% between 2007 and 2008. In large part these decreases can be attributed to declines in manufacturing as a result of the economic downturn and to the roughly 1,500 projects registered in the CDM.[10] In contrast, a recent study by law professor and energy expert at Stanford University's Program on Energy and Sustainable Development, Michael Wara, estimated that accounting for offsets, European emissions were actually about 1% higher in 2008 than they were in 1990.

The real danger here is that many experts agree that offset projects lead to questionable reductions at best. For instance, a November 2008 report by the U.S. General Accountability Office (GAO) examined the CDM and found that its effects on greenhouse gas emissions are uncertain, largely because it is 'nearly impossible' to determine the level of emissions that would have occurred in the absence of each project.[11] A report by Germany's Institute for Applied Ecology found that 40 percent of CDM projects registered by 2007 represented 'unlikely or at least questionable' emissions cuts, partly because many offset projects would happen even without offset funding.[12] Currently, the UN contracts third-party consultants to verify a project's additionality – however, between November 2008 and September 2009, the UN suspended the accreditation of the world's two largest auditors of CDM projects for not properly vetting projects before approving them.[13]

Again, there are many excellent articles detailing these and other reasons why offsets fail to live up to their promises in terms of delivering real greenhouse gas reductions. My purpose here is to look beyond questions of efficacy and explore the *human* costs of offsets, across the globe. In so doing, I call attention to some of the most urgent reasons to oppose the use of offsets. For, mounting evidence shows that as offset projects play out 'on the ground', and as corporations use offsets to continue business as usual, they are having immediate and dire consequences for local communities.

Mount Elgon, Uganda/Appalachian Mountains, United States

Mount Elgon, Uganda offers one of the most well-documented and most violent examples of an offset project gone awry. In 1990, the Dutch Electricity Generating Board vowed to surpass Kyoto Treaty goals and reduce its carbon emissions to 1989 levels by 1994-1995 with a further 3–5% reduction by the year 2000. The board aimed to accomplish its goals by improving energy efficiency at its plants, developing new technologies, and by compensating for emissions through 'cost effective' measures.[14] To implement the last of these strategies, the power board established the Forests Absorbing Carbon Dioxide Emissions (Face) Foundation, a nonprofit corporation dedicated to 'establish[ing], maintain[ing], and/or enhance[ing], forest vegetation' in order to absorb carbon dioxide.[15] The forests also earn carbon credits for the CO_2 they ostensibly sequester.[16] The foundation (which in 2002 spun off from the power companies) could then sell the credits and reinvest the proceeds in further forest projects. In 1994, Face partnered with the Uganda Wildlife Authority (UWA) to plant 25,000 hectares of trees inside Mount Elgon National Park. In exchange for financing the planting of the trees, Face received the rights to the carbon sequestered by those trees – estimated at 2.11 tons of CO_2 over 100 years.[17] While the trees have thrived (especially in areas where agriculture had been encroaching on them), a number of research reports have found that the people surrounding the tree plantations have had the opposite experience.

A year before the Face-UWA project began the Ugandan government declared Mount Elgon a National Park. In so doing, it evicted approximately 6,000 people (some of whom had been living there for 40 years), giving them nine days to vacate their homes.[18] A year later, UWA took over management of the Park, which entailed protecting the biodiversity of the area, managing the carbon plantations and securing the park's borders.[19] Evicted villagers, who were left homeless and without access to land to graze their cattle or grow subsistence crops, attempted to continue using park land. When UWA rangers responded with violence, local villagers organized to regain their land. In 1998, they filed land claims against the UWA and the Ugandan government. Several NGOs and universities heard about the situation on Mount Elgon and launched their own investigations, which corroborated villagers' claims. For instance, a World Rainforest Movement report published in 2006 details villagers' descriptions of UWA rangers committing rape, arson, shootings and other violent acts. According to the report, villagers retaliated by throwing stones, burning trees, and sabotaging rangers' vehicles.[20]

Another reason for local antipathy towards the Face project is that, according to villagers, the forest project has not lived up to its promises of sustainable development. Initially, project leaders promised to employ local people to work in the national park and tree nurseries and as tree planters. However, the World Rainforest Movement quotes local council officials who contend that the project employs very few people and most of the jobs are only available during the planting period.[21] To this day, the UWA continues to prevent local people from using the land, and violence and retaliations continue, despite a 2005 court ruling that an area of the national park should be set aside for villagers to live on and continue farming.[22] To be fair, land disputes on Mount Elgon predated the FACE Foundation's offset project, and the UWA maintains that the offset forest has nothing to do with its conflict with surrounding villagers. At the same time, the funding generated by the project likely provided additional incentives and justifications to administer evictions and violently patrol the area.

If we follow some of that funding and track the carbon credits generated on Mount Elgon, we find a maze of corporations, subsidiaries, and carbon-emitting ventures. Indeed, one of the major criticisms commonly leveled at carbon trading schemes is that they create an opaque web of financial instruments ripe for corruption.[23] For example, the FACE Foundation is a non-profit organization, but the offset reductions generated by its projects are marketed by a Dutch for-profit partner, known as the Climate Neutral Group (CNG). CNG sells credits to over 500 businesses. It also partners with another for-profit company, Green Seat, which sells offsets (including those created on Mt. Elgon) exclusively to individuals and corporations wishing to balance out emissions from airline travel.

After several major news outlets reported on the violence on Mt. Elgon in 2007, Green Seat posted a notice on its website claiming that neither it nor CNG used offsets from Uganda forestry projects any longer. The Face Foundation also claimed to have stopped planting trees in the park and to be disengaging from the project. 'At this stage we don't get any carbon credits for this project', Denis Slieker, director of the Face Foundation, told the *LA Times* in 2007, 'We do not plan to expand anymore in Mount Elgon before these matters are resolved.'[24] Yet a recent visit to the Face Foundation's website describes the Mount Elgon project as 'on going.'

After 2007, it is unclear exactly what kinds of carbon-producing projects the Mount Elgon project offset. It is certain, however, that it has enabled the building of at least several coal-fired power plants. First, the FACE Foundation was initially established to offset emissions from a new 600 MW coal-fired power station in the Netherlands. Second, Climate Neutral Group customer, Enesco is one of the top three energy companies in the Netherlands. Enesco is considered to be a particularly 'green' energy company – in 2008, Greenpeace ranked it the 'cleanest' power company in the Netherlands,[25] and on January 1, 2008, the company proclaimed that its internal business operations were '100% climate-neutral'.[26] Yet, my research revealed that 61.2% of the company's energy supply comes from natural gas, a fossil fuel, and 19.7% – nearly one-fifth – comes from coal.[27] Importantly, one quarter to one third of all carbon dioxide emissions worldwide come from burning coal.[28] In addition to high levels of carbon dioxide, coal plants also produce sulfur dioxide, nitrogen oxide, carbon monoxide, mercury and arsenic (among other pollutants).[29]

Even if the Ugandan project were able offset the climate harm generated by coal-fired power plants, it would not be able to offset their human costs. In fact, in 1974 the Netherlands closed all of its coal mines due to their dangerous conditions; yet, in 2008, the country imported 3.6 million short tons of coal from the U.S. making it one of the world's top coal importers.[30] The global demand for coal has expanded a controversial method of coal extraction, known as mountaintop removal, which uses explosives to blast away a mountain peak and expose coal seams. While coal companies claim the practice is safer and more efficient than traditional shaft mining, critics contend that it has already ruined more than 500 mountains while dumping tons of toxic waste into streams and valleys, and that its blasts are driving nearby residents (those who can afford to move) from their homes. A U.S. Environmental Protection Agency (EPA) study estimated that by 2012, mountaintop removal projects in Appalachia will have destroyed or seriously damaged an area larger than Delaware and buried more than 1,000 miles of mountain streams.[31]

Without foliage and natural layers of soil, the land is rendered unable to retain water. As a result, floods have increased and their waters carry highly toxic debris. For instance West Virginia resident Maria Gunnoe's home sits directly below a 10-story valley fill that contains two toxic ponds of coal mine waste. Before mining began, Gunnoe's property was not prone to flooding, but since the mine became operational, her property has flooded seven times, covering her land with toxic coal sludge. In 2007, Gunnoe and her colleagues at the Ohio Valley Environmental Coalition (OVEC) won a federal lawsuit against the U.S. Army Corps of Engineers that repealed mountaintop removal valley fill permits in southern West Virginia granted without adequate environmental consideration, and banned the issuance of new permits. But the Corps defied the federal judge's orders and granted permits to construct two new valley fills above Gunnoe's community.[32]

Today, the battle over mountaintop removal continues. During his campaign U.S. President Barack Obama expressed concern about mountaintop removal projects. However, in late May 2009, the U.S.EPA stated that it would not block 42 of 48 mine projects under review, including some of the most controversial mountaintop mines.[33] Obama has also been a proponent of so-called 'clean' coal technology, which captures the carbon released by coal-fired power plants. Importantly, however, this technology does not address the immediate dangers of the mining process itself. In the meantime, thanks in part to the publicity surrounding clean coal as a viable climate change solution, the coal industry remains strong. In fact, international coal lobbyists are currently working to establish clean coal projects as certified carbon reduction programs.

For the Mount Elgon community, the ramifications of carbon offsetting are clear – the offset forest intensified existing land disputes and accelerated displacement, violence and impoverishment among local villagers. Then, if we then follow some of the offset credits generated by the project to their buyers, we find Dutch energy companies whose energy portfolios include coal-fired power plants. Pursuing the path of the credits even further, we arrive at the Appalachian region of the U.S., which supplies the Netherlands with most of its coal, at great cost to local communities. In sum, this example demonstrates how the trail of carbon offsets–in this case, from Uganda to Appalachia – is lined with threats to human rights to health, safety and well being.

Rural Sri Lanka/ The Pacific Northwestern United States

Another early offset project similarly exemplifies both the direct and indirect ways in which offsets can violate the rights of everyday people. In the late 1990s, the U.S. state of Oregon instituted groundbreaking laws curtailing carbon emissions. Around that time, the city of Klamath Falls proposed building a 500-megawatt natural gas fired power station. But to comply with the new state laws, the city would need to find a way to offset the greenhouse gases generated by the plant. Eventually, local officials decided to partner with PacifiCorp Power Marketing, Inc., (PPM), a non-regulated affiliate of energy giant, PacifiCorp. PPM promised to spend $3.1 million on off-site carbon mitigation projects. $500,000 of that went into a revolving loan program to equip remote households in India, Sri Lanka and China with photovoltaic systems. Based on solar power, the new systems would replace the carbon-emitting kerosene lamps commonly used by households that are 'off the grid.' Project developers estimated that over thirty years, the solar systems would prevent the release of 1.34 million tons of CO_2. The power plant would then be allowed to emit that same amount of CO_2.[34]

To implement the project, PPM contracted SELCO, a US-based solar electric company. SELCO piloted the project in Sri Lanka, targeting tea plantations. At the time, 90% of the tea workers were without grid-based electricity and 66% were considered illiterate. Ostensibly, workers earned average wages of \$1.58/day; however for most, their wages went to repay debts in a system left-over from British colonial rule. The system of indenture meant that SELCO had to cooperate with the plantation owners in order to launch the project. Those owners, who were struggling to maintain profits under a newly privatized system, offered solar loans exclusively to tea plantation workers, and then further indebted them.[35]

In addition to exacerbating poor economic conditions, the program also exacerbated social tensions. For instance, in one test case, solar loans were first offered to people living in a particular area, most of who were Tamil. This excluded the Sinhalese, a neighboring ethnic group which already had tense relations with Tamils. Furthermore, the solar panel program angered local politicians, who had historically used promises of getting villagers 'on the grid' as a means to win elections. Some began issuing threats to discourage villagers from entering into the loan program. Other villagers feared that, if too many people bought the solar systems, the village would never get on the grid, which in turn would prevent small business and economic development opportunities. In fact, shortly after some families acquired the solar panels, those without them started throwing stones at those who had them. Yet, these problems went largely unpublicized – rather, reports about the project upheld it as a major success story.[36]

Meanwhile back in Oregon, the City of Klamath Falls and PPM completed the cogeneration plant in 2001. The plant came online to much acclaim, winning awards and notoriety for being 'the cleanest fossil-fueled power plant ever constructed in the U.S. in terms of greenhouse gas emissions,' according to the website for PPM's current owner, Spanish energy company, Iberdrola Renewables. Certainly, the plant *did* pioneer advancements in energy efficiency by co-generating electricity and steam, and by relying heavily on wastewater. Overall, the plant is estimated to attain a net energy efficiency of 54% – 20% higher than conventional coal-fired electric generation plants. But beyond those savings, one-quarter of the plant's much-touted emissions reductions come from offsets.[37] In fact, the project's innovative use of offsets is as much hailed as its new technology. As part of this success, Klamath Falls granted plant owners permission to expand. Shortly after it came online, PPM built a secondary, peaking plant and this spring, the city approved a permit to increase the plant's generating capacity.[38]

Although the Klamath Falls cogeneration plant, itself has not directly garnered opposition, natural gas-fired power plants present problems on several levels. First, they emit nitrous oxide, particulate matter and other pollutants, as well as noise pollution. Second, they bring a risk of pipeline explosions. Third, because natural gas is a fossil fuel, these plants contribute to climate change. In this case, the city of Klamath Falls spent millions of dollars (including a $9.4 million upgrade to its waste water treatment plant[39]) to build a fossil-fuel based (albeit efficient) energy source rather than devoting similar resources to the development of renewable energy alternatives.

Ultimately, the plant's copious awards and positive publicity bolstered the value of several companies, illustrating the tangled webs that international power companies weave. In 2000, the British company, ScottishPower acquired PacifiCorp, including PPM. In 2006, Warren Buffet's MidAmerican Energy Holdings Company bought PacifiCorp for $5.1 billion in cash, minus PPM, which remained under the ownership of ScottishPower.[40] That same year, Iberdola announced that it was buying ScottishPower for $22.5 billion, creating one of Europe's largest utilities.[41] All of these companies now lay claim to Klamath Falls' landmark, 'green' cogeneration project, made possible in part by the Sri Lankan solar loan program. In turn, the positive publicity awarded to these companies helps gloss over the fact that each has its own mixed environmental and social records.

For example, although according to its website, PacifiCorp's motto includes 'Responsible environmental management', it has long been embroiled in controversy over its hydroelectric dams. While the company claims that the dams are sources of renewable energy, environmentalists and Native tribes maintain that they have led to the loss of natural fish habitats that offer subsistence to Native tribes and have diminished water quality in the Klamath basin. Moreover, environmentalists claim that these dam reservoirs create high concentrations of toxic algae and (ironically) release 104 million metric tons of methane annually, the single largest source of human-related methane emissions on the planet. Scientists have made preliminary estimates based on water quality conditions on the Klamath, and found that up to 1/8 of the carbon emissions displaced by the Klamath Hydro Project are regained from methane emissions.[42]

PPM's current owner, ScottishPower is one of the world's five largest power companies and has also recently been criticized by environmental groups. ScottishPower owns Longannet, the second largest coal-fired power station in the UK, as well as the Cockenzie power station. Like the Grangemouth plant mentioned at the beginning of this article, both power stations sit on the Firth of Forth. Both were also set to close in 2015 due to their noxious emissions. However, ScottishPower recently made significant upgrades to Longannet, extending its life beyond 2015.[43] It has proposed another major upgrade for Cockenzie, which would transform it into a natural gas-fired power station. However, environmentalists contend that such upgrades would not only prolong a reliance on fossil fuels, but also that such expansions would be unnecessary if more resources were devoted to developing renewable sources such as wind, wave and tidal power. [44] On a more immediate level, ScottishPower came under fire earlier this year for having raised its consumer fees along with increases in energy costs and then reluctantly lowered them again even though energy costs had reduced.[45]

In this case, tracking offsets from rural Sri Lanka to the Pacific Northwestern U.S. illuminates both the short and long-term implications of carbon offsetting for human lives. In the short term, the solar power project in Sri Lanka intensified social and economic tensions for the local villagers it was meant to serve. Offsets also provided a way for Klamath Falls to adhere to state greenhouse gas emission limits *and* build a fossil fuel-based power plant that emits carbon as well as other pollutants. In the longer term, offsets allowed project developers to claim that the plant was 'carbon neutral'. In turn, this positive publicity had a doubly negative effect. First, it boosted the value of the companies, increasing their power and influence over workers, local politicians and local communities. Second, upholding offsets as being able to 'zero out' carbon emissions circumvented investment in more sustainable energy resources. In sum, this case exemplifies the labyrinthine trail that carbon offsets wend – from corporate beneficiary to corporate benefit – paths covered with a 'greenwash' that hides their real environmental and human costs.

Grangemouth, Scotland/Sao do Buriti, Brazil

This essay's last example returns to the case with which it began. This case clearly demonstrates the direct, negative impacts of offset projects on local communities, as well as the ways in which the benefits that offsets offer to industries indirectly lead to human rights abuses. Yet this case also emphasizes that, while offsets have the potential to bring harm to local communities, they also offer an opportunity for new kinds of transnational alliances that can challenge world leaders to find more sustainable solutions to climate change.

To review the case, in 2002 Plantar (see also Chapter 9 in this volume), an iron foundry company with operations in Brazil threatened to switch from burning charcoal to coal in order to increase its capacity. The switch would also significantly increase the foundry's greenhouse emissions, so the World Bank offered to help Plantar expand the eucalyptus plantations that provide its charcoal. The Bank financed the Brazilian project through its Carbon Fund, into which British Petroleum (BP) had invested substantial amounts of money to offset emissions from some of its operations such as the Grangemouth oil refinery on Scotland's east coast. Thus, after receiving carbon credits from the project, BP was able to continue to operate Grangemouth without significantly reducing the plant's emissions.

However, a 2008 report by the Sustainable Energy and Economy Network's co-director, Janet Redman, states that the eucalyptus trees' enormous roots almost immediately began to soak up vast amounts of water in and around their environs. Villagers now had to travel increasingly far to find water, as well as traditional subsistence and medicinal plants. In addition, the tree plantation relied on herbicides and pesticides, which local farmers claim killed crops and poisoned streams. Furthermore, the water shortage destroyed some small businesses that had been in families for generations. Finally, Redman writes, 'Perhaps more seriously, groups allege that Plantar pressured local residents to sign letters of support for the project or forfeit employment at the plantations.'[46] Those residents who did publicly oppose Plantar claim that they and their family members were threatened and/or hired to work at the plantation.[47]

Meanwhile, Grangemouth, which is one of Europe's largest oil refineries, emits sulfur dioxide, nitrogen dioxide and small particulate matter into the air. In addition, officials at the Scottish Environment Protection Agency (SEPA) have cited the refinery as 'one confirmed source' of an oil slick covering several square miles of the Firth of Forth.[48] Grangemouth residents have long complained about high rates of asthma, as well as the smells and noise coming from the plant.

The refinery has a similarly noxious track record on a social level. In late April 2008, the Unite union (Grangemouth's workers' union) became embroiled in a dispute with the refinery's current owner, INEOS, over pension policies. The union accused the company of buying assets and then cutting costs by introducing new working practices, lowering wages, and terminating pension schemes.[49]

INEOS has also come under fire for its involvement in another carbon offset project with its own set of human rights violations. Briefly, in 2005, INEOS partnered with GFL (see also Chapter 12 in this volume), which produces HCFC 22, a refrigerant gas for air conditioning units and refrigerators. GFL wanted to institute a program to capture and recycle HFC 23, a potent greenhouse gas that is a byproduct of producing HCFC 22. INEOS supplied the technology for the program, and both companies received the right to claim the carbon offsets. However, residents of Gujarat, where the GFL factory is located, claim that the factory has made them sick with joint aches, bone pains, unexplained swellings, throat and nerve problems and temporary paralysis. A recent investigation by the UK's *Daily Mail* found 'dangerously high levels of fluoride and chloride – fluoride in the water was more than twice the international acceptable limit. All the water fell well below any safe drinking standards and the soil had worryingly high levels of these chemicals.'[50] But these chemicals do not contribute to global warming. Thus, those monitoring the program considered it successful in so far as it reduced greenhouse gas emissions.

Thanks in part to this myopia, the CDM Executive Board approved the Gujarat project in 2005 and awarded INEOS and GFL an undisclosed number of Certified Emission Reduction units (CERs) over time (INEOS' website predicts that together with a second, similar project in Korea, the Gujarat project will generate 3 million tons of CERs annually). Both companies were then free to sell those credits to industries falling short of their national emissions caps. For instance, in 2006, GFL made news for doubling its sales revenue by selling a record number of carbon credits to Noble Carbon Credits group of Singapore, Rabobank Nederlands and Sumitomo Corporation,[51] most of which resold the credits mainly to large industries needing to comply with Kyoto Protocols, including power and oil companies.[52] According to some reports, some of the proceeds from GFL's sales went to build a Teflon and caustic soda manufacturing facility which uses processes known to be massively polluting.[53]

This example thus illustrates another way in which carbon credit schemes violate human rights – by creating dangerously perverse incentives for polluters to continue to pollute. First, carbon offset schemes reward corporations for lowering greenhouse gas emissions while allowing other highly, and deadly, toxic emissions. Second, GFL's handsome profits from capturing HFC-23 have inspired other HCFC-22 manufacturers to follow suit, drastically lowering its cost. Experts predict that soon, a global over-reliance on the chemical, itself a powerful greenhouse gas, will result.[54]

This case thus well demonstrates the ramifications of carbon offset projects for human rights across the globe. Most directly, offsets allowed the continued pollution of the Grangemouth community, and they introduced new hardships for people in Sao Jose do Buriti. More indirectly, the notion that Grangemouth's emissions were being neutralized made it an attractive asset that increased the profitability of its various owners, enabling them to invest in other toxic projects. In the case of INEOS, I propose that purchasing Grangemouth made it an even more powerful player in the petrochemicals industry which in turn made it better able to fight off opposition from workers or local communities, or perhaps to lobby for the certification of new kinds of CERs. As well, the acquisition may have bolstered the company's ability to continue financing its investments in other offset projects such as the HFC-23 program.

At the same time, this case also demonstrates how awareness about carbon offset projects' trails of tears can connect communities in very tangible ways and catalyze collective action. For instance, in 2003 activists opposed to the Sao Jose do Buriti project attracted the attention of global NGOs, which helped local activists disseminate information about their situation. Eventually, Carbon Trade Watch (a project of the Transnational Institute) initiated a project to connect residents of Sao Jose do Buriti and Grangemouth through the exchange of video diaries. As the resulting documentary film depicts, residents of both communities reacted powerfully to a new awareness of their connected plights and spoke of newfound determination to continue their local struggles. In Scotland, the video diaries inspired one participant first to become an activist with Friends of the Earth and then to run for local office.

Thus, while offsets link communities around the globe in extended chains of emiseration, they also bring new opportunities for transnational alliances and partnerships to challenge market-based solutions to climate change. Fostering such opportunities, though, requires a concerted and well publicized stripping of the green veneer in which offsets are currently washed. Only then can we reveal the ecological and social tarnish hidden beneath and implement alternative solutions.

Conclusions

Certainly, not all offset projects violate human rights in as direct a manner as some of the cases presented here. At the same time, some carefully documented accounts of the ways in which offset projects fail to benefit local communities *are* beginning to emerge. As they do, we find that offset projects have great potential to do more harm than good. By taking such reports a step further and tracking the offsets to the industries they advantage, this essay reveals multiple consequences for human rights and human lives. Yet at the same time, offsets create new opportunities for transnational alliances and partnerships to challenge those consequences. In order to create such opportunities, though, we must further strip the veneer from the greenwash that accompanies offsets and reveal the ecological and social tarnish hidden beneath. As this essay also makes clear, the need for widespread opposition to the human rights violations wrought by offsets is urgent. The examples presented here likely represent the tip of an impending iceberg – carbon markets now trade over US $1 billion annually, and the climate bill currently under debate in the US Congress could send those numbers skyrocketing.[55] This December, world leaders will gather in Copenhagen to revisit and renew elements of the Kyoto Protocol. Under consideration is the certification of new ways to generate carbon offset credits, including biofuels and forest conservation. Although these programs appear environmentally and socially sound, their efficacy at reducing climate change is questionable, especially as they are intended only to zero-out industrial emissions. More immediately, like the programs described here, these initiatives commodify natural resources and thus have the potential to disempower local communities caught in the middle of land-grabs, or to exacerbate inter-group tensions as communities compete for offset dollars.[56] Today we have a narrow but important window of opportunity to redirect the course of climate change mitigation. Myriad non-market based solutions to climate change exist which promote, rather than violate, human rights. Given the human stakes of the system currently in place, we have no choice but to establish more humane and effective alternatives.

Notes

* I would like to extend my thanks to Barbara Rose Johnston, who initially helped me clarify the premise of this article. I am also grateful to Steffen Böhm, Siddhartha Dabhi and Gregory Button for their generous and insightful comments on earlier drafts. A shorter version of this article appeared in the September 9, 2009 issue of /Counterpunch/ and can be found at http://www.counterpunch.org/ checker09092009.html.

1 For more information from 'The Carbon Connection', a documentary produced by *Carbon Trade Watch*. See http://www.carbontradewatch.org/carbonconnection.

2 For the purposes of this article, I define carbon offsets as emissions reductions that are claimed to result in less greenhouse gases accumulating in the atmosphere, such as tree plantations (which are supposed to absorb carbon dioxide emissions) or fuel switches, wind farms and hydroelectric dams (which are argued to reduce or displace fossil energy); Lohmann, L. (2009) 'Uncertainty Markets and Carbon Markets: Variations on Polanyian Themes', *New Political Economy:* p. 12.

3   Lohmann, L. (2008) 'Carbon Trading, Climate Justice and the Production of Ignorance: Ten Examples', *Development*, 51: 359–365.

4   As mentioned, these critiques are plentiful. For some particularly comprehensive examples, see Lohmann, L. (2006) *Carbon Trading: A Critical Conversation on Climate Change, Privatization and Power.* Uppsala: Dag Hammarskjold Foundation; Smith, K. (2007) *The Carbon Neutral Myth: Offset Indulgences for Your Climate Sins.* Amsterdam: Transnational Institute; Fahrenthold, D. (2008) 'There's a Gold Mine in Environmental Guilt', *The Washington Post*, October 6; Lohmann, L. (2008) 'Carbon Trading, Climate Justice and the Production of Ignorance: Ten Examples'; Lohmann, L. (2009) 'Uncertainty Markets and Carbon Markets: Variations on Polanyian Themes'; Faris, S. (2007) 'The Other Side of Carbon Trading', *Fortune Magazine*, August 30. See also white papers from organizations such as Carbon Trade Watch, the Corner House and the World Rainforest Movement.

5   A growing body of literature documents the degree to which climate change disproportionately affects poor and marginalized communities. For some recent examples, see Crate, S. and M. Nuttall (eds) (2008) *Anthropology & Climate Change: From Encounters to Actions.* Walnut Creek, CA: Left Coast Press; Leary, N., C. Conde, J. Kulkarni, A. Nyong and J. Pulhin (eds) (2008) *Climate Change and Vulnerability.* London: Earthscan; Timmons Roberts, J. and B. Parks (2006) *A Climate of Injustice: Global Inequality, North-South Politics, and Climate Policy.* Cambridge, MA: MIT Press.

6   Bullock, S., M. Childs, and T. Picken (2009) 'A Dangerous Distraction', *Friends of the Earth*, http://www.foe.co.uk/resource/briefing_notes/dangerous_distraction.pdf.

7   http://carbon.newenergyfinance.com/.

8   Fahrenthold, D. (2008) 'There's a Gold Mine in Environmental Guilt'.

9   Ball, J. (2008) 'Green Goal of 'Carbon Neutrality' Hits Limit', *Wall Street Journal*, December 30.   http://online.wsj.com/article/SB123059880241541259.html?mod= googlenews_wsj.

10 Kanter, J. (2009) 'Do Carbon Offsets Cause Emissions to Rise?', http://greeninc.blogs.nytimes.com/2009/05/08/do-carbon-offsets-cause-emissions-to-rise/.

11 US GAO. (2008) 'Carbon Offsets: The U.S. Voluntary Market Is Growing, but Quality Assurance Poses Challenges for Market Participants', *GAO*, August 28, http://www.gao.gov/products/GAO-08-1048.

12 Schneider, L. (2007) 'Is the CDM fulfilling its environmental and sustainable development objectives? An evaluation of the CDM and options for improvement', *Institute for Applied Ecology*, Berlin: November 5, http://www.oeko.de/ oekodoc/622/2007-162-en.pdf.

13 Fortson, D., and W. Georgia (2009) 'Carbon Market Hit as UN Suspends Clean Energy Auditor', *The Sunday Times*.

14 Stibbe, W., J. Van der Kooij., J. Verweij., and P. Costa (1995) 'Response to Global Warming: Strategies Of The Dutch Electricity Generating Board', *Report to the 16th Congress of the World Energy Council*, http://www.ecosecurities.com/Assets/3170/Pubs_Response%20to%20global %20warming%20Strategies%20of%20the%20Dutch%20Electricity%20Generating%20Board. pdf.

15 This quotation is taken from the Face Foundation's website, www.stitchingface.org.

16 Whether and to what degree the carbon absorbed by trees is equivalent to carbon emitted by the burning of fossil fuels is a subject of great debate. See for instance recent editorials by journalist Fred Pearce (http://www.guardian.co.uk/environment/2008/dec/17 /carbonoffsetprojectscarbonemissions) and Stanford University climate scientist Ken Caldeira (www.nytimes.com/2007/01/16/opinion/ 16caldeira.htm).

17 House of Commons, Environmental Audit Committee. 'The Voluntary Carbon Offset Market', *Sixth Report of Session 2006-07*. London: TSO. p.30.

18 Himmelfarb, D. (2006) 'Moving People, Moving Boundaries: The Socio-Economic Effects of Protectionist Conservation, Involuntary Resettlement and Tenure Insecurity on the Edge of Mt. Elgon National Park, Uganda', *Agroforestry in Landscape Mosaics Working Paper Series.* World Agroforestry Centre, Tropical Resources Institute of Yale University, and The University of

Georgia, 2006, page 16. Available at: http://www.yale.edu/tri/wkppragrofor.html; According to Himmelfarb, 'the exact number of people left homeless varies depending on the source'. In October 2003, *New Vision* reported that 561 families were left without land as a result of the re-drawn boundary at the Benet Resettlement Area Businge, G. (2003). 'The Benet To Sue Gov't Over Land', *New Vision*, 2 October. http://www.newvision.co.ug/D/8/ 26/309690.

19 Great Britain House of Commons Environmental Audit Committee 2007:30.

20 Lang, C. (2006) 'Uganda: Notes from a Visit to Mount Elgon', *World Rainforest Movement bulletin N° 115*, February, http://www.wrm.org.uy/bulletin/115/ Uganda.html.

21 Lang, C. (2006) 'Uganda: Notes from a Visit to Mount Elgon'.

22 Lang, C., and T. Byakola (2006) '"A funny place to store carbon", UWA-FACE Foundation's tree planting project in Mount Elgon National Park, Uganda', *World Rainforest Movement*, December.

22 Lohmann, L. (2008) 'Carbon Trading, Climate Justice and the Production of Ignorance: Ten Examples'.

23 Lohmann, L. (2008) 'Carbon Trading, Climate Justice and the Production of Ignorance: Ten Examples'.

24 Faris, S. (2007) 'The Other Side of Carbon Trading'.

25 http://somo.nl/news-en/greenpeace-awards-red-and-green-thermometers-to-e-on-and-eneco/.

26 http://corporateuk.eneco.nl/outlook_and_strategy/Business_Operations/Pages/ Default.aspx.

27 Wilde-Ramsing, J., T. Steinweg and M. Kokke (2008) *Sustainability in the Dutch Power Sector*. Amsterdam: Center for Research on Multinational Corporations.

28 These numbers are disputed. See Goodell, J. (2007) *Big Coal: The Dirty Secret Behind America's Energy Future*. Mariner Books for justification of the former and http://www.worldcoal.org/coal-the-environment/climate-change/index.php for justification of the latter.

29 U.S. EPA (2005).

30 U.S. Energy Information Administration (2009) 'U.S. Coal Supply and Demand 2008 Review', http://www.eia.doe.gov/cneaf/coal/page/special/exports_imports.html. The Netherlands also buys 11% of the world's offsets, making it the third largest buyer (tied with Japan). See 'A Dangerous Distraction'.

31 US EPA Region 3, 2005.

32 For more details see http://www.goldmanprize.org/2009/northamerica.

33 Controversial coal mining method gets Obama's OK: Environmentalists decry 'mountaintop removal', Tom Hamburger and Peter Wallsten, Chicago Tribune June 1, 2009, www.chicagotribune.com/news/nationworld/chi-tc-nw-mountaintop-mining 053jun01,0,3998035.story.

34 Lohmann, L. (2006) *Carbon Trading: A Critical Conversation on Climate Change, Privatization and Power*, pp. 43-48.

35 Lohmann, L. (2006) *Carbon Trading: A Critical Conversation on Climate Change, Privatization and Power*.

36 For example, in 2002, SELCO won an Award for Corporate Excellence from the U.S. Dept. of State. See also SELCO founder, Neville Williams' biography, Williams, N. (2005) *Chasing the Sun: Solar Adventures around the World*. New Society Publishers.

37 See http://www.iberdrolarenewables.us/klamath.html.

38 As reported in Business Wire, February 23, 2009. http://www.allbusiness.com/energy-utilities/utilities-industry-electric-power-power/11791359-1.html.

39 American City and County (2002) 'Klamath Falls, Ore' 1 December, http://americancityandcounty.com/mag/government_klamath_falls_ore/.

40 Richardson, K., and R. Smith (2005) 'Berkshire Unit to acquire PacificCorp for $5.1bn', *Wall Street Journal*, 24 May.

41 Mclean, R. and T. Heather (2006) '2nd-Biggest Spanish Power Concern to Buy ScottishPower', *New York Times*, 29November.

42 http://www.bepress.com/cgi/viewcontent.cgi?context=ijeeps&article=1116& date=&mt=MTI0NzA4MDI5MQ.

43 Milner, M. (2009) 'A Giant Chemistry Set on the Firth of Forth', *The Guardian*, 1 January, http://www.guardian.co.uk/business/2009/jan/01/ukcoal-alternativeenergy.

44 Edwards, R. (2009) 'Why what happens to this Scottish power station could make life even harder for this farmer in Malawi', *Sunday Herald*, 20 June. http://www.sundayherald.com/misc/print.php?artid=2515600.

45 Williamson, M. (2009) 'ScottishPower sees revenue soar', *The Herald*, 30 June, http://www.theherald.co.uk/business/news/display.var.2517233.0.scottishpower_sees_revenue_soar.php.

46 Redman, J. (2008) 'World Bank: Climate Profiteer', *Institute for Policy Studies*. Washington, DC, p.24. www.ips-dc.org/getfile.php?id=181.

47 Again more information on the documentary 'The Carbon Connection', produced by *Carbon Trade Watch*, http://www.carbontradewatch.org/carbonconnection.

48 Johnston, I. (2007) 'A Gift from Scotland to Brazil: Drought and Despair', *The Scotsman*, 7 July, http://thescotsman.scotsman.com/world/A-gift-from-Scotland-to.3302061.jp.

49 The Economist (2008) 'Running on Empty', 387: 46.

50 Ghouri, N. (2009) 'The great carbon credit con: Why are we paying the Third World to poison its environment?', *Mail online*, 1 June, http://www.dailymail.co.uk/home/moslive/article-1188937/The-great-carbon-credit-eco-companies-causing-pollution.html.

51 Kalesh, B. (2006) 'Gujarat Fluro sets record on carbon credit sale', *The Times of India*, 10 August, http://www.articlearchives.com/law-legal-system/environmental-law-air-quality-regulation/1765101-1.html.

52 The United Nations Development Programme 'The Clean Development Mechanism: A Users Guide', http://www.undp.org/energy/docs/cdmchapter7.pdf.

53 Ghouri, N. (2009) 'The great carbon credit con: Why are we paying the Third World to poison its environment?'.

54 Schwank, O. (2004) 'Concerns about CDM Projects Based on decomposition of HFC-23 Emissions From 22HCFC Production Sites', October, http://cdm.unfccc.int/public_inputs/inputam0001/Comment_AM0001_Schwank_081004.pdf.

55 Lohmann, L. (2009) 'Uncertainty Markets and Carbon Markets: Variations on Polanyian Themes'.

56 Lohmann, L. (2009) 'Uncertainty Markets and Carbon Markets: Variations on Polanyian Themes'.

4

How Sustainable are Small-Scale Biomass Factories? A Case Study from Thailand*

Tamra Gilbertson

Tell me which industry you can call clean, I have never seen one. (Sunthorn Yensook, Nam Song resident)

Introduction

Carbon offsets are not reductions. For each project that is developed in the South, an equivalent level of pollution from fossil-fuel power stations or heavy industry is permitted to continue in the global North. In addition, the system typically funds the expansion and building of new industrial and power projects which are insensitive to the needs of local communities. The implications of such projects on health, land use and water resources are rarely addressed. In this respect offset projects reinforce an unsustainable development paradigm.

The majority of CDM projects are predominantly run by large, highly-capitalized firms or agencies, since they are the companies best placed to hire expensive carbon consultants and accountants, liaise with officials and pay the fees needed for UN registration. The result is a system that subsidizes some of the most polluting companies in the world.

The CDM is presented as a system that helps the spread of renewable energy. However, the definition of 'renewable' projects does not automatically indicate environmentally sustainable or socially just.

This photo essay is a story of two Thai communities fighting for their livelihoods. It aims to highlight the experiences of two communities that fought for rights to their lands and health and to demonstrate that even small-scale biomass energy projects – which are often seen as among the 'better' offset projects – can be detrimental to the lives and livelihoods of local residents.

A.T. Biopower and the CDM[1]

There are currently 24 registered CDM projects in Thailand and close to 100 projects in various stages of the pipeline. In 2007, the Thai government established the Thailand Greenhouse Gas Organization in order to fast-track CDM projects after investors complained that the Office of Environmental Policy and Planning (the original Designated National Authority (DNA) was too slow, and could thus jeopardize Thailand's opportunity to sell profitable CDM credits.

In 2001, A.T. Biopower put forward a plan to build five rice husk-burning biomass power stations with the objective to bundle them and acquire CDM financing. The A.T. Biopower project was the first CDM project registered in Thailand, and among the first five for which baseline methodologies were approved by the CDM Executive Board.[2] The first power station was built in Pichit near the fertile banks of the Nan River in north-central Thailand. The Pichit station is a 22 megawatt capacity thermal power plant located next to the community of Sa Luang in Hor Krai subdistrict in the province of Pichit, about 200 kilometres north of Bangkok. The plant is located 1 kilometre from the Nan River and has a daily fuel requirement of 500 metric tons and a daily water requirement of approximately 2,200 cubic metres. The station burns in its entirety rice husks. The power station is surrounded by a 12 metre high fence comprised of newly planted eucalyptus and pine trees.[3]

The credits generated by the project are bought by Japan Mitsubishi UFJ Securities, a financial services group, and Chubu Electric, a Japanese power company which is registered in The Netherlands to minimize its corporate tax obligations. Chubu interestingly owns a 34 per cent stake in A.T. Biopower, allowing the company to use the Thai project to avoid its domestic emissions reduction obligations in Japan.

What 'Waste'?

Biomass is often touted as a renewable resource which provides benefits for local communities and reduces the demand on fossil fuels. Defining what a waste product is can be very complicated for local communities because often resources and 'waste' are used and reused in a continual cycle with benefits to the environment. Far too often the waste in question already has a purpose within a local economy.

The staple crop grown in the region is rice which depending on the season and rains will produce two or three crops per year in the fertile soils. The rice is then brought to a de-husking facility to separate the kernels from the husk. The kernels are sold on to vendors or stored by the community. The husks have been used for centuries to absorb animal droppings, mostly from chickens. The resultant product is used as an agricultural fertilizer as well as for brick manufacturing. The rice husk and manure mixture is a natural fertilizer that releases minerals into the soil and builds soil content. Rice husks therefore play a vital role in agriculture.

Local farmers in the region commented that they will have to replace this natural fertilizer with chemical fertilizers now because demand from the power plant has driven up the price of rice husks, meaning they are no longer affordable.[4] Local chicken farms and brick factories have to go further away to source rice husks, destroying a once self-sufficient system.

A truck piled with bags of rice husks on its way to the Pichit power station.

The A.T. Biopower project claims to be replacing power generation which would otherwise require oil, coal and natural gas. It also claims that the resulting ash by-product will be used for cement production, further reducing the environmental impact. No mention is made of existing uses for rice husks, which are presented merely as waste products. This fiction is elaborated on by the project validator, Det Norske Veritas (DNV), which claims that uncontrolled burning or dumping of rice husk, without utilizing it for energy purposes, is the predominant current practice.[5] No supporting evidence is offered to back this up, and the wording is simply copied from a standardized text that DNV applies to all such projects.[6]

By assuming that the burning of rice husks is climate neutral, talking up the 'sustainability' of the project and talking down the local environmental impacts, the project developers are able to maximize the number of Certified Emissions Reductions (CER, the carbon credits issued as part of the CDM) issued to A.T. Biopower. Over a period of seven years, it is projected that 495,405 CER will be issued. When sold on the market, these might plausibly fetch between $10 and $30 each, with each CER claimed to represent a metric ton of carbon emissions.

Heath Risks

The residents near the Pichit power station have complained about respiratory problems and aggravation felt in their skin and lungs. 'While my harvest has nearly returned to normal, health problems from the dust have persisted. Residents, especially children, have developed skin rashes and breathing difficulties, which is why we've closed up our windows and doors', a local resident explains.[7]

Silica (SiO_2) is the main mineral component of RHA (85-90 per cent). It carries serious health risks, particularly to the respiratory system.[8] Silicosis is an irreversible lung disease which is normally found in workers at mining operations or rock quarries, but it can also be caused by inhaling RHA.[9] A few years ago certain villages in Northern Thailand were dubbed 'villages of widows' because of the large number of pestle-and-mortar-making workers who died from silicosis. China reports 24,000 deaths per year due to silicosis.[10] Villagers living next to the Pichit plant keep their doors and windows closed or the ash piles up on everything. These health risks have not been addressed by the company and like most villagers living next to a factory or power station, they fear complaining.

The villagers complained of noise pollution when the power station was being built. The engine was so loud in the first month of operation that residents living opposite the power station complained of having to shout in order to be heard by each other. Instead of slowing operations or modifying the engine, the company responded by offering the villagers ear plugs. Each time the villagers have complained about the station, the standard response has been to offer them gifts to stay quiet. (Local residents asked to remain anonymous.)

Resistance to the A.T. Biopower Station

In 2001, A.T. Biopower was still in planning stages and the company was considering sites in which to build its five factories. One of the sites being considered was located in Nam Song, a river-dependent community in the Phayuha Khiri district, Nakhon Sawan province of Thailand. This community is located about 50 kilometres from the now functioning Pichit plant. It is located on the fertile flood plain of the Chao Phraya River, just downstream from where two tributaries merge at Nakhon Sawah (Heavenly City).

After six years of struggle against the site proposal the Nam Song community successfully deterred the developers from building on their lands.

Across the road from the Pichit plant the company has started dumping the biomass waste next to homes of local residents. According to a local resident near the Pichit plant, they were offered 'as much ash as they wanted for free because the company does not want it.'

Suraphol Pan-ngam shows his neighbor's aquaculture operation located on the banks of the Chao Phraya River. The fish are raised for community consumption and for sale in the local market.

The main source of livelihoods is agriculture, which has been developed in a way that turns the seasonal floods into a resource. When the water subsides in the dry season, the fertile banks are planted with cabbage, broccoli and other seasonal vegetables. When the water is high in the rainy season, it is used to flood rice paddies and aquaculture facilities are assembled on the river's edge.

Nam Song residents were immediately concerned about the impacts a new power plant could have and visited a community impacted by a similar rice-husk burning station in Wat Sing district, Chainat province, about 40 kilometres southwest of Nakhon Sawan, owned by another company. One community leader reasoned that 'The developers only told us positive sides about the power station and we are uneducated so we needed to find out about the negative sides too.' Residents of Nam Song then travelled to Wat Sing, where the local community was living with the affects of a biomass power station. After talking to the residents in Wat Sing the residents and understanding the levels of pollution they live with the residents from Nam Song were committed to form their own opposition.

The community forests are also an important resource, providing food, building materials, high ground for livestock, traditional practices and medicines.

Moving Forward

After months of information gathering, the Nam Song residents experienced a major setback when the local tambon (subdistrict) government illegally agreed to install the power station in Nam Song. The Thai government requires developers to have a public hearing process with residents before proceeding. At the public meeting the local government officials and the company consultants met with the community and asked them to sign their names on a piece of paper labelled 'consultant meeting'. The consultants and local government officials added names of villagers who were not in attendance. The company showed the list of names to the local authority, stating that 88% of the 528 villagers who attended the meeting agreed to the power plant being built. In the meantime, A.T. Biopower placed a deposit on the plot of land they planned to develop in Nam Song.

This incident provoked the villagers to send a grievance letter to the local government. Initially, they were divided over whether the power station should be built, which caused strife in daily life as well as between family members. Eventually, they resolved to end their divisions. The entire community of Nam Song agreed to sign the letter stating their objections to the meeting and to the proposed station. The villagers then created the Nam Song Conservation Club to co-ordinate a full-scale campaign against the project.

Banner reads, 'Stop the Electricity Plant A.T.B'.

Reaching Out

The Nam Song Coservation Club then began gathering research with the aid of other movements and organizations. The villagers sought to show that the rice field was on a flood plain and an inappropriate power plant site, and that building it so close to where they lived constituted a threat to the health of the people and the river. The campaign grew over time including several rallies of over 700 people outside the provincial government headquarters, door-to-door organizing and on-going meetings.

The developers used several tactics that are common in such situations in their attempts to stop the protests. Members of a community in the nearby Pichit province, who were also facing the possibility of a new biomass power station, were sent by the company to bribe the village leaders, offering them compensation to stop protesting. All of the village leaders were told by developers and local government they could be in danger if they continued the campaign. Various threats were made, large bribes were offered, and the villagers were repeatedly lied to in attempts to destroy their unity.

Site of proposed A.T. Biopower plant near Nam Song during the seasonal floods.

Bribery and Coercion

The project developers invested a lot of time and energy in their attempts to persuade the Nam Song community that the project was beneficial, but they were not convinced. 'We do not need factories or development, we live with nature and we like the way things are', stated Jongkol Kerdboonma, a member of the club. Another resident stated 'We knew the power plant was bad because it involved money'. The company promised the community a development fund and a new health fund in an attempt to smooth over relations, but the local leaders remained sceptical. 'Which doctor will tell us that we are sick from the pollution if the doctor is hired by the company?', they asked.[11]

Interestingly, the Nam Song community was never offered any electricity from the power plant, not even at a subsidized rate. Each household pays 300 baht per month to the national grid. The Nam Song Conservation Club states three main reasons for their opposition to the rice husk power plant:

- We have lived self-sufficiently on this river for generations, so why would we want to destroy the land with pollution that would be bad for the people and the environment?

- We already knew they would dump the ash into our river, and that it would pollute the river and the fish.

- Rice husks are not an agricultural waste product to begin with. We use them for the chicken pens, and after they have absorbed the chicken waste we use this as a fertilizer. If the station was built here rice husks would be too expensive to use as a fertilizer, and we would have to switch to other fertilizers.

Essential Roles of Women

The women in the village played an essential role in fundraising, organizing and maintaining trust within the community. The women made handicrafts and sweets to fundraise for the campaign. They sold t-shirts and sweets at meetings, which provided an opportunity to talk with others about the struggle and build trust. They canvassed an area of 10 km² and gathered 4,000 signatures for one of the rallies at the government headquarters.

Women from Nam Song creating handicrafts for fundraising for the campaign.

The success of the women's work was such that they too were targeted and harassed by the project developers. The developers lied to the women, telling them that the men in the village were receiving bribes from the company. They were then further questioned about why they would want to keep supporting the men if they themselves were not receiving money as well. The women's awareness that this tactic was being used in an attempt to derail their organizing confirmed to them the importance of their work for continuing their struggle.

Jongkol Kerdboonma in her garden collecting vegetables for cooking.

Organizing Together

An open and democratic organizing process helped the community maintain its stamina. One resident stated, 'We made all of our decisions together at meetings, which prevented internal conflicts from arising.' The residents acknowledged that there were disagreements and tensions during the difficult phases of the struggle. 'We would scrutinize each other, even watch each other and everyone was very tense'. However, the community continued to organize, reach out for support and demonstrate. They received solidarity and help from other community movements, NGOs and the Assembly of the Poor, the largest grassroots movement in Thailand involving tens of thousands villagers who are affected by unjust policy and development. The Nam Song residents said they 'learned a lot from each others' struggles' and maintained their unity so that no one accepted the bribes or backed down from the threats.

After six years of struggle, and with the help of several outside solidarity organizations, they were able to approach the National Human Rights Commission (NHRC) to request an official investigation. In 2007, the NHRC recommended that the power plant should not be built on the grounds that it was inappropriate to build on the flood plain, and that it would violate human rights by polluting the river and damaging the villagers' livelihoods.

> We can not rely on any laws to protect us, which are no better than a piece of paper, so we had better protect ourselves, stated Soontan Yentosuk.

And while for the moment Nam Song residents are protected from developers they still fear that other power plants will try to build on this particular site. The Nam Song Conservation Club remains committed to protecting their community from development projects because they are aware they may need to build another campaign at any moment.

The Nam Song Conservation Club kept very good records of its campaign, including photos and notes from meetings.

Conclusions

Carbon offset projects follow pre-packaged designs that do not deal with the real complexities and intricacies of communities and livelihoods. In the case of A.T. Biopower, rice husks that were used for agricultural purposes are now burnt, showing a considerable insensitivity to the context in which the project has been developed. In both villages, in fact, the proposed biomass power plants threatened to undermine local struggles and low-carbon livelihoods, and subvert existing practices that have potential to be applied elsewhere as everyday solutions to tackling climate change. This was compounded by bribery and threats on the part of the company, as is so often an accompaniment of infrastructure projects conducted in the name of 'development'. The bribery and coercion perpetuated mistrust and division between the two communities. In the case of Nam Song, a strong and concerted campaign of local organizing was able to resist the advances of the company – benefiting too from solidarity with other local organizations.

In the case of Sa Luang, however, the station was developed – with A.T. Biopower now brushing aside the concerns and health of local residents. CDM project designs are based on the principle of 'additionality', which means that they should prove that they provide a saving in relation to current 'business-as-usual'. As Lambert Schneider of Germany's Öko Institute puts it, 'If you are a good storyteller you get your project approved. If you are not a good storyteller you don't get your project through'.[12]

Members of the Nam Song Environmental Conservation Club

The local population loses considerably. The increased reliance on synthetic fertilizers carries with it increased health risks, as does the production of rice husk ash at the site, which can cause silicosis – a fatal respiratory condition. Despite A.T. Biopower's claim that this ash would be recycled for use in cement production, there is clear evidence at the site that it has been dumped next to the residents' houses. Further health and environmental risks could result from the combustion of the rice husks within the power plant itself, since this process also generates sulphur dioxide, nitrogen oxide, carbon monoxide and other dust particles. There is also a significant risk of local water pollution – along similar lines to those documented in the NHRC recommendation not to build a rice husk biomass power station in Nam Song.

More generally, the legacy of such development projects is that they pit communities against each other, and encourage divides within them too. When encountering local protest, the common response of the company has been to resort to a range of bullying tactics – including threats, lies and bribery. Further, what was deemed as a Human Rights Violation in one village is ignored and not applied to another village 50 kilometres away.

The experience of Nam Song, however, shows that local resistance can be effective when there is a strong basis for unity. An open decision making process, and the central involvement of women in the campaign, were important conditions for this.

Notes

* Field visit conducted in November 2008 by Nantiya Tangwisutijit, Tamra Gilbertson and Ricardo Santos. Additional Research by Nantiya Tangwisutijit and Tamra Gilbertson. Special thanks to Larry Lohmann for contact and editorial support and of course to The Nam Song Conservation Club for their generosity, kindness and inspiration.

1 A.T. Biopower is owned by the following companies: The Netherlands 34%: Chubu Electric Power Company International B.V. ('CEPCOI') registered in the Netherlands, the subsidiary of Chubu Electric Power Company Incorporated (CEPCO), which is the third largest electric company is Japan having generating capacity of 31,735 MW (as of 31 March 2004); Channel Islands 32%: Al Tayyar Energy Ltd. ('ATE') registered in Jersey, Channel Islands, United Kingdom, an Abu Dhabi-based development and investment company that focuses on renewable energy and energy efficiency projects; Finland 27.55%: Private Energy Market Fund L.P. ('PEMF'), registered in Finland, a Helsinki-based million private equity fund that targets opportunities being created by the deregulation and restructuring of the energy sector Worldwide, including renewable energy and energy efficiency projects. Finnish Fund for Industrial Cooperation Ltd. ('Finnfund') , registered in Finland , and 80% Finnish government-owned development finance company the provides long-term risk capital for private projects in developing countries; Malaysia 5%: Flagship Asia Corporation ('FAC'), registered in Labuan Federal Territory , Malaysia , a sustainable infrastructure project development company and the founding sponsor of ATB; England 1.45%: Rolls-Royce Power Ventures headquartered in London, England, has been ranked by a leading industry survey as the top developer in the market between 5 to 150 MW, with experience in developing cost-effective, environmentally aware power solutions.

2 Saner, R. and A. Arquit (2005) 'Exploring the relationship between FDI flows and CDM potential,' *Transnational Corporations*.

3 http://www.atbiopower.co.th/power_plant/power_plant_e.htm.

4 Personal Interview with Nantiya Tangwisutijit and Tamra Gilbertson, 11 November 2008.

5 See 'A.T. Biopower Rice Husk Power Project in Pichit, Thailand' Validation Report, p.10, 27, http://cdm.unfccc.int/UserManagement/FileStorage/OUR7L1SX25WD2 DXB1BHNCAGCR7PPW1.

6 The 'baseline methodology' used by the project is ACM0006 (version 04) 'Consolidated baseline methodology for grid-connected electricity generation from biomass residues'.

7 Tangwisutijit, N. (2009) 'THAILAND: Renewable energy not so clean and green after all?', *IPS*, 23 October.

8 www.ricehuskash.com: This RHA in turn contains around 85% - 90% amorphous silica. Copyright © 2001 Elsevier Science Ltd and Techna S.r.l. All rights reserved. Studies on silica obtained from rice husk. N. Yalçin and V. Sevinç, Sakarya University, Arts and Sciences Faculty, Chemistry Department, Serdivan, 54180 Sakarya, Turkey. Received 14 February 2000; revised 14 March 2000; accepted 19 April 2000. Available online 13 February 2001. Abstract:

The potential and limits of rice husk to prepare relatively pure activated silica were investigated. For the activated silica, rice husk samples were submitted to a chemical pre-and post-treatment using HCl, H_2SO_4 and NaOH solutions. Samples were incinerated at 600°C under static air and flowing atmospheres (air, argon and oxygen). The product was characterized in terms of silica content, particle size distribution and morphology, specific surface area and porosity. The particle size distribution range from 0.030 to 100 □ m. The structure is amorphous. The specific surface area reaches value of 321 m²/g, porosity diameter is 0.0045 □ m, specific pore volume is 4.7297 cm³/g. Purity is 99.66% SiO_2.

9   Liu, S. et al. (1996) 'Silicosis Caused by Rice Husk Ashes', *Journal of Occupational Health*, 38, 257: 62.

10   http://www.who.int/mediacentre/factsheets/fs238/en.

11   '1) Environmental Impact Protection Guarantee Fund which will pay compensation for the damages the power plant has caused to the environment of the community such as excessively over-standard smog emitted from the plant's smokestack 2) Community Development and Environment Fund which will support and develop the education, heath care, occupation for a better standard of living of people in community' (http://www.atbiopower.co.th/power_plant/power_plant_e).

12   Presentation at conference on Review of the EU ETS, Brussels, 15 June 2007, cited in Lohmann, L. (2008) 'Carbon Trading: Solution or Obstacle', *The Impact of Climate Change on India:* 16.

5

Politics of Methane Abatement and CDM Projects based on Industrial Swine Production in Chile[*]

Cristián Alarcón

Introduction

Once the urgent need for facing climate change – which, let's not forget, has been caused by a historical process of capitalist industrialization highly based and dependent on the overconsumption of forest resources and fossil fuels – is established, and once the extremely grave consequences of this process are widely acknowledged at those centres where a lot of global political power is concentrated, the question about what to practically do about climate change and which measures to take remains a tremendous global problem. One of the roots of such a problem is that, even if some countries accept their responsibilities in the global climatic crisis, they try to keep their favorable positions in a world extremely divided in terms of the distribution of consumption and production. Partly because of that there are many reasons to be extremely concerned in relation to the negotiations prior to the United Nations Climate Change Conference in Copenhagen in 2009 (COP15). In fact, at the last G8 summit in Italy it was recognized that 'a temperature increases of 2% ought to be avoided' and that this could be done by 'achieving at least a 50% reduction of global emissions by 2050'.[1] While this reduction of emissions is presented as a big step forward in terms of facing climate change, it nonetheless contradicts scientific evidence telling us about the necessity of much more radical cuts of emissions. That is, the IPCC recommends that greenhouse gas (GHG) emissions must be globally reduced by 80-90% by 2050.[2] Moreover, the IPCC Chairman Rajendra Pachauri highlighted the lack of an agreement in the G8 to peak emissions in 2015.[3] However, right after the G8 meeting Jose Manuel Barroso, representing the EU, reaffirmed the EU's commitment in terms of a 50% reduction of emissions by 2050 declaring that 'this is what science tells us'.[4] As shown above, science actually tells us something else than what Barroso wants us to believe.

So, we are facing a double exposure to both the dangers of climate change and a dangerous global ruling class that in opposition to what climate change science tells us is simply trying to preserve the status quo. This also configures one of the greatest divergences of our time: what should be one of the most important political questions of our time, leading to massive political mobilization, becomes reduced to a question of selecting 'options' and making 'choices' within the logic of a capitalist market, which are then communicated to the public and transformed into possibilities for investments. 'The market will solve it', seems to be the tenor.

This is to a great degree the logic of the Clean Development Mechanism (CDM) established in the Kyoto Protocol (KP), which is aimed at creating investment projects that, and this is the hope, would reduce global GHG emissions and promote sustainable development in the 'developing world' at the same time. However, criticisms of the CDM have been widespread. The Financial Times (FT), for example, started a semi-critical article about the CDM as follows:

> What do Latin American pig farms, Chinese refrigerant factories and explosives makers in South Africa have in common? The answer is that they all qualify for carbon credits under the Kyoto Protocol.[5]

The FT quoted acknowledgments made by the representative of the KP Ivo de Boer about several failures of the CDM. Yet, the article finishes stating that 'The need to keep up, and expand, investment flows to poor countries to help them cut their emissions, is a key reason why Mr de Boer is adamant that any post-2012 agreement must include "a toolbox" by which countries can cut their emissions, and which would include trading in carbon credits'. Two aspects of this statement are interesting to note here: First, the insistence on carbon trading in the post-Kyoto regime, and, second, the discourse on 'investment flows to poor countries'. Despite all the documented problems with the CDM, it seems that carbon markets are here to stay.

The Kyoto Protocol and the Clean Development Mechanism's Multiple Objectives

As we know, the KP and the CDM have two main declared objectives: reductions of GHGs and achievement of sustainable development. Obviously, both objectives are a matter of definition. Within the IPCC, for example, references to the CDM can be found in several parts; for example, the Glossary of the IPCC's report from 2007 says:

> Defined in Article 12 of the Kyoto Protocol, the CDM is intended to meet two objectives: (1) to assist parties not included in Annex I in achieving sustainable development and in contributing to the ultimate objective of the convention; and (2) to assist parties included in Annex I in achieving compliance with their quantified emission limitation and reduction commitments[6]

A third CDM goal is to lower the cost of compliance for Annex I parties.[7] The two main goals of the CDM reproduced above are bounded to two kinds of definitions: while the goal of reducing emissions rests purely on an agreement about how many future emissions of GHGs should be reduced, the second goal of sustainable development makes reference to something rather undefined and even fuzzy. This question about the meaning of sustainable development within the KP and its mechanisms has raised doubts in many places.[8] It is now quite evident that 'sustainable development' must be seen as a kind of 'empty signifier', which has been added to the Kyoto mix out of discursive considerations. But in the end it seems to mostly mean the continuation of 'capitalist development'. Similarly, only to say that GHG emissions have to be reduced without making any reference to what climate change science tells us needs to be done, has been critiqued widely. Hence, it has been argued that the KP has simply failed and it has to be ditched.[9] Other critiques of the CDM can be found in papers by Wara and Victor.[10] These authors argue that,

> We doubt the CDM is an effective means of engaging developing countries for two reasons. First, fundamentally, the CDM works mainly by encouraging countries to avoid broader commitments and thus rewards exactly the opposite behavior that should govern the long-term efforts to build an effective regime for regulating emissions of greenhouse gases. Second, the CDM does not seem to be working well.[11]... the CDM, as currently structured, has serious problems with both the cost-effectiveness of its interventions in developing countries and with the credibility of the reductions these interventions produce.[12]

Another critique has to do with the fact that GHGs considered within the CDM are various, different and caused by a disparity of production and consumption processes. Because of this multiplicity, GHGs are made equivalent to each other by inventing a measure called 'CO2-equivalent'. So, methane emissions, for example, and depending on the approach used, can be conceived as having 21 or 25 times more global warming potential (GWP) than CO2. This fix has been a core element within the technical implementation of the CDM. But here resides also one of the most criticized outcomes of the CDM: the fact that a large portion of Certified Emission Reductions (CERs) via the CDM are generated through the management of GHGs other than CO2. In some cases such non-CO2 gases are possible to manage in cheaper ways than CO2. Wara, for example, shows how the relation between HFC-22 production and HFC-23 abatement provides what he denominates 'perverse economic incentives' through the CDM (see also Chapter 12 in this volume). Wara and Victor therefore doubt that the CDM delivers any significant reduction of emissions.

The failure of the CDM in achieving development goals can be found in many places as well. A detailed analysis and evaluation of how development goals are achieved in CDM projects is offered in a report for the International Institute for Sustainable Development (IISD), which analyzed more than 200 CDM projects.[13] Through applying certain development criteria the study found that the best project gets only 58 points out of a maximum of 100. Hultman et al. (2009) have shown the same fact, and they emphasize the crucial factor of national government deciding sustainable development criteria:

> Sustainable development, being a politically central but poorly executed tenet in the CDM, is currently not incorporated into its core incentive structure. In the current system, any additional, noncarbon specific requirements are likely to increase costs. Similarly, it is logical that private investors focus their efforts on countries that pose lower political and economic risks for their projects, or who impose weak sustainable development criteria, and the CDM is no different in this regard from other forms of foreign investments.[14]

Yet, these criticisms often remain ideologically attached to the logic of 'the market'. Prins and Rayner, for example, still argue that the problem of the KP is its top-down implementation of markets and therefore a bottom-up process of market-creation would be necessary. For Chafe and French, after stating that the carbon market will be a significant feature of the global economy landscape in the years and decades ahead, one of the most important benefits of carbon markets is political: 'They are creating powerful economic constituencies that favor stricter international action to stabilize Earth's climate'.[15]

In this chapter I argue that there is a need to go beyond such 'internal critiques' and offer a more totalizing critique of the CDM, a critique that considers broader aspects of the political ecology and political economy implied by climate change.

Case Study

A focus on CO_2 emission that created by the burning of fossil fuels captures an important portion of the discussions on climate change. Yet, another crucial dimension of the problem is industrial livestock production and its generation of other GHGs in form of byproducts. Industrial livestock relates to a massive production and consumption of food, involving a very lucrative business that is mostly shared by a few large companies and oligopolies.[16] On the other hand, food scarcity is deepened in many countries,[17] and industrial production of food has become associated with several social and environmental conflicts.

A report by the Food Agriculture Organization (FAO), 'Livestock's long shadow: Environmental issues and options',[18] shows that the process of industrial pig production and the management of pigs' manure is a major source of GHGs. 'At a global level, emissions from pig manure represent almost half of total livestock manure emissions. Just over a quarter of the total methane emission from managed manure originates from industrial systems'.[19] And 'Nitrogen concentration is highest in hog manure (76.2 g/N/kg dry weight)'.[20] The report also points out that

> globally, methane emissions from anaerobic decomposition of manure have been estimated to total just over 10 million tonnes, or some 4 percent of global anthropogenic methane emissions (US-EPA, 2005). Although of much lesser magnitude than emissions from enteric fermentation, emissions from manure are much higher than those originating from burning residues and similar to the lower estimate of the badly known emissions originating from rice cultivation. The United States has the highest emission from manure (close to 1.9 million tonnes, United States inventory 2004), followed by the EU. As a species, pig production contributes the largest share, followed by dairy.[21]

Moreover, pig farms are major consumers of water: 'In particular, pigs require a lot of water when kept in 'flushing systems'; in this case service water requirements can be seven times higher than drinking water needs'.[22] Pigs are fed with fishmeal which contributes to the overexploitation of fish stocks. For example, 'nearly 38,000 tons of fishmeal are used in UK pig rations each year. This represents around 14% of the 270,000 tons of fishmeal consumed annually in the UK'.[23] In a global context, FAO's report states that 'demand for fishmeal from the pig production sector continues to increase (from 20 percent of global fishmeal supply in 1988 to 29 percent in 2000)'.[24] A Life Cycle Assessment (LCA) of Danish pig production shows, for example, that: 'The environmental impacts were 3.77 kg CO_2 eq. global warming potential, 319 g NO_3 eq. eutrophication potential, 59 g SO_2 eq. acidification potential, and 1.27 g ethene eq. photochemical smog potential per kg Danish pork delivered to Harwich Harbour'.[25] In addition to all these environmental consequences, we can add that the spreading of diseases, such as human influenza, has become associated to industrial pig production. At a time when swine flu had not reached the pandemic status we are facing today, Mike Davis had already exposed the deep connections between new types of human influenza and industrial pig production.[26]

Pig producers in Latin America can take advantage of the CDM since the abatement of GHGs caused by pig production in a Non-Annex B country can generate carbon credits to be traded within carbon markets. The Chilean company Agrosuper is the world's eighth largest pork producer. Using vertical integration, cutting-edge corporate communication systems and even web pages in Japanese and English, the company is a global player in the food industry.[27] The company has produced and exchanged carbon credits from swine manure management by registering several CDM projects, and it has recently submitted another CDM project. After an evaluation process Agrosuper's CDM projects have been accepted. (One project was rejected by the CDM board, but it has since been re-submitted for approval). Once accepted, Agrosuper's credits have been successfully traded in the carbon market.[28] The first operation was formally signed in the Regional offices of FAO in 2004 with attendance of the Chilean ministers of energy, economics and environment as well as companies' representatives. The national media announced this operation as one of the biggest sales of carbon credits at that time, and a corporate film about this CDM project can even be found on YouTube.[29]

The CDM projects of Agrosuper are labeled either as methane capture and combustion from swine manure treatment or as advanced swine manure treatment.[30] The technologies used in the projects have been developed by Agrosuper and basically consist of the capture and burning of methane, which means that methane is converted into CO_2, a GHG that has a lower GWP. Agrosuper presents its participation in carbon trading as follows: 'Voluntarily decreasing the emission of greenhouse gases, Agrosuper has become the first agro-industrial company to generate Certified Greenhouse Gas Reductions (Carbon bonuses).'[31]

The Japanese Tepco and the Canadian TransAlta Corporation, two companies[32] operating in Annex I countries and needing to offset their emissions to accomplish KP goals in their countries, purchased Agrosuper's credits. TransAlta announced the operation in the following terms: 'With emission trades like this one, TransAlta is able to cost effectively take action now to reduce greenhouse gas emissions'.[33] In the case of Tepco, this carbon offset operation is presented in its 'Sustainability report 2008'[34] as a case of 'international cooperation activities'. Tepco justifies carbon trading by stating that: 'Japan is among the most energy-efficient countries in the world. Compared to other countries, there is little room left for cost-effective domestic measures. Therefore, TEPCO is actively working to achieve its CO2 reduction target, by employing the Kyoto Mechanisms to even more efficiently reduce greenhouse gas emissions while contributing to sustainable development in developing countries'. In other words, Tepco ensures to Japanese consumers that they can continue using energy and keeping their production and consumption levels as usual because reduction of emissions is being achieved among other places in Chile.

What is meant by the reduction of emissions is the amount of methane abated in Chile. We can analyze these CDM projects following KP's two objectives:

Reduction of Emissions

In line with the KP, one of the CDM's goals is fulfilled. In fact, the baselines used in such CDM projects show that without the technique used the methane would have been released to the atmosphere. However, as these emissions are caused by an export oriented pig production, what are the real costs involved here? The fact is that a big part of these emissions are deeply rooted in pig meat trade and export from Chile to other markets. In addition, and as the business goes very well for the company, production increases in order to satisfy demand and so are also emissions. A circle becomes evident here: as the export of pig meat increases, more methane is produced and more carbon credits can be generated. However, as we will see below, this process has both environmental and human dimensions not accounted for in the CDM. And also, a contradictory effect can be noticed: as mentioned earlier, methane has a higher GWP than CO2. On the other hand, CO2 lasts much longer in the atmosphere than methane which has a long-life of 12 years (IPCC). This point is crucial to understand the politics of methane abatement as a way of facing climate change. On a broader and global scale, this can even be seen from a cost effectiveness point of view:

> Given the relatively short life-time of the temperature response of methane reductions and that the target will be met beyond the middle of this century the shadow price of methane will be relatively low compared to what is found in the cost-benefit analysis or to its GWP value calculated over 100 years. The case is different for carbon dioxide since it has an almost irreversible effect on the temperature. Hence, given a cost-effectiveness approach (with a 2 K target) the use GWP overvalues the importance of reducing methane in year 2020, and correspondingly, relatively more economic resources should be devoted to reduce long-lived greenhouse gases such as carbon dioxide.[35]

Sustainable Development

According to the CDM, Agrosuper's projects should contribute to sustainable development goals. What the projects say in order to fulfill such a requirement is for example: 'The project activity can be stated as a relevant improvement for sustainable development, distressing local (odors) and global environmental pressures. This advanced system minimizes the odors related to swine manure management, because organic matter is stabilized through an aerobic process.'[36] The same claims about sustainable development are repeated word by word in all the other CDM projects of Agrosuper. But besides the general negative patterns and consequences of industrial pig production presented above, we can add some local socio-environmental issues connected to these CDM projects:

– The company has had several and serious negative impacts on the local environments where it has been operating. The Chilean environmental assessment system has registered a large number of violations of environmental regulations, which range from illegal discharge of sewage into rivers to illegal overloaded of manure in lagoons; from dust pollution to illegal extraction of water. In 2008 the environmental NGO Terram reported that, according to a study by the Chilean maritime authority, Agrosuper's activities close to a lake caused serious pollution.

– Another grave case took place in the El Yali wetland, which is a Ramsar site into which sewage from Agrosuper was discharged. In addition to the pollution of the wetland caused by Agrosuper, the firm obtains subterranean water from that wetland. Agrosuper is one of the main users of these water sources, which has been estimated to have negative impacts on the wetland.

– In another part of the country characterized by water scarcity, the company has a large project of industrial pig production and also has plans to build a private port for production purposes. Not surprisingly, water issues have become the subject of conflict there.[37] Not surprisingly either, this project is also aimed at producing CERs within the CDM. Table 1 shows some of the violations of environmental regulations attributed to Agrosuper in facilities linked to CDM projects ('fine' shows how little the company would have to pay for these violations.[38] The 'date' shows that the violations are fairly recent and hence occur in plants that have been approved as part of the CDM):

Plants linked to a CDM project	Violations	Fine (Approx. Chilean Pesos)	Date
Corneche	a) unauthorized construction of new lagoon to storage RILES[39] without impermeabilizacion measures b) Unilateral change of point of discharge of sewage	16,362,000	24 July 2007
Huasco	Non-authorized extraction of water	8,667,000	7 Feb 2008
Huasco	Non-compliance in control of dust pollution from its facilities	16,103,000	27 April 2007

Plants linked to a CDM project	Violations	Fine (Approx. Chilean Pesos)	Date
La Manga	Unauthorized construction of a well for disposal waste. A well many times bigger than the authorized	Warning	17 June 2008
Huasco	Illegal construction of two lagoons for water storage (size bigger than 50,000 m3) related to pig farms	17,403,500	30 April 2008
Maintenlahue	Spill over of sewage (Liquid Industrial Residues) to the Maintenlahue marsh, storage of non treated purines[40] in furrows non considered in the project	16,023,000	8 Aug 2006
Maintenlahue	Non reforestation of 5,39 ha required in the management plant Maintenance of untreated purines in a lagoon of 61,000 m^3	9,661,800	16 April 2007
Peralillo	Maintenance of untreated purines Unauthorized construction of a waste lagoon	16,135,500	8 March 2007
La Manga	Unauthorized spill over and discharge of liquid residues; Non notice to authorities of spill over of residues; Unauthorized discharge of sewage into a stream that leads to a wetland (A Ramsar site and also a national natural reservoir)	16,362,000	18 July 2007
Ramirana	Discharge of sewage into a stream not considered for this purposes in the environmental authorization	15,138,500	22 Feb 2005

- Along with this environmental record, an outbreak of listeria was detected in 2009 at one of the company's plants. Listeria was caused by bad sanitary conditions, and the company was forced by the authorities to close the plant down temporarily.

- Furthermore, the company has been involved in serious conflicts with workers at its facilities. Agrosuper employs about 17,000 workers, and there have been a string of bad working conditions, anti-unions practices, low wages and violations of labor rights. In 2007 more than 600 workers were striking at the Rosario and Lo Miranda pig processing plants with the goal of obtaining the right of collective bargaining and better salaries. As a consequence, the company fired the workers en masse. Today, at least 200 workers are still in the courts trying to obtain their rights. Aware of the consequences of sanitary conditions within food processing plants, some workers and unions have been active whistleblowers even at the risk of losing their jobs. This was the case at Agrosuper's Huechuraba plant where bad meat, which was still supposed to be sold on the market, was exposed by workers.

As has been pointed above, a big part of Agrosuper's pig production is aimed at export with a considerable portions going to Japanese consumers. In 2008 the Japanese authorities alerted the public about considerable levels of dioxin contained in pork entering the Japanese market from Chile.[41] A temporal ban on pork imported from Chile was declared. One of the companies affected by the ban was Agrosuper.[42] In other words, the very same industrial pig farming that enabled Japanese consumers to keep up their energy consumption through the CDM was banned because of dioxins found in the pork meat. Interestingly, Tepco, one of the buyers of the CERs created by Agrosuper, has a no less dubious environmental record. In 2002 Tepco was at the center of a scandal related to safety conditions at its nuclear power plants, and it became known that the company had been falsifying security tests at its plants. In this case, also a whistleblower permitted to discover the practices of Tepco.[43]

Gas Commodity Fetishism and Unequal Exchange

Tepco's and TransAlta's clients are probably not thinking about pig manure or the socio-environmental consequences of industrial pig production in Chile when they switch on their electricity. Nor do they think about workers getting fired, trying to better their working conditions. This is not very surprising, given that under capitalism people are alienated from places, resources and producers. The process of exchange of commodities within capitalist markets conceals the social relations underpinning the process of creating exchange value from use values. In this regard, Marx clarifies that it is not only labor which is the source of use value and wealth:

> Labor is not the source of all wealth. Nature is just as much the source of use values (and it is surely of such that material wealth consists!) as labor, which itself is only the manifestation of a force of nature, human labor power.[44]

Harvey points out that commodity fetishism also conceals geographical relations interwoven in the process of capitalist production and exchange.[45] In Marx's analysis of capitalist, production and consumption are dialectically linked, and commodity fetishism is that which conceals the real material dynamics of such a dialectical process of capitalist exchange. Mass consumption (particularly in the North) has been the attempt by capital to keep social control and integrate workers into, and legitimize, its system of accumulation, bearing in mind though that this takes place in parallel with ongoing unequal exchange – both in social and ecological terms – in other parts of the world. For, a part of the workers in countries that have accumulated more capital, to consume cheap commodities depends on production allocated somewhere else. Within this context, sustainable development under capitalism is also a way of continuing the dialectical process of development and underdevelopment associated with differences between countries and within countries. Carbon markets, which supposedly aim to reduce GHG emissions, are one expression of this global dynamic.

Yet, the nature and conceptualization of a market for carbon credits challenges our theoretical imagination. But with this carbon commodity, what is really produced and consumed here?[46] The first thing to realize here is that these markets are very much government driven – so much about the neoliberal ideology of 'free markets'. For a company in a country like Chile, for example, the process of entering into the carbon market is not that difficult, as there is a whole governmental framework to take advantage of the KP and the CDM. That is, companies get support all the way before selling their credits and hence profiting from these markets, not to mention the generous support industrial pig production companies receive in general terms. But, of course, we are dealing here with global neoliberal contexts, involving companies and consumers in Japan and Canada as well. The point is that climate change implies a politics where tales of environmentalism, cooperation and development are rooted in global structures of political ecology and political economy. That is, the attempt to deal with climate change through logics of cost effectiveness must be understood as an expression of the global structures of political ecology and political economy of capitalism. In short, it is historically specific.

Some people have named our era 'anthropocene' because 'we' have had the global social power to change the climate.[47] One should add, though, that the social power to create such an era lies within the dynamics of production and consumption shaped by capitalism worldwide. In fact, to root the economic system to a great extent in fossil fuels is a specific feature of capitalism. And the organization of massive and increasing consumption beyond biophysical limits, based on the overuse of raw materials, ecosystems and foodstuffs, must also be understood as a specific feature of capitalism. Therefore, we cannot miss the fact that climate change is a matter of global social power and its relation to local and global ecosystems. Any attempt, I would argue, to truly understand the core of the ecosystems-society relationship under capitalism must deal with the dialectical relation between production and consumption. Equally, any attempt to think of truly sustainable human-nature relationships needs to radically rethink and transform both the mode of production and the mode of consumption. Tragically, the solutions mostly favored today are those shaped by the ruling capitalist elites. That is, capitalist market mechanisms and cost-based strategies are strongly internalized as the only ways to solve climate change, reinforcing commodity fetishism through carbon trading. It is illustrative to note here that the IPCC's reports have somehow reflected this logic in the rationale of cost-based proposals, which have been favored to environmental targets. Actually, this issue goes back to the First Assessment Report (FAR), as one author points out:

> Although the IPCC involves thousands of participants, there are scientists who are marginalized by core group members of the IPCC. For example, during the drafting process of the FAR, a study that based analysis of response options on environmental targets was excluded by IPCC WG3 as basis for the formation of

emission scenarios as its effects-based logic ran counter to the IPCC's cost-based assessment of strategies.[48]

As we know, the whole process of negotiating a climate change regime is based on the recognition of an environmental crisis. When it comes to the idea of crisis, the critics of capitalism tend to feel as if they are navigating in a fertile territory, and hence quick conclusions are made in terms of hopes and social mobilizations. But hopes for what? Development? What the climate crisis teaches us is that in parallel to the efforts of agreeing the radical measures needed to face climate change properly, we also need to consider development and unequal global social relations, because both are part of the same capitalist coin that has produced the climate crisis in the first place.

Conclusions

The CDM projects critiqued in this paper connect the industrial production of pigs in Chile to electricity consumption in Japan and Canada. This case has unveiled some basic flaws of the CDM logic: the persistence of unequal exchange in both flows of energy and labor terms, and a myth of technology transfer through the global KP regime. The technology in this case was basically developed in Chile, which is inherently connected to the goal of exporting more and more pig meat to international markets. This pig production in Chile can be conceived as a typical example of a process of underdevelopment, enriching the local capitalist class through a process of degradation of local environments and exploitation of local workers. What is new is that this underdevelopment process is now made possible through discourses of climate change mitigation and 'sustainable development', which, in fact, has the only aim of maintaining high levels of energy production and consumption in both Canada and Japan, hence cementing these countries' privileged positions in the world system. Besides, these reductions of emissions can be conceived as illusory, as they are based on a questionable baseline. The logic of the CDM is this: the same company that with one hand affects both environments and people negatively gets money and makes windfall profits for supposedly helping to mitigate climate change with the other.

It is now clear that the ideological construction of neoliberalism is aimed at convincing people that we can face the challenges of climate change within the capitalist systems. In responding radically to climate change, we cannot forget, as David Harvey has insisted on, that the neoliberal project is a class project, and as such carbon trading must be understood within the category of the trans-national capitalist class. In a broader context, human ecologist Alf Hornborg[49] has recently put it this way:

> We are not all sitting in the same boat, as the metaphor goes. We are sitting in at least two different boats, but one is pulling us all toward disaster. There are definitely powerful social groups who have very much to gain – at least within the anticipated time-frame of their own lifetimes – from the current organization of global society.[50]

The CDM is just one expression of such a global pattern, and we have to criticize it as such. But it is not enough to just keep the spirit of critique alive and to produce the best analysis of capitalism if we are not moving into the direction of social changes. Here resides the challenge of a full recovery of the somehow missing link between the critique of capitalism and the performative goal of starting to create a different world here and now. Making stronger the critique of the Kyoto Protocol without imposing a new path is futile today. Thus, any attempt of thinking about the sustainability of ecosystems on earth must be concerned with how production and consumption are materially organized. In my view, the starting point for such a new path would be social, rational, non-profit and truly eco-based planning, as opposed to capitalist markets and capitalist class organization of the world economy.

Notes

* This article is a modified part of a conference paper titled 'On the assumptions and consequences of climate change economics and the Clean Development Mechanism as responses to climate change. Examples from Chile', presented at the conference Climate Change, Power and Poverty in Uppsala, Sweden, October 2009; I would like to express my thanks to Steffen Böhm and Siddhartha Dabhi for their helpful editorial support; Disclaimer: the author is a Chilean lawyer that worked for about 7 years representing workers and trade unions and defending labour rights in Chile. The legal representation of a group of workers referred to in this article is carried out by one of the author's partners in Chile.

1 http://www.g8.utoronto.ca/summit/2009laquila/2009-declaration.pdf.

2 IPCC WGIII chapter 13:776 / WGIII SPM: 15 /WGIII chapter 3:229.

3 http://www.nytimes.com/cwire/2009/07/21/21climatewire-ipcc-chief-raps-g-8-calls-for-global-greenho-41901.html.

4 http://europa.eu/rapid/pressReleasesAction.do?reference=MEMO/09/329&format =HTML&aged=0& language=EN&guiLanguage=n.

5 http://www.ft.com/cms/s/3add4666-7b0e-11dd-b1e2-000077b07658.html.

6 Metz, B., O.R. Davidson, P.R. Bosch, R. Dave, and L.A. Meyer (eds) (2007) *Glossary of the Contribution of Working Group III to the Fourth Assessment Report of the Intergovernmental Panel on Climate Change*. Cambridge, UK and New York, USA: Cambridge University Press. http://www.ipcc.ch/pdf/assessment-report/ar4/wg3/ar4-wg3-annex1.pdf.

7 Wara, M. (2006) 'Measuring the Clean Development Mechanism's Performance and Potential', *P. W. P. #56, Program on Energy and Sustainable Development* at the Center for Environmental Science and Policy Stanford University.

8 Olsen, K. H. (2007) 'The clean development mechanism's contribution to sustainable development: a review of the literature', *Climatic Change, 84*(1); Hultman, N., E. Boyd, et al. (2009) 'How can the Clean Development Mechanism better contribute to sustainable development?', *Ambio*, 38(2): 120-122; Boyd, E., N. E. Hultman, et al. (2007) 'The Clean Development Mechanism: An assessment of current practice and future approaches for policy', *Working Paper 114*, Tyndall Centre for Climate Change Research.

9 Prins, G. and S. Rayner (2007) 'Time to ditch Kyoto', *Nature*, 449: 973-975.

10 Wara, M. (2006) 'Measuring the Clean Development Mechanism's Performance and Potential'; Wara, M. W. and D. G. Victor (2008) 'A Realistic Policy on International Carbon Offsets', *PESD Working Paper #74*, Program on Energy and Sustainable Development at the Center for Environmental Science and Policy Stanford University.

11 Wara, M. and D.G. Victor (2008) 'A Realistic Policy on International Carbon Offsets'.

12 Wara, M. and D.G. Victor (2008) 'A Realistic Policy on International Carbon Offsets'.

13 Cosbey, A., D. Murphy, et al. (2006) 'Making Development Work in the CDM', *Phase II of the Development Dividend Project*, International Institute for Sustainable Development.

14 Hultman, N., E. Boyd, et al. (2009) 'How can the Clean Development Mechanism better contribute to sustainable development?', p.122.

15 Chafe, Z. and H. French (2008). 'Improving Carbon Markets In: State of the World 2008, Innovations for Sustainable development'. *A worldwatch Institute Report on Progress Toward a Sustainable Society*, New York: WW. Norton & Company, p.106.

16 Dicken, P. (2007) *Global Shift. Mapping the Changing contours of The World Economy.* London: Sage.

17 In this context, it is important to note that: 'With the number of chronically hungry people in the world now higher than during the baseline period, the World Food Summit target of reducing that number by half by the year 2015 has become more difficult to reach', ftp://ftp.fao.org/docrep/fao/meeting/014/k3175e.pdf.

18 FAO (2006) 'Livestock's long shadow. environmental issues and options Rome, Food and Agriculture Organization of the United Nations'.

19 FAO (2006) 'Livestock's long shadow. environmental issues and options Rome, Food and Agriculture Organization of the United Nations', p.99.

20 I am emphasizing concrete consequences of pig production. In some cases other livestock production systems have worse environmental performance and consequences than those associated with pigs, as for example is the case of cattle.

21 FAO (2006) 'Livestock's long shadow. environmental issues and options Rome, Food and Agriculture Organization of the United Nations' p.98; the FAO report states that a total annual global emission of methane from manure decomposition could be estimated as being of 17.5 million tonnes of CH4.

22 FAO (2006) 'Livestock's long shadow. environmental issues and options Rome, Food and Agriculture Organization of the United Nations', p.29.

23 http://www.iffo.net/intranet/content/archivos/85.pdf.

24 FAO (2006) 'Livestock's long shadow. environmental issues and options Rome, Food and Agriculture Organization of the United Nations', p.207.

25 Lundshøj, R. (2007) 'The environmental impact of pork production from a life cycle perspective', *Faculty of Agricultural Sciences*, University of Aarhus and Aalborg University, p.38.

26 Davis, M. (2005) *The Monster at Our Door. The global Threat of Avian Flu.* New York: The New Press.

27 http://agrosuper.co.kr/index_new.html.

28 http://www.agrosuper.cl/site_english/?p=noticias;
http://diario.elmercurio.cl/detalle/index.asp?id={b2329ac4-ac5d-40ab-92c9-d3b3ebe09201}.

29 http://www.youtube.com/watch?v=XPRVHJTu1dQ&feature=related.

30 The list of CDM projects includes: Methane capture and combustion from swine manure treatment for Corneche and Los Guindos; Methane capture and combustion from swine manure treatment for Peralillo; Methane capture and combustion from swine manure treatment for Pocillas and La Estrella; Advanced swine manure treatment in Maitenlahue and La Manga; Ramirana Emission Reduction Project of Agrícola Super Limitada and Advanced swine manure treatment for the Huasco Valley Agroindustry.

31 http://www.agrosuper.cl/site_english/?p=respaldo_produccion; Voluntarily decreasing the emission of greenhouse gases, Agrosuper has become the first agro-industrial company to generate Certified Greenhouse Gas Reductions (Carbon bonuses). http://www.agrosuper.cl/site_english/?p=respaldo_compromiso Perfil agrosuper: http://www.sofofa.cl/BIBLIOTECA_Archivos/Eventos/2008/08/05_abascunan.pdf;
UNEP Risoe Center. CDM Pipeline overview. Analysis and Database. Updated 1st July 2009. Accessed: 15 July 2009.

32 Also a third firm, Rabobank from The Netherland, purchased credits that were then sold to a UK based utility company.

33 http://www.transalta.com/transalta/webcms.nsf/AllDoc/8B3302B4EF255
    F6387257157004FC26F?OpenDocument.

34 https://www4.tepco.co.jp/en/useful/pdf-2/08report-e.pdf.

35 Johansson, D. and F. Hedenus (2009) 'A Perspective Paper on Methane Mitigation as a
    Response to Climate Change', *Copenhagen Consensus Center*, p.10; This report is in the line of
    climate change economics, so the argument is valid when discussing within the border of
    capitalist economics.

36 So it is stated in Agrosuper's CDM project Maitenlahue and La Manga, according to the
    project description submitted to the CDM board (p.45).

37 http://www.youtube.com/watch?v=PbtUC4grSEw.

38 1 UTM in December 2007=63 USD. 500 UTM=31.500 USD.

39 *Residuos Industriales Líquidos (RIL)*, Liquid Industrial Residues.

40 http://en.wikipedia.org/wiki/Purine.

41 http://www.thepigsite.com/swinenews/18638/japan-suspends-chilean-pork-imports.

42 Clarification: I am not saying that the company exported the dioxins. The point is that those
    exported dioxins are results of a way of producing pigs in Chile in which Agrosuper is a
    central player and developer.

43 http://www.greenpeace.org/international/news/japanese-nuclear-safety-scandal;
    http://cnic.jp/english/newsletter/nit92/nit92articles/nit92coverup.html.

44 From the 'Critique of the Gotha Programme' (1875). In *Capital*, this point is also made in the
    following terms: 'Capitalist production, therefore, develops technology, and the combining
    together of various processes into a social whole, only by sapping the original sources of all
    wealth - the soil and the labourer'.

45 Harvey, D. (1996) *Justice, Nature & the Geography of Difference*. Oxford: Blackwell.

46 Fine, B. (2008) 'Looking at the Crisis through Marx: Or Is It the Other Way about? Other',
    https://eprints.soas.ac.uk/5921/; Ben Fine tries to explain the bizarreness of carbon offsets:
    'And, in commodity markets, we have futures trading at its most bizarre with carbon offsets.
    Commodity fetishism has surely arrived at perfection when we can buy and sell in a market
    for not producing something in the future (especially when, in fact, carbon trading is about
    allowing that undesirable carbon to be produced for you by someone else as well as yourself
    on the grounds that they might produce less of it than you would if you were producing what
    they produce as well as what you yourself will carry on producing)'.

47 http://en.wikipedia.org/wiki/Anthropocene.

48 Kameyama, Y. (2004) 'IPCC: its roles in international negotiation and in domestic decision
    making on climate change policies', in N. Kanie and P. Haas (eds) *Emerging Forces in
    Environmental Governance*. United Nations University Press, p.148.

49 Hornborg, A. (2009) 'Zero-Sum World: Challenges in Conceptualizing Environmental Load
    Displacement and Ecologically Unequal Exchange in the World-System', *International Journal of
    Comparative Sociology*, 50(3-4): 237-262.

50 Hornborg, A. (2009) 'Zero-Sum World: Challenges in Conceptualizing Environmental Load
    Displacement and Ecologically Unequal Exchange in the World-System', p.238.

6

Paying the Polluter? The Relegation of Local Community Concerns in 'Carbon Credit' Proposals of Oil Corporations in Nigeria

Isaac 'Asume' Osuoka

Introduction

While the reduction in fossil fuels production and usage is central to the discourse on climate change, the neoliberal mechanisms for climate change mitigation have been widely critiqued. In the case of oil producing regions of the Global South, key socio-economic and environmental concerns of 'host' communities to hydrocarbon exploitation remain unaddressed. Carbon trading schemes may in fact worsen the social dynamic of dispossession and disempowerment of local communities by the state/corporation alliance as it occurs in Nigeria. I examine the proposals for 'carbon credit' within the Clean Development Mechanism (CDM) as made by transnational oil and gas corporations involved with the West African Gas Pipeline and Eni-Agip's Kwale gas project. Both projects claim that utilization of natural gas from the Niger Delta for export and marketing to power plants would contribute to reducing the greenhouse gas emissions associated with routine flaring of associated gas. A look at the proposals and the responses of 'stakeholders' such as the government, World Bank Group, and local civil society groups exposes not just the inherent flaws in the proposals, but the CDM itself, and shows how carbon trading may relegate community concerns about local pollution, livelihoods and 'resource control'.

The methodologies of the Clean Development Mechanism (CDM) allow for carbon credits to be awarded to projects aimed at reducing gas flaring. Corporations involved in oil production in Nigeria have made applications in relation to investments for reducing associated gas flaring. However, from the point of view of communities in Nigeria's Niger Delta, the prospect of these corporations making additional profits from selling or gaining carbon credits is tantamount to adding insult to injury. For fifty years the corporations, with the connivance of colonial and independent governments, have organized crude exploitation in ways that have destroyed local livelihoods and turned the Delta area, as indeed the country, into a war zone characterized by massacres of whole communities and widespread dislocation. But now the corporations largely responsible for the problems of oil production, which have failed to stop the dangerous flaring of oil's associated gas, are seeking some extra pay. This they would do simply with the presentation of some paperwork indicating how they hope to solve a bit of the problem they have created. Perhaps unsurprisingly, in the last few decades, we have been told that the market solves all problems, including cleaning up its own mess. But does it?

Gas flared in Nigeria contains high amounts of methane and carbon dioxide – major greenhouse gasses that are contributors to global warming. Gas flaring from the Niger Delta oil fields produces emissions that are more than the combined emissions of the rest of sub-Saharan Africa.[1] While no reliable data have been provided by the government and corporations, one estimate is that oil corporations in Nigeria were flaring 2.5 billion standard cubic feet of associated gas daily in 2004. This represents 70 million metric tons of CO_2 emissions per year.[2] Global outrage over continuing flaring in Nigeria and the acceptance of projects that reduce the practice for credits within the CDM system have resulted in applications from oil corporations.

This chapter concerns how oil corporations operating in countries like Nigeria try to exploit these schemes. We shall examine the character of hydrocarbon exploitation in the country and how the state, from colonial times, has condoned the flaring of associated gas and other acts of pollution. This is in spite of evidence of environmental damage and wastage of national energy resources. A critical look at how associated gas is generated would help expose the inadequacies of flaring data presented by industry and government, which are often used to justify claims that gas infrastructure projects would lead to reduction in flaring. These would explain inherent flaws in the applications by oil corporations for carbon credits; in showing how futile the CDM is, beyond just providing fresh profit opportunities corporations. We shall end by pointing out the way towards alternatives from the actions of the people themselves in resisting fossil fuels exploitation.

Marketing the Atmosphere

The crisis of neoliberal globalization, currently expressed in financial meltdown with socio-economic reverberations, offers us another opportunity to examine the problems that were created by an enforced obsession with the 'market'. In most sectors states were pushed to deregulate so that corporations would self-regulate. Laws and institutions were restructured in a form that tended to weaken citizens while further empowering corporations.[3] We have seen how attempts have been made to replace liability and the idea of 'the polluter pays' with schemes of 'corporate social responsibility'. Here states and multilateral institutions like the United Nations endorsed voluntary mechanisms that suggested a kind of superiority of corporations over national laws and local systems. Some international 'civil society' organizations joined up with corporations to promote schemes and voluntary principles on everything from human rights practice of corporations to how companies may (s)elect to consult with local people.

As the global impact of climate change was increasingly becoming a serious issue for world governments, the market paradigm and the predilection towards corporate voluntarism came to define official solutions. In the area of carbon reduction, such market mechanisms were promoted via carbon trading schemes, epitomized under the Kyoto Protocol that emerged from the third Conference of Parties of the United Nations Framework Convention on Climate Change (UNFCCC) in 1997. Carbon Trading and its sister Clean Development Mechanism (CDM) were part of Kyoto's 'flexible' market based mechanisms which are supposed to help reduce greenhouse gas emissions by allowing the major emitting industrialized countries to be granted emissions permits up to limits accepted by the protocol, calculated in units of carbon dioxide (one ton of carbon dioxide equals one Certified Emission Reduction). The governments could trade these permits onto the industries in their countries. Fuzzy *cap and trade* systems have since emerged that allow companies that have not used up all of their permits or that want to continue polluting to sell or purchase excess carbon credits to or from other companies.

Specifically, the CDM is a system for generating credits. Corporations can present investments in projects in developing countries of the South that will supposedly reduce emissions and thus generate new credits. The companies could use this credit to 'offset' its emissions of greenhouse gases over the permitted levels in their home country. Or sell such credits for profit in the carbon market.

Citizens groups and their networks such as Oilwatch and the Durban Group have since challenged these market based mechanisms for 'instead of materializing the objectives of the Climate Change Convention through concrete and effective measures...by changing patterns of production and reducing fossil fuel consumption, transfer responsibility for emissions reductions onto third parties and create new business opportunities for the polluters.'[4] For Oilwatch, such mechanisms are mere palliatives that allow corporations and countries with the most emissions to do 'too little' and reinforce the marginalization of the peoples of the Global South.

Though the United States was to pull out of Kyoto under George W. Bush, European Union countries championed it with an Emissions Trading Scheme (ETS) launched in January 2005. However, their ratification of the Kyoto Protocol was the beginning of 'scandals and market mishaps' as the whole system was bedevilled with 'intrinsic problem in setting an artificially-generated market price for carbon'.[5] Existing carbon trading systems do not provide any clear and adequate measurements of emissions and 'carbon offsets' nor mechanisms for global enforcement. No wonder the markets are crashing along with the edifices of global neoliberalism.

Efficiency and cost effectiveness became major considerations with neoliberalism. With the CDM, it was assumed that by reducing emissions in a developing country cheaply, instead of concrete actions in the developed countries, corporations can achieve benefits for the global climate in a more cost effective manner. However, as it has been pointed out elsewhere in this volume, the real cost of achieving such reductions in the developing countries in the form of loss of land and livelihoods, human rights abuses of local people etc. are not calculated.

In the examples of the gas flare reduction projects that have been presented for the CDM, sponsors' claims for carbon credit fail to account for all the costs that have been forced on Nigerian communities through all stages of oil and gas exploitation – from exploration to transportation and refining.

Petroleum and Exploitation

Petroleum exploitation is considered as one of the most environmentally damaging of all human activities. When extracted from the tropics, negative impacts commence from the exploration to production stages. In Nigeria's Niger Delta area, communities are confronted by the industry directly at the level of production (seismic exploration, drilling, forest and farmlands clearing for pipeline construction). They face direct loss of land rights by the instrumentality of military era decrees. The absence of enforcement of state regulations have meant that oil corporations pay only lip service to environmental and social standards with the results of pollution of all sorts. Communities continue to face discrimination in the way profits and revenues from extracted resources are used and are subjected to unfair compensation, and they suffer violence on their persons and their communities whenever they sought to protest or demand justice.

Post independence the Nigerian state has tended to retain the character of the colonial; that of facilitating exploitation of (petroleum) resources for export and depending on oil rents. It is not surprising, therefore, that little or no action has been taken to address the problem of associated gas flaring over fifty years after the first oil well was drilled at Oloibiri in the Niger Delta.

Crude Oil and Gas Flaring in Nigeria

Of particular concern to us here is associated gas flaring by the oil corporations, which produces continuous noise, rise in temperature in communities close to flare sites, acid rain and retarded crop yield, corroded roofs, respiratory and skin diseases. In addition, many communities suffer the impact of the 'perpetual light' with the unnatural illumination from gas flares at night.[6]

Gas flares don't just happen. It is a consequence of ill conceived and badly regulated exploitation for crude oil. In Nigeria's Niger Delta area, both onshore oil fields contain very large amounts of 'associated gas' that are mixed with crude oil in an area that has been described as a 'gas colony'. Under British colonial rule, Shell had prospected for and discovered these petroleum deposits in large quantities. In tune with the norm of the colonialist that engaged in resource extraction purely for export and profit expropriation, the corporation and their government at that time considered only the crude oil as valuable, as there was a ready 'market' in Europe. However, in drilling for crude oil, gas that was mixed with the crude in the reservoir also comes out. This is called associated gas in the industry. From 1956, Shell decided to collect the crude, while treating the associated gas as a waste product – burning it off. This is what

89

is referred to as 'gas flaring', which received particular critique with the rise of Ogoni resistance. It is done at flow stations, which are huge installations where crude oil is collected from a network of pipelines (flowlines) that transport oil from the numerous oil wells. At the flow stations, associated gas is burnt off in huge stacks that produce giant flames. Waste water from the oil wells are often emptied into pits near the stations while the treasured crude oil is collected and pumped into pipelines that take oil to the coastal export terminals.

In 1956 other options that could have been considered by the developers to avoid the local pollution of associated gas flaring was to reinject into the reservoir or to build gas powered power plants to provide electricity to the Niger Delta region and shore up the Nigerian grid. But local people did not constitute a profitable market and the extra investments for reinjection were considered too expensive in comparison to the cheaper option of burning into the natural environment. And Shell and the British administration were not oblivious of the consequences of associated gas flaring, as has been exposed in comments made by a British officials a few years after Nigeria gained flag independence. In 1963, Mr. J.S. Sadler, the British Trade Commissioner in Nigeria made the following comments in a memo to British Foreign Office in London:

> Shell/BP's need to continue, probably indefinitely, to flare off a very large proportion of the associated gas they produce will no doubt give rise to a certain amount of difficulty with Nigerian politicians, who will probably be among the last people in the world to realise that it is sometimes desirable not to exploit a country's natural resources and who, being unable to avoid seeing the many gas flares around the oilfields, will tend to accuse Shell/BP of conspicuous waste of Nigeria's 'wealth'. It will be interesting to see the extent to which the oil companies feel it necessary to meet these criticisms by spending money on uneconomic methods of using gas.[7]

To better understand the problem of gas flaring in the Niger Delta area (and expose the lies of later day carbon credit seeking corporations), it is important to indicate that apart from the crude oil fields that contain associated gas, there are also many pure natural gas deposits. This gas is referred to by the industry as non-associated gas. Oil corporations in Nigeria have built large Liquefied Natural Gas (LNG) plants in Bonny while others are planned. These 'gas utilization' projects like Chevron's Escravos Gas plant have been build to be fed with mainly non-associated gas, the production of which does not affect the rate of oil's associated gas production and flaring.[8]

One estimate is that Nigeria has 124 trillion cubic feet (tcf) of proven gas reserves (of which about 50% is non-associated gas), making it the ninth among the countries with the largest proven gas reserves globally.[9] However, there are several of such estimates. One presented by government officials suggest that Nigeria has the seventh largest proven gas reserves with 182 tfc.[10] Yet figures concerning oil and gas production in Nigeria are widely held to be manipulated and exploited for a range of reasons: For multinationals to avoid taxes, in order to justify an increase in Nigeria's OPEC quota (which are allocated on the basis of proven reserves). Other reasons for this are to mask the real volume of associated gas flaring, and most recently, to gain carbon credits.

Indeed, in the light of disclosures that Shell had overstated its proven reserves in Nigeria to fraudulently boost is share value in the US and UK, discrepancies in the figures presented above seem to reflect a more widespread practice – wherein data disclosure on production figures serves business strategy.[11]

Gas Flaring and the CDM

As the local complaints about associated gas flaring became louder, oil corporations and Nigerian National Petroleum Corporation (NNPC) have occasionally presented bizarre claims and figures to show how they are acting to reduce the problem. Most of these involve manipulation of associated and non-associated gas production volumes. For, example, in NNPC's official data for gas production and flaring in 2007, the actual mix of associated and non-associated gas is concealed – leading to a misrepresentation, as shown below.

Volume of	in BSCF
Gas produced in 2007	2,415.65
Gas utilized (for liquefied natural gas)	1,626.10 (67.32%)
Gas flared	789.55 (32.68%)

Table 1: Nigeria Gas Flaring and Utilization (Source: Nigeria National Petroleum Corporation 2007 Annual Statistical Bulletin)[12]

While the above figures released by the government's NNPC suggest that the rate of gas flaring by all the oil companies operating in Nigeria was 32.68% in 2007, data that presents the volume of associated gas produced side by side non-associated gas in a particular period would show a more accurate rate of flaring.

With the dearth of complete data from the industry, my estimation based on available crude oil production figures arrives at different figures. It is the common understanding that every million barrel of crude oil produced in Nigeria contains about a billion standard cubic feet of associated gas. In 2007, the NNPC reports that the combined crude oil production was 803 million barrels. This should contain about 803 billion standard cubic feet of associated gas. Based on this estimate, we have presented an alternative table as a more accurate estimate of volume and percentage of gas flaring in 2007.

	in BSCF (unless otherwise stated)
Crude oil production in 2007	803 mil barrels
Estimated associated gas generated from crude oil production in 2007	803
Total volume of gas produced in 2007 (sum of associated and non-associated gas)	2,415.65

	in BSCF (unless otherwise stated)
Estimated volume of non-associated gas produced in 2007 (obtained by deducting associated gas estimate from the total gas production for 2007)	1,612.65
Volume of gas utilized (for liquefied natural gas)	1,626.10
Volume of gas flared	789.55
Estimate of associated gas flared in 2007	98.33%

Table 2: Independent estimate of volume and percentage of gas flaring in 2007

This expanded table, which takes associated gas into consideration, gives a more accurate view of gas flaring in 2007 – producing estimates of 98%. This percentage figure appears more consistent with oil industry history and continuing practices such as the near total utilization of non-associated gas for LNG and other commercial gas projects. Before non-associated gas fields were developed by oil corporations for LNG and other gas export ventures, over 90% of all associated gas was flared, as shown in official government data. NNPC records reveals that 98.45% and 98.27% of all gas produced was flared respectively in 1969 and 1975. The high percentage of flaring in those years in NNPC data (compared to 32.68% in 2007) is because there were no non-associated gas developments at that time. Later, the non-associated gas fields were developed for utilization in LNG and other gas export projects. By presenting figures of gas utilization out of the sum of associated and non-associated gas, the corporations and the NNPC could show a reduced percentage of flaring. To further buttress this point, we will take another look at NNPC's 2007 'summary of gas production and utilization' of all oil corporations operating in Nigeria. This summary shows that Texaco and Addax flared 97.69% and 86.34% of all the gas they produced during the year. In the same table, Shell and Elf (Total) are reported to have flared 12.58% and 15.24% of all the gas it produced in 2007.[13] My conclusion is that the low flaring figures recorded for Shell and Elf is because of the very high volume of non-associated gas developments for the Bonny LNG. On the other hand, neither Texaco nor Addax was involved in non-associated gas production for LNG, resulting in a more realistic figure for flaring by these corporations.

Rather than developing programmes to capture associated gas, and reduce flaring, oil corporations have continued to develop new non-associated gas fields that, in fact, account for the 10.69% increase in gas production 2007, as compared to 2006 figures provided by the NNPC.[14]

Major oil corporations in Nigeria have used the LNGs and other non-associated gas projects to attempt to justify such claims of reduction of gas flaring. But this would only be true if all gas produced in Nigeria are associated with crude oil. To show how disingenuous the oil companies have been, even the World Bank, their traditional supporter, rejected Shell's earlier claim that the Bonny LNG would contribute to gas flare reduction. The Bank revealed that initial plants (called trains) of the Bonny LNG were designed to utilize unflared non-associated gas. This meant that the project at the early stages did not contribute to the flaring of associated gas as Shell claimed.[15]

The case of the West African Gas Pipeline (WAGP) in a further demonstration of how proposed projects in the CDM have been employed to claim credits for future reductions in gas flaring by multinationals. As one of the region's largest trans-boundary investments the WAGP, which is incorporated in Bermuda, is a 681 km onshore and offshore pipeline meant to transport natural gas from gas fields in the western Niger Delta of Nigeria to selected consumers in Benin Republic, Togo and Ghana. Construction on the WAGP started in 2004 and was 'substantially completed' in 2007. Project promoters, including Chevron and Shell, stated that WAGP will reduce carbon emissions, provide cheaper, more reliable and environmentally friendly energy, and foster economic development and integration in Ghana, Togo, Benin and Nigeria. But the WAGP was linked to a previously existing Escravos-Lagos Pipeline (ELP), which collects gas from Chevron's Escravos Gas Plant that was built to process unflared non-associated gas.

It remains unclear how the WAGP would reduce flaring, when it is intended to process not the 'waste' gas associated with crude oil that has been flared for decades, but rather the newer non-associated gas reserves often developed for export. So the question is how Chevron, Shell, the Nigerian government and other sponsors of the WAGP can explain how to end associated gas flaring by building a pipeline to transport non-associated gas? Interestingly, when this author and other citizen-activists confronted Chris Miller, a Chevron official and project manager of the WAGP face to face with this question, Miller admitted that projections on greenhouse gas emissions reduction from the WAGP were 'theoretical'.[16] But such assumptions did not stop Chevron from continuing to mount a spirited campaign to justify its request for carbon credits on account of the WAGP.

Such theoretical estimations of emissions reductions is commonplace in the industry, and should disqualify carbon credit seeking corporations for failing to meet *measurability* criteria of the CDM. Indeed, the corporations would fail on all other criteria of the CDM including additionality, sustainable development, Environmental Impact Assessment and community consultation.

In Nigeria, gas flaring is also already prohibited and companies like Chevron and Shell have been paying a penalty for non-compliance. So, oil industry projects like the WAGP claiming to reduce gas flaring cannot be said to provide any *additionality*.[17] Also, oil and gas production rarely contributes to *economic development* in developing countries. Fifty years of oil and gas development in Nigeria have resulted in mass impoverishment, as a result of pollution and mismanagement of oil rents. Dependence on rent from oil and gas have resulted in the abandonment of other sectors of the economy like agriculture and manufacturing that contribute more to GDP.

Communities in Nigeria and Ghana protested the inadequate processes and content of Environmental Impact Assessment (EIA) for the WAGP. The draft EIA was not made available to community people for comments. Copies of the draft EIA supposed to be on public display at the Lagos State Ministry of Environment were hidden in the office of the Permanent Secretary, away from the reach of community people, contrary to the mandatory provisions of the Nigerian Environmental Impact Assessment Act No 86 of 1992. Community members expressed serious reservations which culminated in a legal challenge of the project at the Federal High Court of Nigeria.

After the WAGP, other Nigerian oil operators have continued to make carbon credit claims, presenting the same flawed arguments as did Chevron. Significantly, Eni-Agip, an Italian transnational oil corporation, attempts to disapprove Nigerian law in its presentation to the CDM, maintaining that 'whilst the Nigerian Federal High Court recently judged that gas flaring is illegal, it is difficult to envisage a situation where wholesale changes in practice in venting or flaring, or cessation of oil production in order to eliminate flaring will be forthcoming in the near term.'[18] Eni-Agip blames, 'commercial and industrial risks and political uncertainty' for failures to make 'investments' to end associated gas flaring. However, what is lacking in Eni-Agip's argument is the analysis of their own role in promoting instability and local dislocation, or how new gas infrastructure projects would contribute to exacerbating local tensions as issues of resource control and community land rights have not been resolved.

Like the WAGP, Eni-Agip presents an application to the CDM for its projects for the recovery of associated gas that would otherwise be flared at the Kwale oil-gas processing plant by the 'capture and utilization of the majority of associated gas previously sent to flaring at the Kwale plant (Kwale OGPP).' According to the Eni-Agip, 'large portion' of associated gas produced from the five oil fields in Oil Mining Lease 60 (Ahaka, Beniku, Okpai, Kwale, Irri-Isoko) has been flared upon separation from the oil at the Kwale flow station. This practice according to the company is due to the 'absence of any economically viable, commercial or other outlet for this gas.' Eni-Agip claims that its Kwale gas plant 'will not increase GHG emissions in Nigeria'.[19]

But questions that Eni-Agip, like Chevron (on WAGP), cannot answer are the actual mix of associated and non associated gas that is transported by the network of pipelines and flowlines at Oil Mining Lease 60 or that is actually to be processed at the Kwale gas complex. As mentioned above, a gas complex contains gas gathering and processing infrastructure. As oil fields (with associated gas) exist side by side with non-associated gas fields, gas processing plants could collect both associated and non-associated gas. Lack of clarity of the actual mix of associated and non-associated gas makes it impossible to validate claims that projects would lead to reduction of associated gas flaring. In fact, apart from the mention of 170,000 barrels of oil per day produced by Eni-Agip, there are no figures of current gas production and flaring in Ahaka, Beniku, Okpai, Kwale and Irri-Isoko area (OML 60) in Eni-Agip's documentation related to the CDM.

Conclusion

Oil companies and the Nigerian government obfuscate data on associated gas flaring by presenting mixed figures of associated and non-associated gas. While figures provided by the government suggest a reduction of gas flaring, the actual reality on the ground and a closer examination of flaring data, as I have presented, show that flaring volumes have increased over the years. This situation means that there is no credibility to claims by corporations seeking carbon credit through the CDM. Applications by oil corporations operating in Nigeria expose how corporate entities continue to manipulate carbon trading schemes by presenting inaccurate and 'theoretical' assessments of emissions reduction.

In many cases, as in oil and gas exploitation, the emission of greenhouse gasses are significant not only in causing global climate change; but the processes leading to generation of emissions involve major destruction to the local environment, livelihoods and communities. In the case of crude oil's associated gas flaring in Nigeria's Niger Delta (as elsewhere), there has to be an account for air contamination, as well as oil spills and other manifestations of pollution in a fragile tropical wetlands ecosystem. The WAGP and the Kwale gas plant which have been presented as possible CDM projects fail to account for the human rights abuses, the loss of local resource control and negative distortions to socio-economic life experienced at the local and national levels as a result of oil and gas developments. These examples indicate how emissions reductions schemes are being employed to *further* the oil and gas frontier rather than limit it and as a means of further disenfranchising local populations.

Rather than a discussion of how oil corporations and the government will respect their laws and pronouncements of the courts, we are seeing, with this carbon credit trading, how systems are being created to compensate the polluters and maximize their already colossal profits in ways that are reminiscent of corporate administered colonial rule.

Indeed, carbon credit trading reflects one of the worst forms of the neoliberal fanaticism and attempt at re-legitimating corporate rule as experienced in the past decades. In different ways, carbon credit trading attempts to entrench the rights and privileges of big polluters against local communities by providing rich countries and corporations the framework to delay making structural changes towards low-carbon technologies; therefore allowing the worst polluters to secure huge blocks of pollution rights, to buy more rights; providing compensation to large (oil) corporations while not considering issues of justice for victims of local pollution and allowing so much impunity for historical emissions.

As the current moment presents an opportunity for a deeper reflection on the inadequacies of the market mechanisms to addressing climate and the neoliberal paradigm that informs them, we should also take a look at the local and global forces massing to challenge this system. In the Niger Delta, as in other parts of the world, movements of the people are challenging corporate rule in the streets and creeks and are reclaiming spaces of community power. The Ogoni chased Shell away from their territory and over a decade later community members continue to insist on keeping the oil underground. Even as the state is negotiating to have another oil company re-enter the Ogoni area, some groups, like the Ogoni Solidarity Forum (OSF) are mobilizing against renewed extraction. Other citizens groups in Nigeria (and the Department of Petroleum Resources) have demanded the shutting down of all oil fields that continue to flare gas in the country. Rather than giving carbon credit to polluting oil corporations, corporations should be responsible for compensating local victims for fifty years of associated gas flaring and other abuses.

Notes

1 Oil Change International (2009) 'Gas Flaring', http://priceofoil.org/thepriceofoil/ human-rights/gas-flaring/.

2 Strategic Gas Plan for Nigeria, Joint UNDP/World Bank Energy Sector Management Assistance Programme (ESMAP) (February 2004).

3 For insight into the hegemony of neoliberalism and challenges to the system see Harvey, D. (2005) *A Brief History of Neoliberalism.* New York: Oxford University Press; and articles in *Development Dialogue*, 51, January 2009, http://www.dhf.uu.se/pdffiler/DD2009_51_postneoliberalism/Development_Dialogue_51.pdf.

4 Oilwatch's position on Voluntary carbon Markets, UNFCCC COP14, Poznan, December 2008, www.oilwatch.org.

5 Bond, P. (2008) 'Carbon Crash Emissions trading is a climate red herring. But is it now doomed by the financial meltdown, or will Obama rescue it?', *New Internationalist.*

6 Osuji, L. and G. Avwiri, (2005) 'Flared Gases and Other Pollutants Associated with Air Quality in Industrial Areas of Nigeria: An Overview', *Chemistry and Biodiversity*, 2.

7 Quoted in Environmental Rights Action and Climate Justice Programme (2005) 'Gas Flaring in Nigeria: A Human Rights, Environmental and Economic Monstrosity' June; from comments contained in a confidential 'reasonably comprehensive survey of the history, the present position and future prospects of the oil producing industry in Nigeria' provided by Mr. J.S. Sadler, the British Trade Commissioner in Lagos to the Economic Relations Department of the Foreign Office in London on 9th August 1963: 'Development of Oil Resources in West Africa 1963', File 371/167170, UK National Archives.

8 Anna Zalik examined LNG development in Nigerian and Mexico in Zalik, A. (2008) 'Liquefied Natural Gas & Fossil Capitalism', *Monthly Review.*

9 Economides, M.J., A. O. Fasina, and B. Oloyede (n.d.) 'Nigeria Natural Gas: A Transition from Waste to Resource', in *World Energy*, 7(1), http://www.worldenergy source.com/articles/text/economides_WE_v7n1.cfm.

10 In a presentation at an 'Investor Road Show' on Nigeria Gas Master-Plan, David O. Ige, Group General Manager of Nigerian National Petroleum Corporation (NNPC) reported that, as of 2007 Nigeria had the seventh largest proven reserves of natural gas.

11 In 2004, Shell acknowledged that it was doing business with misleading and false information by overstating its proven oil reserves in Nigeria. Independent media reported that that more than 1.5 billion barrels, or 60 percent of its Nigerian reserves, did not meet accounting standards for 'proven reserves'. Authorities in Britain and the United States investigated the fraud and imposed fines of £17m and $120m (£67m) respectively on the corporation for 'market abuse'. Nigerian authorities did nothing; http://www.nytimes.com/2004/03/19/business/shell-withheld-reserves-data-to-aid-nigeria.html.

12 http://www.nnpcgroup.com/performance/index.php.

13 It is interesting to note that the corporation reports a different figure for its flaring in 2007. Shell says that during the year it 64% of associated gas from its crude oil operations, adding that 'recent production losses have contributed to this fall', http://www.shell.com/static/nigeria/downloads/pdfs/brief_notes/shell_nigeria_harnessing_gas.pdf.

14 Nigeria National Petroleum Corporation, http://www.nnpcgroup.com/performance.

15 World Bank (1995) 'Defining an Environmental Development Strategy for the Niger Delta', 25 March.

16 In a meeting with Chevron and Shell officials in Amsterdam ahead of the November 2000 Climate Summit, Chevron's Chris Miller who was then Project Manager of the WAGP agreed that the estimates then being presented for emissions reduction from the WAGP were merely 'theoretical'.

17 On Monday, 14 November 2005, the Federal High Court of Nigeria, in Benin City has ordered companies to stop gas flaring in the Nigeria, as it violates guaranteed constitutional rights to life and dignity.

18 Clean Development Mechanism Project Design Document Form (CDM-PDD) for projects for 'recovery of associated gas that would otherwise be flared at Kwale oil-gas processing plant, Nigeria', http://www.dnv.com/focus/climate_change/upload/final%20pdd-nigeria%20ver.21%20%2023_12_2005.pdf.

19 Clean Development Mechanism Project Design Document Form (CDM-PDD) for projects for 'recovery of associated gas that would otherwise be flared at Kwale oil-gas processing plant, Nigeria', http://www.dnv.com/focus/climate_change/upload/final%20pdd-nigeria%20ver.21%20%2023_12_2005.pdf.

7

Carbon Sink Plantation in Uganda: Evicting People for Making Space for Trees

Ricardo Carrere

The UK-based New Forests Company[1] is establishing tree plantations in Uganda. The company states that 'whilst based on commercial forestry economics, our projects are underwritten by carbon credits ... in compliance with the Clean Development Mechanism'.[2] This means that its profits from the sale of wood will be increased by selling 'carbon credits' to polluting industries in the North. It also means that companies buying these carbon credits should be also held responsible for the impacts of these plantations on local peoples and the environment.

The story starts in 2004, when New Forests Company leased two plots of land from the Ugandan National Forest Authority: an area of 9,000 hectares in Namwasa Central Forest Reserve in Mubende district and 8,000 hectares in Luwunga Forest Reserve in Kiboga district. In both cases, local people (Ugandans) were living in the area and obtained their means of livelihood from its natural resources. However, government officials defined them as 'encroachers' and the managing director of the National Forest Authority, appealed to the government to help them out with the 'encroachers' and declared that once the encroachers were 'dealt with' they [the New Forests Company] would plant trees at Luwunga.[3]

It is interesting to note that the official document presented to the Executive Board of the CDM does not mention the fact that thousands of people would need to be displaced from the area to make way to the company's plantations.[4]

In the case of Kiboga, a 2008 news article[5] reported that at least 2,500 residents received eviction orders from the National Forest Authority (NFA). Kiboga Resident District Commissioner James Sserunjogi described the situation as follows: 'NFA officials have already destroyed crops belonging to people who are living in the forest reserves without informing us area leaders. This is totally wrong because NFA is there to ensure the well being of Ugandans.' He asked: 'Why do they evict them without our knowledge?'

Some of the affected residents claimed that they had lived on the land since 1975 and that it was wrong for NFA to evict them without first getting them alternative land: We should be given somewhere to go. We shall not just go away because we have nowhere to go with our families and cattle, an elderly man who identified himself as Mr William Butera said.

A NFA official in Kiboga, who refused to be named, said they had given the residents enough time to leave the forest reserves but that they refused to comply. 'What can we do now because the land in question was given to a company called New Forest Company to plant pine trees? The company wants to start work,' he said.

The district chairman, Mr Kizito Nkugwa, said over 4,000 residents in Kiboga District were landless due to evictions and that the issue was likely to result into insecurity in the district.[6]

In the case of Mubende, the following quotes from an article published in 2009,[7] provides evidence on how the New Forests Company's activities are impacting on local people. According to the article, residents in the villages of Kyamukasa, Kyato, Kicucula, Kisiita, Mpologoma, and Kanaamire denounced that armed groups were beating people, abducting them and destroying their crops and houses. Such actions were meant 'to subdue them to leave their land, which they have occupied for decades', so that the New Forest Company could plant its trees. 'My banana plantation on three acres has been destroyed by the people who are trying to evict us. They even took 10 bags of maize from me,' Jessica Nyinamatama, a 56-year-old widow, who is taking care of nine orphans, said.

The local land committee chairman, William Mpamira stated that 'Two of our neighbours were abducted by armed people who are trying to evict us', adding that 'Richard Twahirwa was arrested on June 26 and Cyprian Munyagaju was arrested on July 13. Up to now, we don't know their whereabouts.'

According to Mpamira, the population is suffering night attacks and as a result most residents have resorted to sleeping in the bushes. He also added that 'we doubt whether the intention of the company is to plant trees and protect the environment,' because 'since 2005, they have been cutting down trees which we had preserved for commercial timber.'

As a result of the situation they were suffering, the villagers decided to go to Kampala, where they petitioned the lands minister, Omara Atubo, to stop the evictions. In response, the minister vowed to stop the investor from evicting the residents and said: As a ministry in charge of land, we are saddened by what has happened to you. It is important to respect your rights irrespective of whether you occupy the land legally or not. There is no need for your colleagues to disappear, your property to be stolen or crops to be destroyed, Atubo said as the villagers applauded.

The minister said he would summon the resident district commissioner and the company officials to respond to the reports. Atubo also promised to lead a team of investigators to Kitumbi on a fact-finding mission.

'This is an urgent case because it is about life and death. These acts against our citizens should stop immediately. Investment is only good if the residents benefit from it. Human beings are more important than trees,' he stated.

The above should be sufficient to disqualify the company, but there is yet more to be said about its activities, which the company defines as 'sustainable and socially responsible forestry'. The meaning of this is shown clearly in the pictures and short text on its own web site,[8] which show that the 'responsible' process begins with the destruction of local biodiversity in two steps: 1) manual 'bush clearing' 2) 'chemical spraying'. Once the local vegetation – shown in the pictures – has been totally eliminated and the environment polluted with chemical herbicides, it is substituted by two fast-growing alien tree species (eucalyptus and pine) planted as monocultures over large areas of land. These green deserts are the 'New Forests' from where this company takes its name. In its presentation to the CDM, the company even dares to say that the conversion of what it claims to be 'degraded grassland and unproductive or not sustainably used agricultural lands' to pine and eucalyptus plantations will result in 'improving the local water flow' – when it is a well know fact that such plantations deplete water resources.

Evidence about how 'socially responsible' the company can be is also provided in the above mentioned pictures. Two of them show a few women working in very uncomfortable conditions in a makeshift tree nursery. Another photo shows a 16-strong 'clearing team' without appropriate protective clothing for the task. Finally, the 12 workers of the 'chemical spraying' team are shown from too far away to assess if they have been provided with the necessary protective gear and clothing.

Given that the company does not provide any information on the figure of 1800 workers that it says are 'expected' to work in the plantation,[9] one can only guess that most of them will be employed in tree planting and dismissed once that activity is completed.

But even in the impossible case that all the 1800 workers were to be employed on a permanent basis, the company fails to mention that over 10,000 residents of Mubende and at least 2,500 people in Kiboga have been or will be evicted to make way to its plantations. Which means that on balance at least 10,700 people will be in a far worse condition than before the company's arrival.

In spite of the above, on 25 May 2009 the company received Forest Stewardship Council certification which, according to this organization, enables 'consumers and businesses to make purchasing decisions that benefit people and the environment'.[10] Information on this company in the FSC web page is extremely scant, only stating the name of the certification company (SGS), the certificate code (SGS-FM/COC-006224) and the number of certified hectares (12,607).[11]

Given that this company plans to sell carbon credits within the CDM mechanism, the FSC certificate provides it with the necessary credentials for achieving CDM approval. FSC therefore has a two-fold responsibility: it is telling both wood consumers and carbon credit buyers that these plantations 'benefit people and the environment', which is clearly not true.

Additionally, in this case the 'carbon sink' capacity of these plantations is more than dubious, because the displacement of all those people may result in many more emissions than those allegedly 'captured' by the plantations. As Larry Lohmann clearly explained almost a decade ago 'any communities displaced from carbon plantations … would have to have their activities monitored closely for (say) a century, no matter where they had migrated to, to determine precisely to what extent they were encroaching on forests or grasslands elsewhere, and thus releasing the carbon stored in those ecosystems to the atmosphere.'[12]

Given FSC's past record in providing its label to socially and environmentally damaging plantations, we don't expect it to withdraw this certificate, but we at least demand a United Nations body such as the CDM to look seriously into what's happening with this company's plantations in Uganda and to reject its carbon project.

But what matters most is to highlight the suffering of the local people that are being harassed and evicted under the accusation of 'encroaching' on this company's lands. Who is the real 'encroacher'? The local people who need to make a living in their own country or a UK company that wants to make profits out of those peoples' lands? The answer is quite obvious for anyone who agrees that 'human beings are more important than trees'.

Notes

1 New Forests Company web page: http://www.newforestscompany.com/.

2 http://www.newforestscompany.com/about-us/.

3 Lang, C. and T. Byakola (2006) '"A funny place to store carbon:" UWA-FACE Foundation's tree planting project in Mount Elgon National Park, Uganda', *World Rainforest Movement*, http://www.wrm.org.uy/countries/Uganda/book.html.

4 UNFCCC CDM Executive Board (n.d.) 'The Namwasa Forestation Project (draft)'; http://cdm.unfccc.int/UserManagement/FileStorage/SDRN4ZBQ7XGTGK2W2TH9Y83L11ANG7.

5 Muzaale, F. (2008) 'Uganda: Kiboga Forest Residents Face Eviction', *The Monitor*, 5 February, http://allafrica.com/ stories/200802050894.html.

6 http://allafrica.com/stories/200802050894.html.

7 Mulondo, M. (2009) 'Uganda: Mubende Residents Petition Lands Minister Over Eviction, Harassment', *The New Vision*, 20 July; http://allafrica.com/stories/ 200907210016.html.

8 http://www.newforestscompany.com/project_area/uganda.

9 http://www.newforestscompany.com/index.php/project_area/7/.

10 FSC web page. About FSC; http://www.fsc.org/about-fsc.html.

11 FSC wweb page: FSC registered certificates. Uganda; http://www.fsc-info.org/VController.aspx?Path=5e8cddf3-9b09-46c6-8b11-2fbdad9e2d71&NoLayout=true.

12 Lohmann, L. (2000) 'The carbon shop: planting new problems', http://www.wrm.org.uy/plantations/material/ carbon.html.

8

Tree Plantations, Climate Change and Women*

Raquel Nuñez and GenderCC

Introduction

The impacts of monoculture tree plantations in countries around the world have been widely documented for many years now. It's a common trait that wherever they are established, local struggles arise against them.

The large-scale pattern of tree plantations resulting from an aggressive and thorough transformation of the landscape is far from a forest. Usually consisting of thousands or even millions of trees of the same species – often exotic – bred for rapid growth, uniformity and high yield of raw material and planted in even-aged stands, they require intensive preparation of the soil, fertilization, planting with regular spacing, selection of seedlings, weeding using machines or herbicides, use of pesticides, thinning, mechanized harvesting, and in some cases pruning. They typically occupy areas already being used in various ways by local people where they compete for the water and soil nutrients needed by other crops or by livestock. They can also cut off sunlight to crops planted in or near plantations.[1]

The high and huge blocks of fast-growing trees can boost species of either mammals, birds and insects, fungi and viruses species whose numbers had previously remained small but can rapidly become pests when large monocultural plantations are introduced. All these impacts have catastrophic effects in the environment and livelihood of rural communities.

Though the evidence of the pervasive impacts of tree plantations is now overwhelming, they continue to be promoted with a string of false claims: that 'plantations are forests,' that 'they protect forests', that 'they create jobs,' that they bring about 'development for local communities'.[2]

Tree Plantations and Climate Change

Climate change has become the most threatening issue for humankind. However, the leaders of the present world business era have been smart and irresponsible enough to turn it into a commercial opportunity.

The huge volumes of carbon emissions caused by wasteful consumption – that are at the root of climate change – can be compensated, say the modern carbon neutral 'magicians'. So the 'carbon offset' was born. The idea is: you emit CO_2, we store it and we charge you for the service. How do we store it? By, for example, planting trees. As if the carbon released through the use of fossil fuels – which has not been part of the functioning of the biosphere for millions of years – could be returned to its original underground storage place by performing other activities such as tree planting![3]

The Kyoto Protocol adopted in December 1997 as part of the United Nations Convention on Climate Change, enables industrialized countries to 'compensate' their carbon dioxide emissions, for example with the establishment of tree plantations in low-industrialized countries under the Clean Development Mechanism. The Kyoto Protocol has thus become another important actor in the promotion of large-scale tree plantations, endorsing the creation of an international emissions millionaire trading market for 'carbon credits' – an additional subsidy for the promotion of tree plantations.

Tree plantations are being publicized as carbon sinks. However, they have yet to prove this role. For one thing, many forests and other ecosystems are destroyed to make place for tree plantations, thereby releasing more carbon than that which the growing plantation can capture, even in the long run.[4]

And yet another crucial issue: will these plantations be harvested or not? If harvested, then they would at best be no more than temporary sinks, releasing most of the captured carbon as the paper or other products from the plantation are destroyed. If not harvested, then large scale tree plantations would be occupying millions of hectares of land at the expense of food production or of other ecosystems that contribute to the general environmental balance of the earth. Tree plantations also displace local communities and deprive them of their livelihoods, thus increasing social disintegration and the numbers of the hungry.[5]

The agrofuel business is yet another turn of the screw in the promotion of industrial tree plantations, creating another market outlet for oil palm as raw material for agrodiesel and likely to span other tree plantations, such as eucalyptus – including transgenic trees – for the production of cellulosic ethanol.[6]

Another new drive in the UNFCCC process is the proposed scheme on Reducing Emissions from Deforestation (and possibly degradation) in Developing Countries, known as REDD. It aims at operating as a financial incentive for reducing deforestation rates by paying Southern countries with tropical forests for doing so.

Apart from the many dangerous implications of this proposal, there is one that has to do with tree plantations: the definition of forests used by the UNFCCC – following FAO's – includes tree plantations. Thus, REDD monies may be used for the expansion of industrial tree plantations disguised as forests.[7]

The carbon offset market and other mechanisms that include tree plantations are false solutions to climate change. They just serve those who try to avoid implementing real changes to a high-carbon consuming economy that is socially inequitable and has led to the present environmental destruction.

Planting millions of hectares of trees in Southern countries not only is by no means a substitute for cutting emissions at the source but also has negative impacts on local communities, with differentiated effects on women and men.

Impacts of Industrial Tree Plantations on Women

When large agroindustrial operations such as monoculture tree plantations destroy the natural base of traditional communities, everybody is subject to material and cultural losses that lead to changes in their roles and status but women are generally left in a more vulnerable position than men.

Within indigenous, rural and peasant communities, women play a fundamental role as caretakers and food providers for their families and communities. Subsistence farming, fetching water and fire, keeping and exchanging seeds, collecting herbs and taking care of the children and the elders as well as cooking, washing and doing the cleaning are among the several tasks carried out by women.[8]

The arrival of large scale tree plantations deprive women from the access to the natural resources on which their livelihoods depend whether by destroying or depriving them from accessing to land, water, food, forest. Women are no longer able to fulfil their families' and own needs. The destruction of subsistence economies increases women's work overload as they have to work inside and outside the house for cash.

They usually end-up working as labourers for plantation companies in a male-dominated system occupying a marginalized position. They are forced into a hard survival struggle. Industrial tree plantations usually mean rural displacement, family and community break up, unemployment, low salaries and economic slavery.

What follows are some cases from around the world that reflect the impacts of large-scale monoculture tree plantations on women.

In Ecuador, the Dutch FACE Programme for Forestation in Ecuador S.A., or PROFAFOR project – promoted under the slogan of: 'Let us save the climate!' – attempts to 'sequester' carbon with pine plantations grown in the Andean Paramo region. A research on the impacts of those monoculture tree plantations on indigenous and peasant communities[9] shows that the Paramo soils store a great quantity of carbon in a thick layer. The loss of organic matter caused by a change in land use, such as the establishment of plantations of fast-growing species – pine and eucalyptus – is not compensated by an input of new litter: 'There is concern that because of its rapid growth it [the plantation] will need much water and therefore dry out the soil. With a drier soil some of the organic matter will disappear, not to be compensated by litter fall, because it is cuticulous, homogenous and foreign to the soil fauna. Thus there is a fixation of carbon by the trees above the soil, but a loss of carbon in the soil'.[10]

The FACE PROFAFOR forestry scheme has been established through forestation contracts signed between the company and land holders or local indigenous communities in exchange for the offer of an income from forestry activities. Under the contract the communities provided free labour force, went into debt to purchase seedlings and land and assumed all of the risks involved in the creation and maintenance of the plantations. The field research has shown that 'incentives provided by the company are insufficient to cover expenses that the communities have to incur to complete the establishment of the plantations. This means that less than six years after signing the contract the Paramo communities have already had to devote their own productive activities and institutions – such as grazing and social resources such as the Minga – to the service of PROFAFOR'.[11]

The disappearance of sources of water in the area was soon felt. Agricultural production, biodiversity, and the daily lives of the people, particularly women were affected. Before the arrival of the pines, water flowed from the springs to drink and to cook, give to the animals and use for the crops. Now, everything has changed for the worse. A field research on the impacts of tree plantations on women[12] gathered the testimony of a woman from the Sierra Paramo:

> We women are the ones who have to give water to the animals at noon and in the evening. We have to search for water to take to the cows because the spring has dried up, and sometimes the river is 40 or 50 minutes away. When we prepare food we have to fetch water and we have to take the kids with us, to get water from the streams, or else we have to dig deep down with a hoe where there have been no pines.

According to research carried out in Brazil[13] 'the impacts of monoculture plantations in Rio Grande do Sul are already visible: the serious drought in the south of the state, where eucalyptus production is most prevalent; the abrupt changes in temperature; the disappearance of the Pampa or temperate grassland biome, leading to the loss of extraordinary biodiversity; the decrease in food production; the drying up of water sources; the pollution and reduced water level in rivers; and the reduced fertility of the soil. Some cities have had to begin rationing water to make up for the shortage'.

In Brazil, where there are currently 5.3 million hectares of monoculture tree plantations, women refer to the river as a space to socialize. It has been a place for meeting, resting, chatting, sharing experiences and knowledge, reinforcing friendships and community ties while washing their bundles of clothes. Women from the Tupinikim village of Pau-Brasil, in Espirito Santo, Brazil, recall what the river used to be for them: it was like a party on the riverbank, all of them washing clothes. It was mostly on Saturdays, and for those who had time, during the week. It was one less chore, because there was all of that water in the river, and everything was easier, says Maria Helena. And Maridéia adds: 'We washed clothes, we collected water for drinking, for cooking… You could catch fish, you could scoop them up with a sieve. All those women… there would be so many there together! It was the place to wash clothes. You would finish washing clothes, then take a swim and leave, you know?'[14]

All those images are just memories after large-scale eucalyptus plantations – some 600,000 hectares – were established in the State of Espirito Santo by Aracruz Celulose doing away with watercourses that played an essential part in the lives of indigenous peoples like the Guaxindiba and Sahy, and the

Quilombola communities. Some rivers and streams practically disappeared while other became clogged with sediment and/or polluted.[15]

In Ecuador, large-scale pine and eucalyptus plantations were established on highland plains. The ENDESA-BOTROSA forest company came to destroy the native forests and replace them with plantations of fast-growing species. The research above mentioned,[16] report several testimonies of the impacts of the plantations on water resources:

> Now that the pine trees have grown the community is feeling the effects, because the water available for consumption and irrigation has gradually diminished. The community has a reservoir, but it is shrinking.

The destruction of primary forests, as in the case of oil palm, or the replacement of food-systems such as agriculture or grazing to give way to large scale tree plantations bring about food scarcity. Food is no longer available except in 'deep forest' areas where only men can go, thus increasing dependence of women on men to collect vegetables from the forest.

In Papua New Guinea, export-oriented oil palm plantations have cleared extensive areas of tropical forests. They have been established under the 'Nucleus Estate Smallholder Scheme' which means that a central company having its own plantation also contracts small farmers to supply it with oil palm fruit. The oil palm companies only pay the men of the family, although women also work in harvesting the oil palm fruit even long hours doing back-breaking work for little reward within oil palm plantations.[17]

Fertilizers and highly hazardous pesticides (such as Glufosinate ammonium, Glyphosate, cypermethrin, carbofuran, Paraquat, Diuron, Metsulfuron), insecticides (such as monocrotophos, methamidophos, carbofuran), and fungicides (such as chlorothalonil and maneb) are used in tree plantations. They are groundwater contaminants and possible endocrine disruptors and produce several health problems both chronic and acute on local populations. In Indonesia, women have reported[18] that often they had no idea about the possible effects of the pesticides they used, especially during the early stages of pregnancy. Women who were weeding were sometimes accidentally contaminated with sprays used by other workers nearby. Pesticides and fertilizers stored in people's homes presented hazards, particularly to women and children who could not read or understand the labels. Empty pesticide containers were occasionally used for domestic purposes and pesticides stored in containers such as old water bottles.[19]

Occupation of territories by industrial tree plantations change communities' livelihoods and thereby women's role. The Brazilian report demostrates 'the impacts of eucalyptus monocultures on women in the light of the experience of Tupinikim and Guaraní indigenous women and Quilombola women in the northern region of the state of Espírito Santo, Brazil'. The study shows how 'the disappearance of the majority of indigenous villages and Quilombola communities led a part of these populations to group together in the small bits of territory remaining of the surviving villages. Others sought out nearby regions to start their lives over'. 'But I would like it if we had the lands to have something better to offer to our grandchildren. To have our own land, to grow things, to raise animals, to have more space to live in... that would be good, right? Because living crowded together like this is very bad', said Rosa, from the Tupinikim village of Irajá. [20]

In Indonesia, oil palm expansion typically implies occupation of customary lands for deforestation and later establishment of oil palm plantations. Local communities displaced by oil palm companies usually have to work in the very plantations that displaced them or end up in urban slums. The important role of women in traditional societies – managing natural resources and maintaining sustainable livelihoods – is lost once plantations replace the forests and agricultural land.[21]

Those profound changes faced by women in the sexual division of labour and in the roles they play in the family and the community, have further exacerbated their subordinate status. In Brazil, according to the quoted research, 'The reduction of the large farms of the past significantly changed women's domestic activities. Women used to take care of their houses and gardens, grow herbs for domestic use, and raise small animals, which were also a source of food. Their children had enough room to play. Faced with the reduction of their territory, many people had to leave the places where they lived to look for work. As a result, many women became domestic workers, babysitters, day labourers, and washerwomen, among other tasks, often facing racial discrimination. In most cases, they work for officials from Aracruz or its subcontractors'. The new tasks affect their role as mothers, forcing them to give up breastfeeding their children at a very young age or to leave them with others while they are still infants, in order to look after the children of urban women.[22]

In Ecuador, 'We used to be able to grow really nice gardens, all of the crops came out really good. But now the forests have been destroyed and the land is drying up. The people let themselves get talked into the plantations. They told us that these wood trees were good, that they would help us, but after the plantations the land's capacity for production dried up, the trees suck up all of the nutrients, even the crops planted far away don't produce anymore', reports a woman from Azuay. Today women have to buy food products sold usually by men who receive the money, and so women are now more dependent on their husbands, some of whom 'spend all the money they get.'[23]

The social prestige of women in their role as food providers and caretakers holding an array of knowledges regarding how to treat diseases with herbs is lost as long as the base of their system is eradicated by tree plantations. Women lose independence and become disempowered.

While some jobs are offered by plantation companies, 'the labour conditions of women workers have much in common with those of men – low salaries, bad working and living conditions, seasonal work, outsourcing – but some degree of differentiation may be established with relation to their work in tree nurseries. In the nurseries of two large forestry companies in Minas Gerais, a large quantity of reiterated injuries caused by making great efforts have been observed, in spite of which women continue to work, many of them with swollen or bandaged hands. They also suffer from rheumatic diseases, probably caused by their constant exposure to cold water in the nurseries and to a generally cold environment in the wintertime', are some of the findings of the Brazilian report.[24]

Women working in oil palm plantations usually help their husbands to meet demanding production quotas, often doing unpaid work. They still have to take care of the children, elaborate the food and collect firewood and water, which now are farther away due to destruction of the forest by the oil palm plantations. In case women work on a hired basis, they often receive lower wages than men. The abrupt dissapearance of the ecosystems that sustained the ways of life of traditional peoples and local communities often disturbs men's role within the family and community/village. Many of them find themselves unemployed which in turn leads to rising rates of alcoholism and domestic violence bore by women and children, as well as exodus of men from rural areas. Quite frequently women have to deal with their home all by themselves.

Increasing prostitution is a common trait in areas where monoculture plantations are most prevalent. Foreign workers drawn by the companies' advertising campaigns and promises of job creation come to the region forming groups of workers without families, often unemployed, which has spurred the emergence of brothels around agroindustrial operations.

The life previous to the establishment of tree plantations was felt better for many women. For Eni, from the Quilombola community of São Domingos, Brazil, the company Aracruz Celulose

> destroyed a part of our life, of our freedom and our culture, our daily life, our health. We were happy, but not now, now our lives are unhappy, we have to fight for what is ours, for our territories, for everything they took from us, and when they took it we lost everything, everything that was ours, and all that's left for us is to protest, right? For us, and for the whole community.[25]

Women: From Victims to Champions

The expansion of monoculture tree plantations has brought about major social changes that serve to disempower women even more in relation to men when it comes to decision-making at the community level and even within the home.

However, such disempowerment is becoming a catalyst for a new empowerment of women. Once invisible members of the community, they are now finding their own voice, and making it heard increasingly louder.

The 1995 Beijing Conference was the forum where women's organizations from Southern countries criticized the hegemonic development model and stressed the responsibility of the North as the leading agent of environmental destruction.

In Brazil, grassroots women's organizations have emerged and given political and organizational response to policies and businesses that have environmental impacts on their lives. Since 8 March 2006 – when near 2 thousand peasant women from Via Campesina destroyed greenhouses and nearly 8 million eucalyptus saplings belonging to the pulp mill company Aracruz Celulose –

International Women's Day has become a day for action against land occupation by monoculture eucalyptus plantations carried out at the expense of food sovereignty and to the detriment of peasant production. GenderCC – Women For Climate Justice – believes that

> the challenges of climate change and gender injustice resemble each other – they require whole system change: not just gender mainstreaming but transforming gender relations and societal structures. Not just some technical amendments to reduce emissions, but real mitigation through awareness and change of unsustainable life-styles and the current ideology and practice of unlimited economic growth. Not the perpetuation of the current division of resources and labour but a responsible cooperative approach to achieving sustainable and equitable societies.[26]

The urgent need is 'to stop promoting false solutions that allow the rich to avoid the major changes they need to make; which help corporations to increase their profits and which have negative knock-on effects on the world's poor and the planet's ecosystem. We need a just transition to a low-carbon society that protects people's rights, jobs and well-being. Natural resources must be conserved for the common good, not privatized and unsustainably exploited. Local communities' sovereignty over land, energy, forests and water must be upheld and reclaimed.'[27]

Paraphrasing Vandana Shiva, the monocultures of the mind must be overcome in the search of a diverse, ecosystemic and loving world. 'We need to question the dominant perspective focusing mainly on technologies and markets, and put caring and justice in the centre of the measures and mechanisms', said Ulrike Roehr, of Gender CC.[28]

The knowledge, systems and networks of indigenous women for protection of biodiversity may well be inspiring. As Anna Pinto, from CORE, India, and part of Gender CC said:

> The relationship between fertility and regeneration, between female spirituality and the sacredness of the earth and its diversity, between sustainability and trusteeship rather than ownership and exploitation is the essence of indigenous culture, the essence of the significance of womanhood and women in indigenous society. It may also be the only ethic that can preserve and conserve our world for the future, any future at all.[29]

Notes

* This chapter was compiled by Raquel Nuñez of GenderCC from Women for Climate Justice's and World Rainforest Movement's books, briefings and position papers.

1 Carrere, R. and L. Lohmann (1996) *Pulping the South:Industrial Tree Plantations in the World Paper Economy.* Zed Books, http://www.wrm.org.uy/plantations/material/pulping.html.

2 'Ten replies to ten lies', *WRM*, http://www.wrm.org.uy/plantations/material/lies.html.

3 WRM (2008) 'Carbon Neutral Magicians', *WRM Briefing*, November 2008, http://www.wrm. org.uy/publications/briefings/ Carbon_neutral.pdf.

4 'Climate Change Convention: Sinks that stink', *WRM*, http://www.wrm.org.uy/actors/CCC/ sinks.html.

5 'Climate Change Convention: Sinks that stink'

6 WRM (2006) 'Oil Palm. From Cosmetics to Biodiesel. Colonization Lives On', http://www.wrm.org.uy/plantations/ material/BookOilPalm2.html.

7 For more reference on the REDD issue, see Chris Lang's chapter in this volume.

8 'Gender and Climate Change Network – Women for Climate Justice Position Paper', UNFCCC COP 13, Bali, Indonesia, Dec 2007, genderCC Network – Women for Climate Justice, http://www.gendercc.net/fileadmin/inhalte/Dokumente/UNFCCC_conferences/gender-cc-forest-final.pdf.

9 Granda, P. and A. Ecológica (2005) 'Impacts of the Dutch FACE-PROFAFOR monoculture tree plantations' project on indigenous and peasant communities'. http://www.wrm.org.uy/countries/Ecuador/face.pdf.

10 Granda, P. and A. Ecológica (2005) 'Impacts of the Dutch FACE-PROFAFOR monoculture tree plantations' project on indigenous and peasant communities'.

11 Granda, P. and A. Ecológica (2005) 'Impacts of the Dutch FACE-PROFAFOR monoculture tree plantations' project on indigenous and peasant communities'.

12 Ramos, I. and N. Bonilla (2008) 'Women, Communities and Plantations in Ecuador. Testimonials on a socially and environmentally destructive forestry model', http://www.wrm.org.uy/countries/Ecuador/Women_Ecuador.html.

13 Barcellos, G.H. and S.M. Ferreira (2008) 'Women and Eucalyptus. Stories of Life and Resistance. Impacts of Eucalyptus Monocultures on Indigenous and Quilombola Women in the State of Espírito Santo', http://www.wrm.org.uy/countries/ Brazil/Book_Women.pdf.

14 Barcellos, G.H. and S.M. Ferreira (2008) 'Women and Eucalyptus. Stories of Life and Resistance. Impacts of Eucalyptus Monocultures on Indigenous and Quilombola Women in the State of Espírito Santo'.

15 Barcellos, G.H. and S.M. Ferreira (2008) 'Women and Eucalyptus. Stories of Life and Resistance. Impacts of Eucalyptus Monocultures on Indigenous and Quilombola Women in the State of Espírito Santo'.

16 Ramos, I. and N. Bonilla (2008) 'Women, Communities and Plantations in Ecuador. Testimonials on a socially and environmentally destructive forestry model'.

17 'Papua New Guinea: Women most affected by oil palm plantations', *WRM's bulletin Nº 120*, http://www.wrm.org.uy/bulletin/120/Papua.html; 'Papua New Guinea: Life can be hard for women in oil palm plantations', *WRM's bulletin Nº 121*, http://www.wrm.org.uy/bulletin/121/Papua.html; 'Papua New Guinea: Women against further expansion of oil palm', *WRM's bulletin Nº 140*, http://www.wrm.org.uy/bulletin/140/Papua_New_Guinea.html.

18 WRM (2007) 'Indonesia: The impacts of oil palm plantations on women, Working Conditions and Health Impacts of Industrial Tree Monocultures', *WRM's Bulletin Nº 121* August 2007 http://www.wrm.org.uy/bulletin/121/viewpoint.html#Indonesia.

19 WRM (2007) 'Indonesia: The impacts of oil palm plantations on women, Working Conditions and Health Impacts of Industrial Tree Monocultures'.

20 Ramos, I. and N. Bonilla (2008) 'Women, Communities and Plantations in Ecuador. Testimonials on a socially and environmentally destructive forestry model'.

21 'Indonesia: Harsh conditions for women workers in oil palm plantations', *WRM's Bulletin Nº 134*, http://www.wrm.org.uy/bulletin/134/Indonesia.html.

22 Ramos, I. and N. Bonilla (2008). 'Women, Communities and Plantations in Ecuador. Testimonials on a socially and environmentally destructive forestry model'.

23 Barcellos, G.H. and S.M. Ferreira (2008) 'Women and Eucalyptus. Stories of Life and Resistance. Impacts of Eucalyptus Monocultures on Indigenous and Quilombola Women in the State of Espírito Santo'.

24 Ramos, I. and N. Bonilla (2008) 'Women, Communities and Plantations in Ecuador. Testimonials on a socially and environmentally destructive forestry model'.

25 Ramos, I. and N. Bonilla (2008) 'Women, Communities and Plantations in Ecuador. Testimonials on a socially and environmentally destructive forestry model'.

26 GenderCC Overview and Strategic Plan, July 2009.

27 GenderCC Overview and Strategic Plan, July 2009.

28 'Gender issues and climate change', *WRM's bulletin Nº 125*, http://www.wrm.org.uy/bulletin/125/Gender_issues.html.

29 WRM (2005) 'Women, Forests and Plantations. The Gender Dimension' *WRM*. http://www.wrm.org.uy/subjects/ women/text.pdf.

9

Shall We Still Keep Our Eyes *Cerrados*[1]?

Rafael Kruter Flores, Fabio Silva and Pedro Volkmann

Introduction

This chapter condemns the effects of one of the biggest forestry enterprise in Brazil. Plantar is one of the main companies trading carbon in Brazil, but it is also known for devastating the local environment and economy. The chapter will present two versions of the Plantar Project: the version of the Plantar Group, which says that the enterprise is helping to save the planet through the removal of greenhouse gases (GHG); and the version of the World Rainforest Movement,[2] which is engaged in uncovering the real effects of this activity, and thus shows that the Plantar Project is in fact destroying lives in many ways.

The *Cerrado*

Brazil is the fifth largest country in the World. It is characterized by a rich biodiversity spread across six different biomes: the Amazon forest; the *Cerrado* (savannah); the Atlantica forest; the *Pantanal* (swamp-land); the *Caatinga* (scrub-land); and the *Pampa* (prairie).

The Plantar project, object of this chapter, is located in the *Cerrado* region. This biome is a grassland ecosystem characterized by small trees that are widely spaced. The open canopy allows sufficient light to reach the ground to support an unbroken herbaceous layer consisting primarily of grass.

The Plantar Project

Plantar S.A. is a holding that brings together companies in the steel area (Plantar Siderúrgica S/A, founded in 1985) and eucalyptus monoculture (Plantar S/A Reflorestamentos, created in 1967), both located in the state of Minas Gerais, in the southeast of Brazil.

Initially dedicated to environmental engineering projects, in the 1980s Plantar S.A. expanded its production activities and became a supplier of iron ingots to both internal and foreign market. Today Plantar's productive activities are concentrated on the cultivation and exploitation of eucalyptus, planted in an area of approximately 230 Km². Eucalyptus serves as an energy source for the production of iron ingots, and it is also sold to companies that use cellulose or wood.

In 2000 Plantar S.A. developed the Plantar Project through a partnership with the Prototype Carbon Fund, operated by the World Bank (PCF).[3] The company sells part of its carbon credits to the PCF through a financial transaction partially subsidized by Rabobank International.[4] The Project aims at avoiding the emission of almost 13 million tons of GHG in the atmosphere in 28 years. The financing plan of the project was agreed upon three counterparts: US\$ 5 million from the Global Environment Fund; US\$ 30 million from various financial institutions and intermediary institutions; US\$ 16 million from the Plantar Group.[5] The project ensures the use of renewable fuel (charcoal from planted forests) instead of fossil fuel (coal) or non-renewable energy (charcoal from native forests) in the iron industry. It is based on Article 12 of the Kyoto Protocol.

The Plantar Project is the first Brazilian project to mitigate GHG approved by the World Bank, responsible for managing the PCF. It is also the first global financial transaction based on carbon credits for reducing CO2 emissions. The Plantar Group operates its businesses through two mechanisms of the Kyoto Protocol: Emissions Trading and the Clean Development Mechanisms. The trade works as follows:

– all countries have quotas or emission permission;

– industrialized countries and their firms that don't want to cut emissions or cannot cut them efficiently buy quotas from less polluting countries (as the case of Brazil) through the PCF;

– the quotas are traded through removal units and reduction units of greenhouse gases. While the removal units relate to conservation of native forest or environmental reforestation (which include the cultivation of eucalyptus and pine), the reduction units refers to the implementation of the Clean Development Mechanism.

According to the Plantar Group, the Plantar Project involves four activities, partially integrated:[6]

– *Forestry Activity:* CO2 removal and storage on 23.100 ha of sustainable eucalyptus plantations, established in place of pastureland – started in 2001.

– *Carbonization Activity:* Mitigation of CH4 emissions in the charcoal production process (wood carbonization) – started in 2004. Scientific research, specifically developed for the project, enabled methane mitigation by improving the process efficiency.

– *Iron Ingots Production Activity:* Avoidance of CO2 emissions in the iron ingots manufacturing process, through using renewable charcoal (carbon neutral) instead of coke or non-renewable biomass – started in 2007/2008.

– *Cerrado Restoration Activity:* Induced regeneration of 400 ha of native *cerrado* vegetation in non-forested land, beyond legal requirements.

The Plantar Project increases the Plantar Group's profits in two integrated ways: the commercialization of iron and wood and the sale of carbon credits. The first is a traditional way of profiting: the sale of the products produced in forestation activity and in the manufacturing of iron. The second is a new way of making money: the removal of GHG from the atmosphere made possible by the reduction of GHG emissions by using charcoal from planted forests for the iron manufacturing process.

The Plantar Project and its Effects

The Plantar Project is known for its innovation in the production of iron ingots and its efforts for reducing GHG emissions – particularly by those who are enthusiastic supporters of the Kyoto Protocol and carbon markets. However, the company is equally well known – particularly by those who are actively studying, analyzing and condemning what really occurs behind the discourse of the international hegemonic model of reducing the global warming – for the damages it has imposed on the local biodiversity and economy, and its contribution to exhausting of natural resources.

Many authors have critiqued the effects of the Plantar Project, for the burning of trees;[7] the expulsion of farm workers;[8] and the contamination of the soil and water by herbicides and pesticides.[9]

This chapter focuses on the research carried out by the World Rainforest Movement (WRM), which has been fighting against the destruction of rainforests and the scam of carbon trading with forest for some time.[10]

Most forestry companies in Brazil were established in the 1960s and 1970s during the military dictatorship which allowed attractive tax incentives. Plantar S.A. started to operate in 1967. At that moment, according to the Brazilian law, the company could not buy such huge extension of land property from the State. Hence, the negotiations were based on fraudulent methods and leasing contracts to occupy thousands of hectares of *Cerrado*. The immediate consequence of the introduction of eucalyptus plantations was the land eviction of thousands of farmers and Tupinikim and Guarani, the traditional Afro-descendent communities. Besides that, the plantations have resulted in increasing unemployment and the despair of local populations, who were left without land, biodiversity and water that enabled them to subsist.

Plantar S.A. is located at the city of Curvelo. In this area the eucalyptus plantations have dried up the rivers and contaminated the local fauna with toxic substances used for forestry management. In the state of Minas Gerais nearly two million hectares were planted with eucalyptus at the expense of burning part of the *Atlantica* forest and the *Cerrado*.

In 2000 the plantation of a tree nursery resulted in the deviation of almost 5 km of a road traditionally used by numerous inhabitants of the area, in order to avoid the 'dust' from the road affecting the eucalyptus seedlings being produced in the nursery. This had grave effects on students, teachers and the community in general, who use this route for walking to school and other places. Additionally, to supply its nursery with water, the company built three dams on the Boa Morte River, deviating the water consumed by the surrounding population and affecting its quality.

Furthermore, the labor conditions of the company in the production of charcoal and eucalyptus include illegal sub-contracting and slave and child labor.

The workers have had accidents and health problems, and there have even been cases of deaths.

Also, the new usage of the *Cerrado* has contributed to a crisis in the local economy, which is based on products from that native vegetation. Various food factories in Curvelo closed down due to lack of raw material, increasing the already high unemployment in the area.

Despite all of these practices, the company received FSC (Forest Stewardship Council) certification in 1998, but only for 4.8% of its land with eucalyptus plantations. This was sufficient for Plantar to sell the so-called 'carbon credits'. The certification enabled Plantar to submit a project to the PCF in which they stated they could help to stop climate change through the 'sustainable' plantation of eucalyptus trees. As a result, the Plantar Project started in 2000, ignoring all the ill practices listed above.

In March of 2003 representatives of citizen movements, churches, parliamentarians, city councilors and citizens of the state of Minas Gerais and its neighbor states, Espirito Santo, Bahia and Rio de Janeiro, sent a letter to the PCF directors stating their concerns over the expansion of large-scale monoculture eucalyptus plantations, which has caused a series of negative social, economic, environmental and cultural impacts. In April of that same year the company also sent a letter to the PCF refuting the concerns of the population by stating they had 'lack of knowledge'. The company then invited the main NGOs engaged in the campaign against the Plantar Project, FASE[11] and WRM, to a meeting at their office. The NGOs representatives stated they wanted to meet at the eucalyptus plantations, but this was denied by the company's sustainable development manager who alleged lack of time for that.

Given the situation, FASE and WRM turned down the company's invitation and visited the area to meet with local people. Here we reproduce their testimony on the visit:

> The overall impact of the company's operations were summarized by a local woman who simply said: 'Plantar finished with all we had.' The meaning of that was made very clear to us by the local people that showed us around the area. Within the plantations, the only thing green were the eucalyptus saplings and trees. The rest was brown, resulting from the widespread application of the herbicide glyphosate (Round-up). The water had either dried up or had been contaminated with agrochemicals, thus depriving local people with the fish they used to catch and eat. Local fauna – which constituted an important element for people's livelihoods – had also disappeared, making the 'hunting and fishing prohibited' sign posts a mockery. Hunt and fish what – said an angry local man – if the company has killed everything?[12]

Conclusions

Climate change is one of the main and most urgent issues we face. It has been at the forefront of the concerns of environmentalists and activists since at least the 1960s, but it was the Kyoto Protocol that finally gave visibility to it at a world stage. Despite all the talk of 'sustainable development' in the Kyoto Protocol, it does not consider the economic inequalities between North and South,[13] and between classes and social groups in national states of the South.

The Plantar Project shows a perverse confluence of two international movements. The first one started in the 1950s with the implementation of large scale monocultures of eucalyptus and pine plantations in Brazil and in other countries of the South, as a response to the needs of large industrial companies which were exhausting their traditional sources of raw material located in the North, mainly in Scandinavian countries, Canada and part of the USA. Hence we have seen a dramatic increase in large scale tree monocultures in many countries of Latin America, where the natural resources and climate offers good conditions for the growing of trees. The other movement started with the approval of the Kyoto Protocol. Today the trade of carbon credits allows countries from the North to keep polluting the air while supporting large scale monoculture in the South. Both moments of this perverse confluence are supported by a network of international and national agents, particularly multinational companies and governmental and inter-governmental institutions.

According to Misoczky, the large scale monoculture model was, in the 1950s, part of the so-called 'green revolution'. The Food Agriculture Organization (FAO) was a key player behind the implementation of the model in the South, and in subsequent decades a series of international agents also became participants in this process, such as the World Bank, the International Finance Corporation and many other international agents. At the national level, the state plays an important role, being ultimately responsible for paving the way for the implementation of the model.

> The first step is usually to carry out viability studies; a second and crucial step is to create or reform legislation to promote direct and indirect incentives in order to make the activity more profitable. Usually, Forestry Departments are created in order to implement national policies, to manage funds and to disseminate the wonders of the model in order to counter criticism from environmental and social organizations.[14]

The implementation of a foreign model of development is not something new in countries of the South. Let us just remember the history of colonization. Then and today colonialization demands an economic and political elite that is supported (and sometimes directed) from the outside. In our case the adoption of the large monoculture model was made possible because of a military dictatorship – partially supported from the outside – resulting in a variety of repressions of people and classes. Today, the justification of this kind of destructive production demands a powerful ideological apparatus, which, combined with economic coercion, provides the mechanisms by which the national elites reproduce themselves.

The Kyoto Protocol, with its carbon trade mechanisms, follows the same logic, and reinforces it. It is supported by a network of international agents, and at the national level it operates through national elites who benefit from destroying local economies and apropriating the common goods, such as the soil and water. The destruction of small farms, the expulsion of indigenous and afro-descendant people, the deviation of rivers, the sub-contrating of labour – all of these actions contribute to keeping local economies in a critical situation of poverty and 'underdevelopment'. At the same time, the economies of the North benefit from their allowances to pollute, by keeping their industries, which has resulted in their 'development' in the first place. The formula, 'development of underdevelopment', conceived by Andre Gunder Frank, may be applied here. For him,

> most studies of development and underdevelopment fail to take account of the economic and other relations between the metropolis and its economic colonies throughout the history of the world-wide expansion and development of the mercantilist and capitalist system. Consequently, most of our theory fails to explain the structure and development of the capitalist system as a whole and to account for its simultaneous generation of underdevelopment in some of its parts and economic development in others.[15]

The mechanisms of the Kyoto Protocol, following basic market principles, subordinate ecology to the capitalist economy. The Clean Development Mechanism allows the Plantar Group to affirm itself as an ecologically friendly company, when in fact it is responsible for destroying ecosystems and the life of people. In this sense, carbon markets perpetuate the logic of the development of the North and the underdevelopment of the South.

Notes

1   *Cerrado* in Spanish means 'closed'. In Portuguese, as we show in this chapter, it is the name of a local biome.

2   The World Rainforest Movement is an international network of citizens' groups of North and South involved in efforts to defend the world's rainforests. It was established in 1986, and it is headquartered in Montevideo, Uruguay; www.wrm.org.uy.

3   The objective of the PCF is to 'pioneer the market for project-based greenhouse gas emission reductions while promoting sustainable development and offering a learning-by-doing opportunity to its stakeholders.' The fund has U$ 180 million and involves seventeen companies and six governments. The aim is to 'pilot production of Emission Reductions within the framework of Joint Implementation (JI) and the Clean Development Mechanism (CDM). The PCF will invest contributions made by companies and governments in projects designed to produce Emission Reductions fully consistent with the Kyoto Protocol and the emerging framework for JI and the CDM' (www.worldbank.org).

4   Rabobank International is a Dutch multinational financial institution located in 43 countries and has a very positive evaluation by the financial market (considered the 4th safest bank in the world by Global Finance). Its market segment is in the area of agribusiness. www.rabobank.com.br.

5   www.worldbank.org.

6   www.plantar.com.br.

7   Reddy, T. (2005) 'Durban's perfume rods, plastic covers and', *Carbon Trade Watch*, http://www.carbontradewatch.org/index.php?option=com_content&task=view&id=180&Itemid=36; Johnston, I. (2007) 'A gift from Scotland to Brazil: drought and despair', *Carbon Trade Watch*, http://www.carbontradewatch.org/index.php?option=com_content&task=view&id=53&Itemid=36; Hartill, T. (2008) 'Durban Group Discusses Carbon Trading and LNG in Astoria', *Carbon Trade Watch*, http://www.carbontradewatch.org/index.php?option=com_content&task=view&id=119&Itemid=36.

8   Verde, P. (2006) 'Kyoto y los Vendedores de Humo', *Carbon Trade Watch*, http://www.carbontradewatch.org/index.php?option=com_content&task=view&id=167&It emid=36.

9   Checker, M. (2009) 'Double Jeopardy: Carbon Offsets and Human Rights Abuses', *Carbon Trade Watch*, http://www.carbontradewatch.org/index.php?option=com_content&task= view&id=307&Itemid=168.

10   The sources of the information used in this chapter are the World Rainforest Movement Bulletins number 60, 65 and 72; see www.wrm.org.uy.

11   FASE (Social and Educational Assistance Federation) is an NGO founded in 1961. Its main office is in Rio de Janeiro, but it is active all around Brazil. Its main objective is to contribute to a democratic society through a sustainable development alternative; www.fase.org.br.

12   http://www.wrm.org.uy/bulletin/70/Brazil.html.

13   According to Misoczky, M. (2009) 'Green deserts expansion in the South of Latin America: the role of international agencies and national states', in A. Guedes and A. Faria (eds.) *International Management and International Relations: A Critical Perspective from Latin America.* London: Routledge. pp. 149-167, North-South issues generally emphasize differences between the industrialized, 'developed' countries of the northern hemisphere and the less developed countries of Africa, Asia, and Latin America. It rests on the fact that the entire world's industrially developed countries (with the exception of Australia and New Zealand) lie to the North.

14   Misoczky, M. (2009) 'Green deserts expansion in the South of Latin America: the role of international agencies and national states'.

15   Frank, A.G. (1966) 'The development of underdevelopment', *Monthly Review*, 18, http://findarticles.com/p/articles/mi_m1132/is_n2_v41/ai_7659725.

10

Clean Conscience Mechanism: A Case from Uruguay*

Steffen Böhm

Introduction

In 2007, Eurostar, the Anglo-French high-speed rail company, announced that it was 'proud to offer carbon neutral journeys.' As part of its 'Tread Lightly' initiative, Eurostar explained:

> We have made a commitment to reduce carbon dioxide emissions by a further 25% per traveller journey by 2012. Consequently, we will be making changes across all areas of our business, from the big things like energy efficiency, paperless ticketing and waste management, through supply chain selection to smaller cultural changes like recycling in our offices... Any remaining emissions will be offset, at no cost to the traveller, meaning that from November 14th 2007, the opening day of St.Pancras International, Eurostar is proud to offer carbon neutral journeys.[1]

The 'Tread Lightly' initiative is supported by Friends of the Earth (FoE) UK, whose 'Big Ask Climate Change' campaign is, in turn, endorsed by Eurostar. FoE's then Executive Director, Tony Juniper, says: 'Eurostar is leading the way by making a real reduction' in carbon emissions.

Meanwhile, seven thousand miles away, in the Pampas region in South America, local landowners and a handful of multinational pulp and paper companies have discovered that the area is suited for growing huge eucalyptus tree plantations, which provide the raw material for the production of pulp and paper. There are already a number of existing pulp and paper mills in this area, and new ones are currently being constructed, turning the Pampas region into a growth area for the global wood pulp industry.

One of these new mills, constructed and operated by Botnia, the Finnish multinational pulp and paper company, is currently starting production in Fray Bentos, a small Uruguayan town on the banks of the River Uruguay, famous in Britain for corned beef and steak pies. The International Finance Corporation (IFC), part of the World Bank, which helps to finance this project, says that this mill 'will help the country [Uruguay] move up the value chain beyond the export of raw materials, while generating some 2,500 much needed local jobs... The plant will generate value added equivalent to 2 percent of Uruguay's entire GDP.'[2] Additionally, Botnia is planning to generate environmentally friendly electricity from biomass in the power plant which is part of the pulp mill. The

IFC claims that the electricity sold by the Botnia mill to the national grid 'can be called green power because it is produced using biomass, which is a renewable resource.'

What links Eurostar and Botnia is the Clean Development Mechanism (CDM), an arrangement under the Kyoto Protocol which allows industrialized countries to invest in projects that reduce emissions in developing countries as an alternative to making more expensive emissions reduction in their own countries.

When Eurostar says that it will 'offset' all those carbon emissions that it cannot avoid itself, and when it claims that all Eurostar train journeys are now 'carbon neutral', it means that the company purchases so-called 'carbon credits' in a number of emerging carbon trading schemes, of which CDM is by far the largest. Eurostar (and hence its passengers) finance carbon reduction projects in developing countries, such as Botnia's biomass electricity generation project, in the hope that this will reduce the planet's overall carbon emissions.

The connection between Eurostar and Botnia's mill in Uruguay is not direct. Capitalist markets are always impersonal: the links between buyers and sellers are hidden, as the commodity ('carbon' in our case) can be traded from one place to the other, concealing the labour that has produced it in the first place – as Marx explained in *Das Kapital*. So, I'm not claiming here that Eurostar directly finances a pulp and paper mill in Uruguay. Indeed, this is one of the problems with the emerging carbon markets. Often one cannot trace what one's carbon offsetting money is really doing to distant communities around the world; one cannot make direct links between carbon sellers and buyers, which means that one cannot scrutinize the carbon reduction claims made.

Nonetheless we have a duty to open the black box of these carbon markets. The money that Northern companies, such as Eurostar, spend on being 'green' and 'carbon neutral' can all too easily be used to prop up industries in the South which are run by neo-colonial Northern companies such as Botnia, whose practices, taken as a whole, may actually increase global greenhouse gas emissions.

Three-Legged Profit Machine

Botnia's offsetting project at Fray Bentos financed by the CDM consists of a 32 Megawatt biomass-based electricity generation plant. Electricity will be generated in the pulp mill's power plant on mill site. About 270 Gigawatt hours will be generated annually – enough to supply all the electricity consumed by 150,000 Uruguayan homes. The project is designed to use black liquor (renewable biomass material derived from the pulping process) for steam and electricity generation in the recovery boiler. Botnia claims that burning eucalyptus to generate electricity emits less greenhouse gas than traditional oil and gas-based electricity generation; and that it will sell the surplus electricity to Uruguay's national grid, thus offsetting 68,000 tons of carbon dioxide a year.[3]

At first, this sounds like a beneficial arrangement for all concerned. Uruguay's economy is boosted, its consumers get electricity, and the environment is spared. It is also good news for Botnia, which stands to profit three times over from its eucalyptus plantations: first, by turning pulp wood into paper; second, by selling electricity to the Uruguayan grid, and third, by selling carbon credits to polluting countries and companies in the North. It seems to prove the point that green and sustainable development is indeed possible, and that companies which do 'good' can still make a healthy profit. But, as is often the case with such mega-developments, all is not quite what it is made out to be.

Green Soldiers

The first major problem with the Fray Bentos scheme is that its main raw material, eucalyptus, is mass-produced in very harmful ways, leading to an array of negative impacts on local communities. Eucalyptus plantations are just as problematic as other biofuels grown across the developing world at the moment, leading to shortages in many core food categories, not to mention the neo-colonial landgrabbing that is a hallmark of large agribusiness operations throughout the so called developing world.

Botnia, for example, through its subsidiary Forestal Oriental SA (FOSA), currently has 89,000 hectares of eucalyptus planted, and a further 103,500 acres available for future use. The aim is to provide the Fray Bentos mill with 3.5 million cubic meters of wood annually, 70 per cent of which will come from its own plantations and the remaining 30 per cent from farmers contracted to Botnia.[4] The company, as well as the IFC's impact studies, claim that these tree plantations are fully sustainable, no adverse environmental effects result from them, and they create employment for rural people in Uruguay. Additionally, Botnia boasts that 'all of Forestal Oriental's plantations have received FSC certification'.[5] That's alright then.

However, even if the tree plantations are fully certified, these control mechanisms don't provide a full picture of what is happening on the ground. Nor do they explain why eucalyptus plantations are universally detested by those who live near them. According to one writer:

> In Brazil, plantations are referred to as 'green deserts' due to their reputation for destroying biological diversity. In South Africa they are known as 'green cancer' because of the tendency of the eucalyptus in the plantations to spread wildly into other areas. In Chile plantations are called 'green soldiers' because they are destructive, stand in straight lines, and steadily advance forward.[6]

Eucalyptus trees originate from Australia where they thrive in a dry climate, developing very deep roots to access water. Plantations have spread around the world because they are fast-growing (on the pampas the trees are mature in about six to seven years) and eucalyptus is fast growing because it is greedy. Each tree consumes up to 100 litres of water per day, so a whole plantation can lower the water table which local people rely on. A World Rainforest Movement (WRM) study on the impact of monocultures in the backyard of the new Botnia plant in Uruguay reported a host of complaints from local residents. A farmer in Guichón, whose land is now surrounded by FOSA plantations, complained that as a result the Boyado stream, which runs though his farm, had completely dried up. In an area called Paraje Pence in the department of Sorianoto one local man stated: 'All the people here have been left with no water; I have a little bit but the well is dirty. Close to here where my father lives there's no water at all.' Another villager told WRM: 'I've lived here my whole life, and we never had any problems with water until they established all these plantations around eight years ago. Now we depend on the local government to bring us water.'[7]

Eucalyptus plantations are also called 'green deserts' because they allow nothing else to grow within them; and plantation managers use herbicides and pesticides to ensure that their tree plantations remain monocultures. 'From a biological perspective, eucalyptus forests are inferior to other types of reforestation, due to their homogeneity and low biodiversity. In this sense, the use of the term 'forest' for these plantations is misleading, but it continues to be manipulated as an ideological tool by the cellulose-producing companies.'[8] In regions with large-scale eucalyptus plantations 'the rivers have been degraded by pollution caused by wide-spread use of pesticides and a process of desiccation, compromising fishing and the quality and quantity of drinking water.'[9]

Eucalyptus plantations are likely to become even more artificial if current proposals to plant genetically modified trees with reduced levels of lignin become a reality. Lignin is a natural glue-like substance that holds wood cells together and makes trees strong and inedible. Because lignin causes yellowing of paper, any lignin remaining has to be bleached away, so paper made from low-lignin trees would be less polluting. However, trees with reduced lignin are more susceptible to viral infections and pest attack, and therefore require increased pesticide use; and there is a risk that reduced-lignin GM trees might cross-fertilize with other trees and spread these characteristics into the wider forest environment.[10]

Pulp Affliction

The renewable electricity generated by Botnia's mill is only made possible because of the pulp processing industry. Thousands of pages of reports, commissioned by the IFC and other governmental and extra-governmental bodies, have concluded that no adverse social and environmental impacts are produced by the new Botnia mill[11] – but many local residents and some environmental groups from within Uruguay have consistently argued the exact opposite. Pulp processing has been labelled 'one of the three most polluting industries of the planet', because of the following problems, all examined in a study on pulp mills carried out by the World Rainforest Movement.[12]

– Size and scale: Today's pulp mills are mega-factories and their very size makes them a risk. The effluents from a large 600,000 metric ton plant are approximately 1000 litres per second.' In an industrial process using so many toxic chemicals, any small release is magnified because of the scale of the factory. Toxic chemical releases may be small compared to the volumes processed, yet more than an ecosystem can support.

– Smell and other emissions: Emissions into the air by pulp mills contain carcinogenic chemical compounds causing hormone imbalance, and reduced sulphur compounds which give off a 'rotten egg' smell that becomes a problem for the surrounding inhabitants'.

– Bleaching agents: To produce white pulp and paper, bleaching agents are needed. 'Many chemical bleaches are reactive and dangerous to transport and for this reason must be made in situ or nearby. This is the case for 21 chlorine dioxide, an extremely reactive greenish yellow gas that explodes easily, representing a major threat to the workers and the neighbouring inhabitants in the event of an accident. Another agent used, elemental chlorine, is very toxic. It is a greenish gas that is corrosive in the presence of dampness.'

– Effluents and water pollution: 'The enormous demand for water in pulp mills may reduce the level of water and the effluents may increase the temperature, a critical issue for the river ecosystem. Generally, mills are installed near a watercourse with a good flow where they can get their supply (at a lower cost) and also discharge their effluents. Chemical and organic residues can combine to produce pollutants that may reduce the oxygen levels in the watercourses where they are released and prove lethal to fish. Studies have revealed genetic damage, hormone changes, liver alterations, cell function problems, changes in blood composition, skin and brachia lesions and reactions by the fishes' immunological system.'

– Chlorines: The pulp industry is the world's second largest consumer of chlorine and the greatest source of toxic organochlorines in watercourses. Some effluents produced in pulp production may combine to form dioxins, furans or other organochlorines which biodegrade slowly and can accumulate in the tissues of humans or other living creatures.

Incidents of contamination have frequently occurred at other locations and continue to do so. In Valdivia, Chile, for example, CELCO (the forestry subsidiary of the Angelini group) opened its new pulp mill in 2004, five years behind schedule because of protests. 'Less than a month later, the nearby communities began complaining about the unbearable smell from the mill.' But bad smells were not the only problem. 'Faced with repeated complaints, environmental and health authorities began to set up inquiries… They found categorical evidence establishing that the company had no system for emissions abatement, control and monitoring.'[13] Serious water contamination from the mill, registered in the nearby Nature Sanctuary Carlos Anwandter at the Rio Cruces, was linked with the death and sickness of dozens black-necked swans, an endangered migratory bird.

The CELCO plant is designed to produce 550,000 tons of bleached pulp annually. When Botnia's mill is in full production it will have an annual output of 1 million tons of bleached eucalyptus pulp, one of the biggest mills of its type in the world. To gain an idea of the volume, consider that a factory of this size needs to be serviced by over 200,000 HGV journeys a year, or one every 2.5 minutes, 24 hours per day, every day. The environmental impacts of pulp production are likely to get worse with the large scale of plants being built today.

It is mainly because of the pollution that the people of the Argentinean town of Gualeguaychú, which overlooks Fray Bentos from the other side of the river, have been up in arms protesting against the project. The town is an important tourist destination, famous for its annual carnival, which draws thousands of visitors to the city every year. Tourists also come to Gualeguaychú to enjoy its tranquil river shores, fishing and water sports. No one whose livelihood depends largely on tourism or agriculture, wants to have a giant pulp mill constructed in their back yard.

Development for the Overdeveloped

However, while the people of Gualeguaychú look on the construction of the Botnia Mill with foreboding, many in people in Fray Bentos and Uruguay welcome the investment into the Botnia pulp mill, which constitutes the country's largest foreign direct investment in its history, and will establish the country as one of the world's major pulp exporters. The project is expected to generate revenues equivalent to two percent of the country's GDP, and to create 2,500 jobs, of which 300 will be in the mill and the rest in ancillary forestry and transport. The project fits in with the World Bank Group's long-term strategy for the development of Uruguay, which recommends investments in forestry and in the diversification of the country's export base to increase its competitiveness globally.[14]

Whether Uruguayans will actually benefit from these revenues is another matter. The plant is being built in a Zona Franca – one of the many Free Trade Zones installed in developing countries over the past decades. These designated areas provide easy investment opportunities for multinational companies without burdening them with national taxes and other unwanted costs. The land for the pulp mill was rented to Botnia for $20,000 for 30 years – enough to rent a flat in London for a year. Botnia does not have to pay any customs duty on machinery and equipment imports, most of which is manufactured in Finland, nor does it have to pay income tax under the free trade area contract. The profits will mainly be given in the form of dividends to foreign shareholders and thus exported out of the country; that is, back to Finland. So, it's actually a development of the Finnish economy. Furthermore, the government has:

> provided forestry companies with generous subsidies, soft credits, and tax exemptions. Over 12 years, the Uruguayan government's support for this sector exceeded $500 million in tax exemptions and direct disbursements, an amount representing almost 4 percent of the country's annual GDP. To facilitate the transportation and export of the wood, the governments of the day made further investments in new ports, bridges, roads, and railway lines.[15]

While this could all be seen as long-term investment in the economic growth of the country, the 'problem is that future investors will certainly demand equal treatment from the state, and the companies will continue to avoid paying taxes. A factory of this size, representing Finnish interests is a powerful economic agent in a country like Uruguay; granting tax exemptions to encourage foreign investment means that this power is transferred to foreign companies'.[16]

These arrangements maintain the unequal power relationships between North and South that have been in place ever since the colonization of South America 500 years ago. Virtually all of the production of the Botnia mill is for export, serving the Northerners' wasteful consumption of ever more pulp and paper. People in the European 'knowledge economy' consume up to 430 kg per head per year, on everything from junk mail to government reports, compared to only about 40-50 kg in the Pampas region.[17] This means that the jobs that are being created in the South are dependent on the wasteful over-consumption in industrialized countries – and will disappear if ever we in the North put our house in order.

What is not taken into account by the IFC and other development institutions is the amount of jobs that are being destroyed. Brazil's Landless Workers Movement (MST) says that a corporation such as the huge pulp firm Aracruz 'creates only one job for each 185 hectares planted, while a small farm property creates one job per hectare.' A Via Campesina poster even claims 5 jobs for every hectare.[18]

In effect the eucalyptus plantations perpetuate the South American tradition of large latifundia, estates covering vast areas of fertile lands, which originally were violently expropriated from indigenous people. As Eduardo Galeano has described so vividly in *The Open Veins of Latin America*, ever since the European invasion, Latin America's lands have served to provide goods in demand in Europe. First it was sugar, then coffee, cacao and cotton; today it is soya, maize, and eucalyptus. These monocultures were made possible because local elites and foreign proprietors owned vast estates, while peasants, forced off the land, have been driven into cities such as Sao Paulo, Rio de Janeiro and Buenos Aires.

Resistance

The plantations have therefore become a focus of resistance for social movements such as the peasant organization Via Campesina, and Brazil's Landless Workers Movement (MST). The MST has been engaged in the fight against the Aracruz' eucalyptus plantations in the Brazilian state of Espírito Santo, where indigenous communities have been struggling to reclaim thousands of hectares of land stolen from them under the Brazilian dictatorship in the 1970s. In August 2007 the Tupinikims and Guarani indigenous people declared victory when the Brazilian government decided that Aracruz should return to them 14,227 hectares of illegally occupied land.[19] On 8 March 2006, on International Women's Day, about 2000 women from Via Campesina occupied an Aracruz plantation in Rio Grande do Sul, 'denouncing the social and environmental impacts of the growing green desert created by eucalyptus monocultures. 'These social movements campaign for real development, by the locals for the locals, where '100% of production [is] destined for the tables of Brazilian workers'.[20]

Resistance against pulp mills and eucalyptus plantations has also been inspired by the long struggle of The Citizens' Environmental Assembly of the Argentinian city of Gualeguaychú. The campaign initiated by environmentalists grew to represent a wide cross-section of town's population from university teachers and business professionals to pensioners and farmers.' They organized road blockades, internet campaigns, legal challenges against Uruguay, and other more clandestine actions, such as the imitation of the corporate website of Botnia.[21] Their slogan, 'No a la papeleras, Si a la vida!' (No to the cellulose plants, Yes to life!) can be seen everywhere in the city: on cars, in shops, in restaurants, and on billboards. Their campaign made national news over three years and although they didn't manage to stop the construction of the Botnia plant, they have succeeded in delaying the construction of another pulp mill planned by ENCE, the Spanish multinational, right next to the Botnia factory. ENCE is now looking to build the plant further down the river.

A Global Scam

Botnia and its financiers want us to believe that an industry with a long track record of pollution, land rights violations and other negative impacts is sustainable, and are using the electricity generation side of the project to give it a green gloss. But even the claim that the electricity is carbon neutral is spurious, because the releases generated by the project as a whole are not taken into account. Besides the emissions arising from the construction of the factory, there are all the carbon releases resulting from project operation: the emissions from the factories producing chemicals associated to pulp production; the consumption of fuel by forestry machinery; timber transportation by trucks to the factory; port movements; and fuel consumption by ships taking pulp to paper factories in Finland and China, etc. A full life cycle analysis of all these energy costs would almost certainly show that 'total releases of greenhouse effect gases by Botnia will be higher than those that would have occurred in the country without its presence'.[22] And the sole purpose of this carbon expenditure is to ensure that we in the North can continue to consume ten times as much paper as people in Uruguay.

This kind of greenwashing is not unique to Botnia. Celulose Irani was the first Brazilian pulp and paper company to sell carbon credits under the CDM, when, in 2006, it sold US$1.2 million worth of credits to Shell, which will use them to continue exploring, drilling, flaring, spilling and polluting.[23]

Nor is this scam unique to the pulp industry. The single largest project type applying for the CDM is hydropower, with more than 400 large dams in China alone applying for credits, while biomass power plants like that at Botnia are the second biggest project type.[24] Like pulp mills, hydro schemes are riddled with environmental problems, and are responsible for displacing hundreds of thousands of peasants from their land. And like the Botnia power plant, many hydro-electric dams are 'non-additional' – that is to say that they would have been constructed anyway, even if there had been no finance through the CDM, so in effect carbon credits are not reducing carbon emissions at all, but simply subsidizing 'business as usual'.

In the last ten years, carbon has become a new commodity. Carbon trading and offsetting is an industry that grows at an alarming rate without any serious checks and balances in place to monitor the real progress in reducing carbon emissions worldwide. Already there is overwhelming evidence that the carbon markets do not work, in terms of the objective they were created for: reducing greenhouse gas emissions. Not only have they failed to introduce significant carbon reductions; in the case of Botnia, and many similar projects, they have the very opposite effect of what they were intended to achieve: they legitimize a further increase in greenhouse gas emissions and prolong the introduction of the measures that will force the Northern countries that have caused climate change to significantly reduce their emissions.

Notes

* This paper was first published in *The Land*, Issue 7, Summer 2009, pp. 6-11.

1   http://www.stpancras.com; see also

http://www.eurostar.com/UK/uk/leisure/about_eurostar/environment/tread _ lightly.jsp.

2   IFC (2006) 'IFC and MIGA Board Approves Orion Pulp Mill in Uruguay 2,500 Jobs to Be Created, No Environmental Harm', Press Release, 21 Nov, http://www.ifc.org/ifcext/media.nsf/content/SelectedPressRelease?OpenDocument& UNID=F76F15A5FE7735918525722D0058F472).

3   IFC (2006) 'IFC and MIGA Board Approves Orion Pulp Mill in Uruguay 2,500 Jobs to Be Created, No Environmental Harm'; and Botnia (2006) 'CDM Project', http://www.botnia.com/en/default.asp?path=204,1490,1494,1373, accessed 17 Nov. 07.

4   http://www.botniauruguay.com.uy/index.php?option=com_content&task= view&id=72&Itemid=96, accessed 19 Nov. 07.

5   Quoted by Chris Lang, http://chrislang.org/2007/05/24/subsidies-and-thebotnia-pulp-mill; FSC stands for Forest Stewardship Council (http://www.fsc.org). The organization sets and controls 'international standards for responsible forest management'. It is, however, not without its critics; see http://www.fsc-watch.org.

6   Petermann, A. and O. Langelle (2006) 'Plantations, Indigenous Rights, & GE Trees', *Z Magazine*, 19(3). http://zmagsite.zmag.org/Mar2006/langelle0306.html.

7   World Rainforest Movement (2003) *Bulletin* N° 72, http://www.wrm.org.uy/bulletin/72/Brazil.html; and Chris Lang (2004) 'Social and environmental impacts of industrial tree plantations', Presentation at WRM/WALHI Southeast Asia Regional Meeting on Oil Palm and Pulpwood Plantations, 29 November, http://chrislang.org/2004/12/02/social-and-environmental-impacts-of-industrial-tree-plantations/.

8   FASE/Green Desert Alert Network (2002) 'Economic, Social, Cultural and Environmental Rights Violations in Eucalyptus Monoculture: Aracruz Cellulose and the State of Espírito Santo', p.27, http://www2.fase.org.br/ downloads/2004/09/553_relat_desc_es_ing.pdf.

9   World Rainforest Movement (2003) *Bulletin N° 72*.

10 World Rainforest Movement (2005) 'Pulp Mills: From Monocultures to Industrial Pollution'. *Montevideo: WRM*, http://www.wrm.org.uy/plantations/Celulose_text.pdf.

11 http://www.ifc.org/ifcext/lac.nsf/content/Uruguay_Pulp_Mills.

12 The information in this section of the article is taken from: World Rainforest Movement (2005) 'Pulp Mills: From Monocultures to Industrial Pollution'. *Montevideo: WRM*, http://www.wrm.org.uy/plantations/Celulose_text.pdf.

13 WRM (2004) 'There's something stinking in southern Chile', http://www.wrm.org. uy/bulletin/83/Chile.html.

14 IFC (n.d.) 'FAQs: Uruguayan Pulp Mills', http://www.ifc.org/ifcext/lac.nsf/ Content/Uruguay_PulpMills_FAQ.

15 Pierri, R. (2006) 'Pulp Factions: Uruguay's Environmentalists v. Big Paper', *Corporate Watch*, http://www.corpwatch.org/article.php?id=13111.

16 http://chrislang.org/2007/06/30/banks-pulp-people-part-1.

17 World Rainforest Movement (2004) Bulletin, 83, http://www.wrm.org.uy/bulletin/83/scenario.html.

18 MST (2006) 'Women of La Via Campesina Take Action Against the Green Desert on International Women's Day', http://www.mstbrazil.org/?q= viacampesinaaction march8, accessed 3 Dec. 07; Via Campesina advert; http://viacampesina.org/main_en/images/stories/pdf/panfleto_ dverde_ing.pdf.

19 MST (2007) 'Brazilian Government: The Lands Occupied by Aracruz Should be Returned to the Indigenous', http://www.mstbrazil.org/?q=aracruzdefeatedbrazil2007.

20 Via Campesina (2006) 'Solidarity action with the Women of La Via Campesina in Rio Grande do Sur (Brazil)', http://www.viacampesina.org/main_en/index.php?option= com_content&task=view&id=129&Itemid=37, accessed 3 Dec. 07; MST (2006) 'Women of La Via Campesina Take Action Against the Green Desert on International Women's Day', http://www.mstbrazil.org/?q=viacampesinaactionmarch8, accessed 3 Dec. 07.

21 Cerri, C.M. (2006) 'El Corte de Gualeguaychú. Buenos Aires: Dunken', p. 13. Bogus website is http://www. botnia.com.ar.

22 WRM (2006) Bulletin, 109, http://www.wrm.org.uy/bulletin/109/viewpoint.html #Uruguay.

23 Lang, C. (2006) 'Paper production + oil = clean development', http://pulpinc.wordpress.com/2006/09/12/paper-production-oil-clean development.

24 International Rivers and Rainforest Action Network, Initial Analysis of Offsets Provisions in the Draft of the American Clean Energy and Security Act of 2009 (ACESA), 15 April 2009.

11

India's 'Clean Development'*

Soumitra Ghosh and Hadida Yasmin

Introduction

After carbon trading was conceptualized in the Kyoto Protocol, India seems to have been the busiest country to put the concept into action. By early August 2008, India had 355 CDM projects accounting for about 31% of the world's total of 1136 projects registered with the CDM Executive Board of the UNFCCC. India's share is highest among all countries, with China standing second with 250 projects. About 2700 million CERs (certified emission reductions) are expected to be generated by 2012, if all these hostcountry-approved projects in India go on stream. By 8 August 2008, a total of 179,888,442 CERs had been issued to projects worldwide, with India accounting for 25.83% and China for the maximum 35.56%.

If we observe the distribution of all UNFCCC-registered projects by scale, there are 611 large-scale projects and 525 small-scale ones, with the energy industry (renewable and non-renewable sources) predominating with 796 projects. Taking into account all projects in various stages – such as those already registered, requested for registration, and waiting for validation – the total number of CDM projects in India comes to be 1021. These include both unilateral and bilateral projects. Projects with involvement of a third party (any Annex-I country) are called bilateral. Countries that are financing most of the bilateral projects in India are the United Kingdom, Switzerland, the Netherlands, and Japan. Other countries involved are Sweden, Germany, Spain, Italy, Austria, France, Canada, Denmark, Finland, and so on.

Project Status (Including bilateral ones)	Number of Projects	kCO2/ Year*	2012 kCO2/ Year**	kCERs Issued***
Validation	620	41,127	190,132	
Registered	355	31,471	214,572	45,385
Registration Requested	46	2,303	11,290	
Total	1021	74,901	415,994 (It will be 808,537 by 2020)	45,385 (from 151 registered projects)

Table 1: Overview of CDM Projects in India (as on 9 August, 2008)[1]
*Annual reduction claimed in 1000-tons of CO2-equivalent per year
**Total reduction to be claimed in 1000-tons of CO2-equivalent by 2012
***Saleable CERs, in 1000-tons of CO2-equivalent, officially issued by the UNFCCC so far

By 2012, all these projects are expected to generate a total of 415,994 kCERs after registration. The 355 registered projects in India, and the 46 more that have requested for registration, are expected to generate 214,572 kCERs by 2012. Among the registered projects, 151 have been issued 45,385 kCERs (Table 1).

Sector-wise Distribution

Most CDM projects in India come primarily under four sectors: biomass (293 projects), energy efficiency (239), wind (208), and hydro (103). Other sectors include fossil fuel switch (42), biogas (31), cement (21), landfill gas (17), and HFC (6). Though HFC comprises the least number of projects, it is expected to generate the maximum quantity of kCERs (78,566 by 2012). Out of the six HFC projects, four are registered and they have already been issued 28,814 kCERs. Table 2 gives an account of sector-wise emission reductions to be achieved by CDM projects in India by 2012.

State-wise Distribution

With 153 CDM projects, Maharashtra's share is the maximum in the country in terms of number. Out of these 150-plus projects, 66 are wind energy projects. If all of them get registered, they would generate 8548 kCERs by 2012 Registered projects across sectors in Maharashtra have already been issued a total of 1245 kCERs, of which 545 kCERs are from wind and 295 kCERs from hydro, while 205 kCERs come from cement and 143 kCERs from biogass. Maharashtra with its 153 registered projects will generate 5852 kCERs annually and 32,623 kCERs by 2012.

Tamil Nadu comes second in terms of number with 141 CDM projects. At 6511 kCERs per year, the state is expected to generate 40,810 kCERs by 2012. The state now has 36 registered projects, and 14 of them have been issued a total of 1444 kCERs. Tamil Nadu has the country's maximum number of wind projects – 69, of which 17 are registered – with 368 kCERs issued and a potential to generate 18,459 kCERs by 2012. Five of its nine registered biomass projects have been issued a total of 399 kCERs. Tamil Nadu comes third in terms of number of biomass projects it hosts (31), after Karnataka (35) and Andhra Pradesh (48).

Gujarat – though it does not have as many CDM projects as Maharashtra, Tamil Nadu, and Karnataka – tops the list in terms of CERs issued (18,772 kCERs) and is also expected to generate the maximum quantity of CERs by 2012 (97,673 kCERs). This is because of its two HFC projects, which have already been issued 17,955 kCERs and are expected to yield 40,459 kCERs by 2012. Out of the total 108 CDM projects in Gujarat, EE (energy efficiency) projects account for 30 and fossil-fuel switch projects for 19.

Very interestingly, Rajasthan with its 53 CDM projects stands second in terms of CERs 'issued (11,456 kCERs from its 14 registered projects (total projects 23) and is expected to be issued another 45,504 kCERs by 2012. Here also, just a single HFC project accounts for 10,518 kCERs. In Rajasthan, 50% of the CDM projects are wind energy projects; out of which seven have been issued a total of 352 kCERs.

Sector	Number of CDM Projects	2012 kCERs*	kCERs issued**
Biogas	31	6,549	304
Biomass Energy	293	67,395	4,215
Cement	21	19,599	923
Energy Efficiency (EE) Energy Distribution	1	234	0
EE households	4	860	0
EE industry	133	18,935	527
EE own generation	106	64,962	6,486
EE service	7	287	2
EE supply side	16	10,456	159
Total of EE		95,734	7,174
Fossil Fuel Switch	42	47,720	794
Fugitive Emissions	12	4,451	0
HFCs	6	78,566	28,814
Hydro Energy	103	34,296	965
Landfill Gas (Waste energy)	17	4,895	76
Reforestation	5	1,018	0
Solar Energy	5	1,280	0
Transport	2	288	0
Wind Energy	208	45,848	2,120
Others	9	8,208	0
Total	1021	415,994	45,385

Table 2: Sector-wise emission reductions by 2012[2]
*Total reduction to be claimed in 1000-tons of CO2-equivalent by 2012
** Saleable CERs, in 1000-tons of CO2-equivalent, officially issued by the UNFCCC so far

Among hydro projects, both Karnataka and Himachal Pradesh host 25 each, Out of the total of 103 in India, Karnataka holds 14 registered hydro projects, with 307 issued kCERs. More than 80% of CDM projects in Himachal Pradesh are hydro projects.

The highest number of biomass projects in India is in Andhra Pradesh, with 48 projects. Out of 24 registered biomass projects in the state, 17 have already been issued 2027 kCERs. Karnataka, with 35 biomass projects is expected to generate 11,283 kCERs by 2012.

In West Bengal, out of 44 approved projects, 33 (75%) are EE projects, which have been issued 278 kCERs. These EE projects are expected to generate 9332 kCERs by 2012.

Emissions Reduction or Business Expansion!

With the Kyoto Protocol turning emissions reduction – arguably, the most important responsibility on humankind today – into profitable 'business', corporations could not have asked for more. Looking at India's CDM scenario in terms of corporate participation, we find that the energy sector, including HFC, is generating the maximum CERs.

The unfortunate fact is that big corporations such as Tata, ITC, Reliance, Ambuja, Birla, Bajaj, GFL, HFL, NFIL, and many others, who keep on emitting millions of tons of carbon dioxide into the biosphere are earning handsome returns in the name of 'clean development mechanism' (Table 3). The current market price of a ton of CO_2 reduced and sold in form of CERs in the global market is generally between 15 and 20 euros, whereas the most optimists of carbon consultants would not have given more than 3.5 euros in 2005! While society gains nothing, corporations reap huge benefits from the business of a new kind.

More than 98% kCERs of CDM energy projects are run by big corporations. More than 50% of the total kCERs (45,386) issued to India went to its four HFC projects by Gujarat Flurochemicals Ltd., Chemplast Sanmar Ltd, Navin Fluorine International Ltd, and SRF Ltd. HFC projects will be issued another 76,212 kCERs by 2012, promising huge monetary returns.

Out of the total 3770 kCERs issued to India's biomass projects (up to 6 June 2008), 3726 kCERs went to the corporate sector. Big corporations also own most of the CDM wind projects in India; corporate-owned projects account for 5824 out of the total 6960 kCERs being generated annually from all wind projects in India. In case of fossil-fuel switch projects, all 788 kCERs issued went to big corporations. So, as usual, the corporate sector has made new fortunes from the CDM regime.

Some of the profit figures for companies engaged in the carbon trade are astounding. Till early 2008, the Jindal group made 11 billion rupees (and perhaps more) from selling supposedly 'reduced emissions' (1.3 million CERs) at their steel plant in Karnataka. Tata Motors sold 163,784 CERs from clean wind projects at 15.7 euros/CER in 2007. Tatas' sponge iron projects in Orissa are set to yield 31,762 CERs every year. Reliance publicly boasts of its CDM Kitty – with seven projects registered with 88,448 CERs per year (till 2007 December), four more CDM projects under validation with 149,533 CERs per year, and seven more potential CDM projects to generate about 400,000 CERs per year. In 2006/7 alone, the GFCL group's earning from carbon money was twice its total corporate assets.

The point is not why they are earning so much! The disturbing fact is that their PDDs are full of half-truths and lies: claiming something, doing something else, and, in the end, showing yet another picture about what they have achieved. Most of the CDM projects we studied in Maharashtra are as polluting as any other industrial project, besides exhibiting barefaced violations over the mandatory social commitments and environmental norms.

Then, how is it possible to pass off these projects as clean ones? Well, in India, the emerging economic superpower, everything is clean; even if you discover layers of fly ash in the food you are about eat, it is clean – especially if it has emanated from a nearby CDM project run by some big corporation! Or else, how do they even get *green* prizes – the bigger the corporation, the more prestigious the prize!

RIL (Reliance Industries Ltd), India's largest private sector entity with businesses in the energy and materials value chain, whose group's annual revenues are in excess of US\$ 34 billion, has recently received a coveted *green award*. RIL's Hazira manufacturing division has bagged the *Golden Peacock Award for Combating Climate Change–2008*). According to the jury, headed by former chief justice of India and UN Human Rights Commission member P N Bhagwati, Reliance grabbed this award for promoting 'energy efficiency' as much as 'controlling greenhouse gases' by initiating various CDM projects.

Owner of CDM Project	Sector	Number of Projects	kco₂/ Year*	2012 kco₂/ Year**	kCERs issued***
Tata	Wind	3	133	836	167
	EE Own Generation	4	663	3521	106
	EE Industry	7	49.7	400	4
	Biomass Energy	1	24	115	
	Biogas	1	7.2	61	19
Birla	EE Industry	4	18.5	178	
	Cement	1	43	436	18
Reliance	Wind	2	108	538	
	EE Industry	6	208	1267	115
	Fossil Fuel Switch	1	1169	6041	
REI Agro	Wind	3	78	434	
Synergy Global Pvt. Ltd	Wind	11	415	2138	
Loyal Textile Mills Ltd.	Wind	3	74	378	
Jindal	Wind	1	15	71	
	EE Own Generation	3	533	3338	
	EE Industry	1	50	265	
	Biomass Energy	1	33	162	
Bannari Amman Sugars Ltd.	Wind	1	19	102	
	Biomass Energy	3	253	1481	
Enercon	Wind	18	1206	6792	349
Essar Power Ltd.	EE Own Generation	3	885	6010	
	Fossil Fuel Switch	3	885	6010	
	EE Industry	1	136	656	
Shri Bajrang Power & Ispat	EE Own Generation	1	108	789	182
	EE Industry	1	9	42	
	Biomass Energy	1	34	172	
Satia Paper Mills Ltd.	EE Industry	1	15	80	
	Biomass	2	55	324	

	Energy				
Indo Rama Synthetics	EE Industry	2	9.5	59	
	Fossil Fuel Switch	1	11	67	
	Cement	1	43	248	
Aditya Birla	EE Industry	6	75	429	12
Haldia Petrochem	EE Industry	1	34	135	
	Fossil Fuel Switch	1	131	657	
H&R Johnson (India) Ltd.	EE Industry	1	35	167	
	Biomass Energy	2	21.5	103	
GACL	EE Industry	1	4.6	46	
	Cement	3	345	3131	
	Fossil Fuel Switch	2	108	1046	
Grasim	EE Industry	2	43	252	
	Biomass Energy	1	52	402	22
Mawana Sugars Ltd.	Biomass Energy	5	190	1118	
BAJAJ	Biomass Energy	6	180	1070	
JCT	Biomass Energy	3	102	592	86
Dwarkesh Sugar Ltd.	Biomass Energy	2	102	530	
	Biogas	1	40	192	
Birla Corporation Ltd.	Cement	2	69	700	78
	EE Industry	2	18.5	178	
ITC	EE Supply Side	1	4.0	40	13
	Reforestation	1	49	470	
	EE Industry	5	108	929	201
	Biomass Energy	2	138	1000	
Chemplast Sanmar Ltd.	HFCs	1	539	5392	342
Gujarat Fluoro-chemicals Ltd.	HFCs	1	3393	51778	12948
	Wind	1	52	243	
Hindustan Fluorocarbon Ltd	HFCs	1	464	4644	
Navin Fluorine Int. Ltd.	HFCs	1	2802	28022	1215
SRF Ltd.	HFCs	1	3834	38336	9624
Acme Tele Power Ltd.	HFCs	1	25	109	
NEG Micon (I) Pvt Ltd	Wind	4	118	821	

Table 3: Big Indian Corporations and their CDM Revenues[3]
*Annual reduction claimed in 1000-tons of CO2-equivalent per year
**Total reduction to be claimed in 1000-tons of CO2-equivalent by 2012
*** Saleable CERs, in 1000-tons of CO2-equivalent, officially issued by the UNFCCC so far

With India's unprecedented thrust on industrialization during the past two decades, big companies are increasing their manufacturing process by the day, thus increasing their turnover. And while doing so they are adding greenhouse gases to the atmosphere like never before. The irony is that they are also making bucketful of money simply by putting a so-called 'clean development' tag to some of their dirtiest projects.

How the Carbon Markets Work

All clean development mechanism or CDM projects need to get themselves registered with the CDM Executive Board of the UNFCCC. Registration does not, however, mean that the projects can go to the market immediately and sell their CERs. A project can only sell its CERs 'officially' when the UNFCCC issues those. Such officially issued CERs fetch the maximum price in the carbon market, because it is assumed that the 'delivery' is guaranteed, or, in other words, the projects are really, beyond any doubt, reducing emissions. Projects without UNFCCC issuance certificates (and even without registration) can still go to the unregulated offset market, and sell VERs (verified emission reductions), which means the validating agency has certified that such projects are promoting 'clean development'. This does not get the same price as a UNFCCC-issued CER is known as 'secondary CERs' in the market, while CERs from a registered project – but not officially issued – are known as 'primary CERs'. Depending on the ability of the broker – and the nature of the marketplace – a VER can get anything between 5 to 10 euros. In comparison, while CERs can fetch as high as 26 euros (the price peaked in last July), the last one-year average stands at around 19 euros.

Unless there is a prior and direct ERPA (emission reduction purchase agreement) with a particular brokerage concern, consultant, or, rarely, an European buyer, secondary CERs are usually sold through various climate/carbon exchanges in Europe and America, though, of late, Asian exchanges have come up, one of them in India. The endbuyer for Indian CERs is usually untraceable, and the exchanges give only bulk sales figures and, that too, not always. Similarly, unless and until a project declares its CER revenues, there is no 'public' way to know how much money a particular project makes, and whether the figures given by the project-proponents in their red-herring prospectus and annual reports are at all correct.

Most Indian projects are unilateral, which means they do not have a specific buyer lined up at the time of registration. Though this apparently increases marketing risk, the arrangement seems to suit most Indian companies, who are in this game simply for more money. Being tied with no specific buyers gives them good bargaining opportunities, and further, to indulge in speculation. Indian projects have been repeatedly reported to hoard CERs for higher prices! This tendency of hoarding, of course is, not confined to unilateral projects. Going by the fact that most of the Indian projects to which CERs have been issued so far are bilateral (which means it declared an 'other party' from Annex-I countries at the time of registration), it is evident that all Indian projects, small and big, unilateral and bilateral, are out for a kill.

Status	Number of Registered Projects	Number of Registered Projects with CERs issued to	kCERs issued*
Bilateral	171	134	46,135
Unilateral	185	20	619
Total	356	154	46,754

Table 4: Unilateral and Bilateral projects in India with CERs issued to[4]
* Saleable CERs, in 1000-tons of CO2-equivalent, officially issued by the UNFCCC so far

One thing has to be said, though: the CDM Executive Board of the UNFCCC has so far been consistently niggardly in issuing CERs to unilateral projects. Indian unilateral projects have only been issued a paltry 619,000 CERs (up to 26 August 2008), whereas the bilateral projects got a whopping 46.15 million! Many of the bilateral projects, especially the HFCs, have been issued many times, whereas only one unilateral project (0112: Nagda Hills Wind Energy Project) was issued twice, rest only once. Does it happen because the UNFCCC considers such projects to be cleaner? Does the 'other-party' involvement in the bilateral project have any influence in making the issuance process faster? Why the discrimination then, when both unilateral and bilateral projects show a characteristic disregard for the declared principles and guidelines of the CDM?

The lure of easy money has led to a muster of vultures in the carbon market; all kinds of speculators, consultants, self-professed carbon gurus, and now the hedge funds and private equity funds have set up their own shops in India. Futures trading in CERs/VERs has picked up in recent months, which means that CDM projects are entering into secure deals with traders who now carry the project's risk burden (the greatest risk is it being rejected by the CDM Executive Board, which seldom happens) in lieu of the larger share of sales profits. It is quite possible that we will see increased financing of new CDM projects by both hedge and private equity players, and given the essentially unregulated, shady, and non-transparent nature of their operations, such projects will continue to be dirtier and more fraudulent. Already the larger parts of the issued credits from Indian projects are being purchased by new carbon finance companies, private equities, and banks. A look at the credit buyer section in the UNEP CDM Pipeline confirms the presence of big names in the field: Meryl Lynch, BNP-Paribas, ABN-AMRO, and so on. CDM is a big money game, and big players have arrived (Table 5).

Fraud? Yes, one must clearly use the word, talking about carbon trading in general, and Indian CDM projects in particular. The main problem with these projects' tall – and immensely profitable – claims of reducing greenhouse gas emissions is that there is no credible and definite way to verify these claims. The validating agency is an organization paid by the project – not for 'validating' the project, but precisely for 'establishing' what the project is claiming is true. Though it ritually invites comments on projects it validates, such comments are, as a rule, ignored. The result is that dirty and utterly ineligible projects sail through, and make money, without bothering to clean up their acts.

Credit Buyers	Number of Projects	Annual kCO_2*	kCO_2**	kCERs issued***
Germany (KfW)	14	4243	35328	15
UK (Cantor Fitzgerald Europe)	10	769	3383	15
Sweden (Carbon Asset Management)	17	1350	12981	2058
Switzerland (Ecoinvest Carbon)	9	298	2400	489
UK (ABN AMRO Bank)	9	226.5	2318	581
UK (Agrinergy)	52	3268	21879	598
UK (EcoSecurities)	9	379	2255	187
UK (Nobel Carbon)	20	225	20130	7554
UK (EcoSecurities)	9	379	2255	187
UK (Merrill Lynch)	3	69	624	146
France (BNP Paribas)	1	35	271	46

Table 5: Major Credit Buyers in Indian Carbon Market[5]
*Annual emissions reduction claimed in 1000 tons of CO2-equivalent per year
**Total reduction to be claimed in 1000-tons of CO2-equivalent by 2012
*** Saleable CERs, in 1000-tons of CO2-equivalent, officially issued by the UNFCCC so far

The biggest instance of this is the waste-heat-based energy projects, mostly located in various sponge iron plants. These projects are legally required to operate ESPs (electrostatic precipitators) to ensure that the smoke emitted by the plants remain reasonably clean. Because an ESP is an expensive machine to run, the plants mostly do not operate it. And, because the ESP remains inoperative most of the time, the waste heat project, which is technically dependant on continuous running of the machine, does not work. That the ESPs do not run is known to everybody – the State Pollution Control Boards, the villagers near the plants, and the workers. Yet, the Indian government approves these projects' CDM claims, the validating agency validates, and the UNFCCC registers and issues CERs. Quality-wise, there is no difference between VERs and CERs from such a project; the pollution caused by it continues all the same. The UNFCCC certification means a few more wads of paper from the validating agency, and the occasional methodological explanations offered by the project proponent. Contrary to the popular belief, such papers, however well-written and convincing, prove nothing, least of all, the emissions reduction claims.

Notes

* This article was first published in *Mausam*, 1(1): 19-26.
1 www.unfccc.org; www.cdmpipeline.org.
2 www.cdmpipeline.org.
3 www.cdmindia.nic.in; www.unfccc.org; www.cdmpipeline.org; www.iges.or.jp.
4 www.unfccc.org.
5 www.cdmpipeline.org.

12

Where is Climate Justice in India's First CDM Project?*

Siddhartha Dabhi

Introduction

December 1997 saw a historic step taken in the direction of climate change mitigation with the adoption of the Kyoto Protocol under the leadership of the UN Framework Convention on Climate Change (UNFCCC), which set binding emissions reduction targets for 37 industrialized countries. The Kyoto Protocol which has been framed in the shadows of neoliberalism, established a series of market mechanisms under the so called label of 'carbon markets' to combat global warming.[1] The basic idea behind carbon markets is to convert carbon emissions reductions into a commodity, which could then be traded between countries, corporations and even individuals, assuming that the trade leads to emissions reduction at the lowest possible cost.[2]

In this paper I critically engage with these carbon markets, specifically by analyzing the Clean Development Mechanism (CDM) project of the Indian company Gujarat Fluorochemicals Limited (GFL), which was India's first and one of the biggest.

Today, more than a decade after the adoption of the Kyoto Protocol, the CDM is growing in size with more than 4200 projects in the pipeline including 1822 projects already registered which will be producing more than 2,900,000,000 CERs (Certified Emission Reduction units) which can be traded internationally.[3] But with this growth in size, there has also been a growth of critiques of the CDM, some of which are collected in this book. This chapter adds to this critical literature by providing new evidence showing the inequalities created by the GFL CDM project.[4]

GFL's CDM Project

Gujarat Fluorochemicals Limited (GFL) was founded in 1987, and production at the plant started in 1989-90. GFL is located in the state of Gujarat, which is in western India. In Gujarat the GFL plant is located in the village of Ranjitnagar in the Panchmahal District. It is engaged in the production of refrigerant gases like R11, R12, HCFC22, etc. Recently, the company has added another product to its repertoire: CERs, which stands for Certified Emission Reduction Units, the new commodity created out of the UN controlled Clean

Development Mechanism (CDM). This is how the company introduces itself and its carbon trading product on its website:

> GFL is amongst India's largest refrigerant gas manufacturing company. In the course of manufacture of HCFC22 (a coolant widely used in air-conditioning and refrigeration applications), HFC23 is generated as a waste product, which is a potent greenhouse gas, with a global warming potential equivalent to 11700 MT of carbon dioxide. As a part of GFL's larger business plan to create a sustainable future, it is one of the few companies in India involved in Carbon Trading. Of the 15 projects approved by the United Nations Framework of Climate Change Convention (UNFCCC) so far, four are Indian and GFL is one of them. It has today the technology in place to bring down the emission levels of greenhouse gases and sell certified emission reduction credits (CERs) to developed countries. GFL is setting up a project for Greenhouse Gas Emission Reduction by Thermal Oxidation of HFC 23, at Gujarat in India. This project has been registered by the Executive Board of the Clean Development Mechanism (CDM), established under the Kyoto Protocol. Apart from being the largest project in India, it is also the first Indian & third in the world to be registered as a CDM... GFL expects to generate more than 3 million tones of CERs annually, which is expected to go up in the future as HCFC22 production grows. These CERs can be traded internationally and can be used as a compliance tool under the Kyoto Protocol as well as several other trading markets like the EU Emissions Trading Scheme. Trade in compliance grade emission reductions is expected to grow to € 10 billion per year by 2008, according to industry estimates.[5]

GFL's CDM project is thus to do with the destruction of HFC23, which is a very potent greenhouse gas, using the process of thermal oxidation. The technology for the thermal oxidation of HFC23 has been imported from a UK based company called Ineous Fluor Limited. The other key players involved in the CDM project are: Cooperatieve Centrale Raiffeisen Boerenleenbank B.A. (Rabobank), Netherlands, Sumitomo Corporation, Japan, and the Government of India. Rabobank is the mediator between GFL and the Government of the Netherlands for the purchase of CERs produced from the project. Sumitomo Corporation provides operations and maintenance assistance through Daikin Industries (Japan), and it also facilitates the sale of CERs in Japan. The Government of India acts as the Designated National Authority (DNA). The Project Design Document (PDD) has been prepared by the Mumbai office of PricewaterhouseCoopers (PwC). The PDD is the most important publicly available document outlining the CDM project's aims and objectives, including its technical design, contribution to emission reductions as well as its environmental impact, socioeconomic benefits and contribution to sustainable development.

As far as sustainable development is concerned this is what the PDD has to say about the project's contribution to sustainable development:

> GFL has expressed its strong commitment to the sustainable development activities by committing a total fund of Rs. 7 Crores (Euro 1.375 Million) approximately for the life of the entire CDM project out of the revenues received if the project is approved and once there is revenue stream from sale of

CERs. These funds will be used for selected community development activities such as education; vocational training; employment; agriculture; sanitation, hygiene & environment; water management; medical and animal health, which will contribute significantly to the well being of the local population and poverty alleviation. Towards water management, for example GFL has estimated that under the Panam River Basin Plan (Sardar Patel Jal Plan) the community contribution for check dam construction shall work out to Rs. 50 lacs half of which is proposed to be provided as a catalytic fund by the project promoters.[6]

On paper, then, this looks like a great project: HFC 23, which is one of the most potent greenhouse gases, is eliminated from the production; technology and capital is transferred from the North to the South; sustainable production and employment in the South is safeguarded and even enhanced; overall greenhouse gas emissions in the world is reduced. The problem with this official story is that it is only a 'half-truth'. That is, the official, often very technical, information presented on the official UNFCCC/CDM website, which is the result of a very bureaucratic decision and control mechanism, hides a lot of facts and doubts of a range of people about GFL's CDM project.

As we know, the devil lies in the details! The problem with this superficial analysis provided by the official literature is that it cuts out a lot of information about the real environmental, social and economic impact of the GFL plant in the local surroundings. Let us therefore present some of this data and information that we feel is important to understand the true impact of the GFL production and its approved CDM project.

Socio-economic Geography

The GFL plant is located in Ranjitnagar village, with Halol, the nearest town, 16 kilometres away. The villages surrounding the plant are Kankodakoi, Nathkuva, Jitpura, Chandranagar, Tarkheda, Arad, and Ranjitnagar. But the villages that are mainly affected by GFL are Ranjitnagar, Nathkuva, Kankodakoi and Jitpura.

To be able to better understand the GFL CDM project and its implication it becomes very important to understand the geographic and socio-economic constitution of the villages surrounding the plant.

- The four most affected villages by GFL's production of potent greenhouse gas like HFC23 are within a 2 kilometre radius of the plant.
- Out of the four villages most affected by the GFL plant, Ranjitnagar, is located 1.5 kilometres north west of the plant, which is in the opposite direction of the prevailing winds and also opposite to the ground water direction, hence it is the least affected by the plant in comparison to the other three villages. Ranjitnagar comprises mainly of Patel community farmers, which are considered to be rich farmers. Nathkuva, Jitpura and Kankodakoi are located within 2 kilometres south west and south east, which directly get hit by the emissions coming from the GFL plant as they fall in the wind direction as well as in the ground water direction. Amongst the four villages Kankodakoi and Nathkuva are the worst hit. Kankodakoi

constitutes of one of the most vulnerable communities known as the Schedule Tribes.

- The 'Pavagadh-Champaner' World Heritage Site is located less than 10 kilometres away from the GFL plant.

The area where the GFL plant is located is considered to be an industrially backward area. The villagers are mostly subsistence farmers, with the rest of the employed population working as factory workers in GFL or other factories located more than 20 kilometres away from their villages. The villages mostly consist of the economically backward Kshatriya caste and the Schedule Tribes. The population in the four worst hit villages is approximately 1200 people per village, and each village constitutes of almost 800 families. The per-capita income of most of the families is below the poverty line and the villages lack basic amenities like pure water for drinking and irrigation, sanitation facilities, energy, proper infrastructure like roads and lighting. The literacy levels are very low in the villages. The health services in the region are also not well developed.

As far as employment is concerned, the main source of employment is agriculture. Most of the farmers are subsistence farmers. The main crops they cultivate are corn, cotton, rice and maize. They also grow vegetables like onion, green peas, etc. Thus the farmers in this area cultivate both, Rabi as well as Kharif crops. The average investment per season is around Rupees 5000. The average income from agriculture in Jitpura, Nathkuva and Ranjitnagar is approximately Rupees 15,000-20,000. The average production level per season is around 1500-2000 kilograms. But the villagers of Kankodakoi village, who are basically tribal, are subsistence farmers, and hardly earn anything from agriculture and thus have to migrate to nearby town or city in search of employment. The average land-holdings of these farmers is, around half an acre.

The other major source of employment is labour in factories. The wages that the labourers get in GFL and in the industries in the nearby cities are very low. Most of the villagers (90%) are employed as unskilled labourers and are daily-wagers. The repairing staffs usually get Rupees 75 per day which is less that US$ 2 per day. Casual labourers earn Rupees 80 per day which again is around US$ 2 per day. The contract casual labourers earn Rupees 104 per day which is around US$ 2.5 per day and the labourers on permanent contract earn Rupees 175 per day which is around US$ 4 per day. The labourers are not covered under any insurance and there are no health and safety standards practiced in the plants.

Sustainability?

The first thing to realize is that even within the CDM process serious doubts has been expressed regarding the methodology approved by the CDM Executive Board, suggesting that it has serious drawbacks of incomplete environmental impact assessment.[7]

There has been plenty of evidence that the CDM project of GFL does not contribute to sustainable development in the region. Studies by Sutter[8] and investigations by CSE[9] and also by Nandene Ghouri[10] provide testimony that the GFL CDM project does not contribute to sustainable development.

The PDD of GFL claims that the project will lead to employment generation of 30-40 direct employees of which 90% is unskilled labour. The first point to note is that, on the basis of the data I collected through data survey and interviews of villagers, including those working at the GFL plant, it is evident that quite a lot of labourers employed at the plant come from other Indian states, like Uttar Pradesh and Bihar, leading to forced migrations for the local villagers. Is this to be considered a sustainable practice? Also, the wages that I have reported on above, are they too considered sustainable?

GFL claims that in its efforts to meet the goals of sustainable development it provides the farmers with agricultural assistance every year during the sowing season. On further enquiry on this issue, I discovered that GFL provides each farmer with 2 bags of urea and 1 bag of DAP every monsoon for farming. But this is a very unsustainable practice as the usage of urea and DAP eventually make the soil infertile and the use of urea and DAP has to be increased every year to maintain the production level. Instead of this highly unsustainable practice, GFL could probably have educated the local farmers with techniques of producing organic manure using earthworms, as this a very highly sustainable farming practice, which helps to increase agricultural produce and moreover it could also create self-employment for certain villagers who could engage themselves in the production of such organic manure. GFL claimed in its CDM project to help improve sanitation services in the villages, but it has been more than three years now since GFL started its CDM project, but yet there is no evidence of any improvement in the sanitation facilities.

As far as agriculture, cattle and human health are concerned, they have been adversely affected by the industrial activities of GFL and the pollution caused by this company. Agriculture is worst hit in the monsoon season, which actually is the prime season for agriculture. From the interviews with the farmers I learned that the pollution has led to reduced agricultural yield as well as poor quality of agricultural produce. The main reason behind this is firstly the air pollution. Secondly, there are very high fluorine deposits in the soil and water. Fluorine is very damaging for the crops. The vegetation around the GFL plant shows abnormal growth. Fruits on the trees in the surrounding region have been found to be affected by the fluorine content of the soil and water. The high fluorine content in the soil has not just affected agriculture, but it has adversely affected human health and cattle. The villagers complain of chronic problems of irritation in the eyes, burning of eyes, rashes on skin, skin pigmentation and acute pain in the joints. There have been no official records of these complaints as far as human health is concerned, as villagers usually end up going to private doctors instead of government hospitals – given the poor state of public hospitals. Health ailments have also been observed in the cattle in the villages surrounding GFL. On 15 September 2004 the villagers undertook an independent veterinary test for the cattle from Nathkuva, Ranjitnagar, Jitpura, Devpura, Kharkhadi and Kankodakoi. The cattle were diagnosed with ailments like reduced milk production, black fever, deformed skeletal system, weakening of bones, watering of eyes, tumours in the body, walking ailments, tumours in the rib cage and diarrhoea.[11] All of these issues have not been considered in the slightest by the official PDD of the GFL CDM project. Hence, the claims of sustainable development cannot be trusted.

Resistance

There has been a fairly long history of explicit resistance by the people from the villages surrounding GFL, which mainly included villagers from Ranjitnagar, Jitpura, Nathkuva, and Kankodakoi as well as activists of NGOs like Paryavaran Mitra. For instance, the villagers filed a Public Interest Litigation (PIL) on 20 April 1996, accusing GFL of causing air and water pollution under the provisions of Water Act 1974 and Air Pollution and Environmental Protection Act 1986. The Gujarat Pollution Control Board filed an affidavit with a report suggesting that the fluorine contents in the water samples has been high, but this does not evidently prove that GFL is responsible for it. Moreover, GFL has argued that in the Effluent Treatment Plant, the effluents are vaporized and hence there is no chance of any effluent disposal in underground water sources, which would lead to higher fluorine levels due company's activities. GFL also argues that the fluorine deposits in the region were very erratic, and hence GFL can not be held responsible for the fluorine deposits in the region. There have been no consistently documented records of hazards to human health due to pollution, as the sample test done by the villagers is too small to reach to any strong conclusion that would stand up in the courts. Regarding agricultural production, reports suggest that agricultural production has been rising consistently. Hence on the basis of the evidence put forth, the High Court of Gujarat passed the following decision on 3rd July 1996:

> As observed by us earlier, the officer of the Board has visited the plant and surrounding places as asked by this court is of the opinion that 'long time scientific study on chronic impact of pollution from the industrial plant on plantation, agricultural crop yields, animal and human health in the vicinity of these areas are required to be carried out by the experts in these fields'. The experts suggested by the petitioners in their report have also stated that they are of the opinion that 'to arrive at definite conclusion, a more detailed study of the area with monitoring of certain parameters is essential'. Under the circumstances we direct the State Government to appoint a committee of experts who shall carry out necessary investigation and shall submit its report to the State Government, on the basis of which the State Government shall take appropriate action in this matter.[12]

Despite losing the High Court case, the local villagers continued to resist GFL, as became apparent to me in a number of interviews. One of my interviewees, for example, is a resident of Jitpura village. He dropped out of school early, so he cannot be regarded as well educated or part of the local elite. After school he started working at GFL as an electrician. But then he left his job and started working independently, with the support of local NGOs, for developmental and environmental issues of the villagers. He has been very active in fighting for the poor village peasants who have been affected by the pollution caused by GFL. Together with other villagers he has tried to unite the residents of Jitpura, Nathkuva, Kankodakoi and Ranjitnagar, organizing many protests against the company. But all these efforts failed because of a lack of unity among the four villages, and because GFL was strong enough in breaking down the movements against it with the help of political and police pressure. Whenever there was a movement, GFL bribed some of the weak villagers and misused their economic helplessness to break down the revolt.

Ever since the company started production at the plant, the villagers have been very sceptical about its safety, as the chemicals used in the plant are very dangerous, and in an accident they could prove fatal for the villagers. In fact, the villagers feel that if there is an accident at the GFL plant, it could have impacts similar to the 'Bhopal Gas Tragedy'. So in 2005, when there was a small blast in the plant, the villagers turned violent against the company. The villagers attacked the plant and an office inside the plant, protesting against the management of the company. Immediately the managers called for the police to tackle the situation, and nearly 20 villagers were arrested and put behind the bars.

In addition to dedicated local villagers, a number of NGOs have also been active in the resistance against GFL. Paryavaran Mitra is an NGO that works for environmental concerns in Gujarat with its office in Ahmedabad. During the High Court case it helped the villagers to get veterinary and water tests done, and it also worked with local and national media outlets to drum up support for the villagers. As a consequence of the failed High Court case, Paryavaran Mitra then started to research the CDM process in India in general, and it realized that no monitoring of these projects was in place. So, the NGO wrote to the UNFCCC asking them about who monitors these CDM projects, and the UNFCCC replied that it is the duty of the government of the host country. So, the NGO then wrote to the Government of India asking them about the monitoring of these projects, and they replied that it was the UNFCCC's responsibility. Hence after many more correspondences and representations, the Government of Gujarat was forced to form a committee to monitor the CDM projects in Gujarat. This committee is called Gujarat Cleaner Production Centre, but I am told that this committee does not function the way it is supposed to, and it merely exists for the sake of formality rather than actual monitoring. As a result, Paryavaran Mitra has reported to the UNFCCC the poor and corrupt implementation of the CDM projects in Gujarat, yet nothing seems to be done about it.

The Centre for Science and Environment (CSE), an NGO that is widely known in India and based in Delhi, has also studied the GFL case in some detail. For example, it has tried to investigate the price at which GFL sells the carbon credits and also who are the major buyers of these credits. However, GFL has never disclosed this information under the pretext that it would be a breach of contract. This shows the complete lack of transparency at the heart of the CDM process. CSE has also studied the Project Design Documents (PDD) of GFL and another company called Navin Fluorochemicals Limited, both of which were prepared by PricewaterhouseCoopers. Very interestingly, CSE has observed that the PDDs of both CDM projects seem to be identical in some places, as in both documents the questions raised by stakeholders are the same and the answers to them are also the same, even though the two projects are located in two different states of India. Is this a coincidence or further proof of the CDM scam going on here?

Money Making Machine

As Schwank[13] argues, CDM projects involving the destruction of HFC23 create perverse money making incentives rather than being a real 'green' initiative. The cost for CER generation in a HFC23 destruction project is roughly around 0.5 USD/CO2 which is very much less compared to the current market prices of CERs. So the question really arises if CER stands for 'certified emissions reduction' or 'certified emissions revenue'. This is evident from the fact that GFL earned € 27 million in the last quarter of 2006, which would have helped GFL expand its 'normal', that is polluting, industrial activities.[14] And mind you these are just the official figures. My interview with a GFL employee revealed that except for Deepak Asher, who spearheads the CDM division of the company and a few other close associates of his within the company, no one else in the company has any knowledge of the CDM project and the revenues generated from it, not even most of the employees in the accounts department. The question is why all this secrecy around the CDM when supposedly it is something noble. The perverse incentives arising from the CDM has led to an increase in the number of companies in India and China trying to seek registration for HFC23 destruction CDM projects – given the high amount of CERs generated through this process. This in turn also gives rise to the perverse incentive of increasing HCFC22 production (and also setting up new HCFC22 production plants), which would counteract the Montreal Protocol which calls for the phasing out of HCFC22 gas. Another normal day for carbon markets then.

Conclusions

It is very important to understand that climate change is not merely an issue of carbon emissions but one that depends on and involves a wide variety of factors. Reducing carbon emissions is not simply a numbers game. Climate change, as mentioned in the introduction and also repeatedly in other chapters of this book, is a crisis which encompasses a wide range of social, economic, political and environmental issues, because emission reductions require a structural change of our life styles. So this brings us back to the key issue of the title of this chapter: 'climate justice'. The main conclusion that can be drawn from the case of GFL and also from other CDM cases from all around the world is that it is not just a matter of carbon emissions, but it is a matter of justice and more importantly of climate justice.

As far as the case of GFL is concerned, for a moment let us assume that the thermal oxidation process actually does reduce greenhouse gas emissions (which have been questioned by many environmentalists). As this case has shown, there are a wide range of other negative factors, such as land and water pollution and fluorosis disorders in humans and animals, which this CDM project does nothing about. That is, there is no sustainable development, although this is what the CDM is supposedly about. Instead, the profit generated by the sale of CERs simply props up very polluting business practices, which has had negative impacts on the local population ever since its inception. Should this money, which is generated through carbon markets, not be used for the benefit of local people, rather than propping up local and national Indian elites?

Instead of the noble claims about 'sustainable development', put forward by the GFL CDM PDD, what seems to be really happening on ground is that GFL is accumulating enormous profits from the sale of CERs and in this process is dispossessing the poor villagers from their rights to clean water, clean air and productive soil.

The conclusions that can be drawn from the CDM project of GFL is that the use of market mechanisms for climate change mitigation has been a failure, as far as the CDM and its impacts on local people are concerned. What the CDM seems to be doing is to provide an incentive to sustain polluting production practices, rather than invest into real sustainable development.

There is next to no control at the local, regional and national level in terms of the implementation of CDM projects. Hence, all sorts of spurious claims can be made by companies and consultancies which only seem to want to profit from carbon markets rather than invest in real clean technologies. As a result, industrial dinosaurs, which already have a long track record of pollution, are propped up alongside local elites that profit from this global money making scheme. But can we really save the climate through these mechanisms? Can we really better the lives of those living in poverty, as the CDM claims to do?

In my view, the solution to the global climate crisis is not some market mechanism which commodifies carbon. This seems to simply provide an incentive to create more carbon and extent existing polluting practices – in both North and South. Instead, the solution lies in finding a green and sustainable alternative to fossil fuels and keeping climate and social justice at the centre of any climate change mitigation mechanism.

Notes

* I would firstly like to thank Mahesh Pandya and Hiral Mehta of Paryavaran Mitra and Nayan Patel of Yuva Shakti and Neeta Hardikar for helping me find local contacts in the villages for data collection and interviews and also thank the villagers for their tremendous support and the courage they showed by helping me with the data collection and for their cooperation during the interviews. I would also like to thank Robert Williams from Sahyog for his help and support by providing the much needed contacts. I would also like to thank Ketan Francis, Prof. Kiran Vaghela, Ankit Parmar and Alex Macwan for helping me with data collection and conducting the interviews and last but not the least I would like to thank Steffen Böhm for his unending guidance and support.

1 http://unfccc.int/kyoto_protocol/items/2830.php.

2 Lohmann, L. (2006) *Carbon Trading: a critical conversation on climate chance, privatisation and power.* Uppsala: Dag Hammarskjold Foundation, www.thecornerhouse.org.uk/pdf/document/carbonDDlow.pdf.

3 http://cdm.unfccc.int/Statistics/index.html.

4 The CDM project of GFL has also been brilliantly investigated by the Centre for Science and Environment (CSE) (http://downtoearth.org.in/cover.asp?foldername= 20051115&filename =anal&sid=1&page=2&sec_id=7&p=2) as well as by Nandene Ghouri (http://www.carbon tradewatch.org/index.php?option=com_content&task=view&id=294&Itemid=36)   and Mahesh Pandya and Hiral Mehta of Paryavaran Mitra, which is an NGO based in Gujarat, India.

5 http://www.gfl.co.in/carbon_trading_pro.htm.

6 The quote has been taken from the Project Design Document prepared by PwC (Mumbai Office).

7 Schwank, O. (2004) 'Concerns about CDM Projects Based on decomposition of HFC-23 Emissions   From   22HCFC   Production   Sites',   October,   2004. http://cdm.unfccc.int/public_inputs/inputam0001/Comment_AM0001_Schwank_081004.p df.

8 Sutter, C. (2003) 'Sustainability Check-Up for CDM Projects', http://e-collection.ethbib.ethz.ch/eserv/eth:27027/eth-27027-02.pdf.

9 http://downtoearth.org.in/cover.asp?foldername=20051115&filename=anal&sid=1& page=3&sec_id=7&p=3.

10 Ghouri, N. (2009) 'The great carbon credit con: Why are we paying the Third World to poison   its   environment?',   http://www.carbontradewatch.org/index.php?option=com_ content&task=view&id=294&Itemid=36.

11 Centre for Science and Environment has also come up with similar findings. Please see http://downtoearth.org.in/cover.asp?foldername=20051115&filename=anal&sid=1&page=2 &sec_id=7&p=2; Also please see Ghouri, N. (2009) 'The great carbon credit con: Why are we paying the Third World to poison its environment?', http://www.carbontradewatch.org/ index.php?option=com_content&task=view&id=294&Itemid=36 .

12 The paragraph has been quoted from the Gujarat High Court judgement.

13 Schwank. O (2004) 'Concerns about CDM projects based on decomposition of HFC-23 emissions   from   22   HCFC   production   sites',   http://cdm.unfccc.int/ public_inputs/inputam0001/Comment_AM0001_Schwank_081004.pdf.

14 Ghouri, N. (2009) 'The great carbon credit con: Why are we paying the Third World to poison its environment?'.

13

The Jindal CDM Projects in Karnataka[*]

Nishant Mate and Soumitra Ghosh

Location

The JSW Energy Ltd is located adjacent to its parent concern, the JSW Steel Ltd, two kilometres from Torangallu village of Bellary district in Karnataka. The plant site situated between Bellary and Hospet falls on the state highway connecting Bellary and Sandur.

What the PDD says

Project Overview

JSW Energy has been commissioned to generate electricity using imported coal and waste gas. The electricity generated is supplied to JSW and the Karnataka state grid, the KPTCL (Karnataka Power Transmission Corporation Limited). The input fuel to the JSW Energy power plant is sourced from JSW Steel, which is generating corex gas and other waste gases from its process and sourcing imported coal. The project is supposed to reduce GHG emissions by increasing the proportion of waste gas in the fuel configuration for power generation.

During the initial operation period, the project faced uncertainties about the availability and steadiness of supply of the corex gas and other waste gases from JSW. Because of these, JSW Energy dropped the plan of utilizing waste gases, and had accordingly applied for and obtained the requisite approval from the KSPCB (Karnataka State Pollution Control Board) to combust coal exclusively. Subsequently, during March 2001, JSW Energy management decided to go for the current project activity so that the use of waste gas is maximized in the fuel configuration and emissions of GHG is reduced. This decision has seriously internalized the potential benefits of CDM. Besides the potential CDM benefits, there is no other incentive for JSW Energy to maximize the use of waste gases for power generation. The project activity involved additional investments to the tune of 240 million rupees (to the investment in power generation using coal) to achieve a steady supply of the waste gas.

The JSW Steel Ltd has also a CDM project operational, that is, generation of electricity through combustion of waste gases from the blast furnace and corex units at its steel plant (in JPL unit 1) at Torangallu in Karnataka. With the advent of the separate JSW Energy, the total amount of corex gas supplied to by the JSW Steel Ltd and the JSW Energy Ltd is metered separately. Also, the JSW Steel Ltd and the JSW Energy Ltd are two separate legal entities.

148

Sustainable Development

The PDD of JSW Energy highlights the company's contention that they have satisfied all the four indicators for sustainable development – social, economic, environmental, and technological well-being – as stipulated by the MoEF (Ministry of Environment and Forests), Government of India, in the interim approval guidelines for CDM projects.

Social Well-Being

– The project demonstrates harnessing power from waste gas sources, which will encourage replication of such project in future across the region.

– The project has built up a knowledge base about the operation of the waste-gas-based power generation and has built up a skill set for such kind of operation.

Environmental Well-Being

– The project activity involves generation of electricity using waste gas, thus replacing a certain amount of fossil fuel used for electricity generation. This has resulted in reduced GHG intensity per unit of electricity generation for the state grid; and, in effect, the total carbon intensity of the Karnataka state has been reduced.

– The project has reduced the local air pollutants and environmental impacts due to increased share in the use of waste gas in the fuel configuration.

Economic Well-Being

– This project will demonstrate the use of new financial mechanism – that is, CDM – in raising finance for power generation from waste gases.

What the Field Study Reveals

A visit to Toranagallu village and discussions with some panchayat office bearers and villagers – Mr Shankar, who is a bill collector at the gram panchayat; Mr Govind, a gram panchayat member; Mr Shivkumar, gram panchayat member; and many other members – revealed that though the JSW Energy Ltd is a very big industry, established on an area covering 250 acres of land, the area is not declared as an industrial area by the government. The main products in this industrial set-up are steel and iron, where production of energy constitutes only a small proportion. JSW has got the land from the government at a throw-away price of 10,000 to 15,000 rupees per acre.

The company, before starting the project, had promised that they would adopt the entire area for all-round development and provide all kinds of civic amenities. However, after acquiring people's lands, the company did not do a single social or developmental activity, and neither do they have any plans for doing so. They blatantly backed out on all their commitments about electricity supply, road construction, health facilities, employment benefits, and so on. The JSW Energy has only constructed a few bus-stop sheds and two roads, which are mainly used by the company.

Very few local people are engaged in the company as workers; most workers are from distant areas and even from other states. The village population is 8,000; but with the number of people coming from other areas to work in the plants, the population touches 80,000. All the workers are contract labourers who are not allowed to work here for more than 2/3 years. Anyone who dares raise any question against the company's work ethics does not get further work.

The plant is indiscriminately releasing toxic waste water into the canal that passes through the village, not only polluting the canal water but also the village pond (which is linked to the canal) and the groundwater. Even the water collected from tube wells is found to be toxic. Besides harming farm produce and activities, water pollution by the industry has made it difficult for the villagers to access safe drinking water and has given rise to incidences of several water-borne diseases in the village. On the other hand, unabated air pollution from the industry has compounded the problems for the local populace. Diseases, such as skin ailments, asthma, and tuberculosis, which were not prevalent in the area, are now common.

The company has also turned back on improving health services in the area. The local government health centre is in a dismal condition, depriving villagers of the requisite health services. However, the JSW Energy has opened a modern hospital named *Sanjeevani*, which only caters to its employees and discourages the local villagers to avail its health facilities by forcing them to pay unaffordable fees. Therefore, people have demanded that the JSW Energy develop the existing health centre, to which the company did not pay any heed.

JSW has also illegally occupied the 600 acre village commons where there were plans of constructing schools, college, and an ITI (industrial training institute).

The villagers complained that before starting a new unit or project, JSW never even bothers to get the consent of the villagers. They call only a few panchayat members for a meeting and bribe them with good food and sweets, and then take several photographs so that a fabricated story about people's consent can be published in the newspapers.

The CDM Hoax

Sustainability Criteria

– The field study clearly suggests that all the indicators of a CDM project have been grossly violated by the JSW Energy in the Torangallu region. No initiative on sustainable development of the region has been taken up by the company, and nor have the local people been involved in any decision-making or project activities. The PDD claims that the project has built up a knowledge base about the operation of the waste-gas-based power generation and has built up a skill-set for such kind of operation. However, there has been no such initiative to even make people aware about a waste-gas-based power system, let alone building any skill-set for such kind of operation.

- The economic development promised in the PDD has also turned out to be a hoax, as most people are now rendered unemployed after losing their land to the project.
- In terms of environmental well-being of the region, the project has been an unmitigated disaster from the beginning. Instead of cleaning up the air and the atmosphere, it literally damned the local populace and their environment with unprecedented air, water, and sound pollution, which, in turn, brought a host of strange diseases to the village.
- JSW Energy has even made no attempts to make people aware about the CDM component of the project, such as GHG emissions, clean mechanism, carbon trading, or even global warming. However, they have claimed in the PDD that the local people have a direct hand in the reduction of GHG emissions in the project and thereby contributing to the mitigation of climate crisis.

Additionality

In case of the Jindal projects, the CDM fraud goes deeper than violations of the sustainability criteria. Every new CDM project has to prove its 'additionality', which means that the project would not have been possible without the CDM benefits – monetary and otherwise. The additionality of the Jindal waste gas projects in Karnataka has been suspect from the very beginning; among others, the noted carbon market expert Dr Axel Michaelowa called the projects 'clearly non-additional', because the projects could have come up irrespective of the CDM money. Dr Michaelowa's submitted a public comment challenging the project developers' claim regarding the timing of when the plant was going to use waste gas for electricity generation: essentially showing that this decision had been made long before the company had applied for CDM funding, and that it, therefore, failed the additionality test, that is, the JTPCL had already decided (and had an incentive) to implement it without the CDM. Dr Michaelowa writes:

> I made a public comment on the first project questioning its additionality. In my view, these projects could become a key precedent for allowing large non-additional energy-efficiency projects into the CDM. My comment read as follows: 'This project is non-additional. Its claim that a decision to use waste gases to generate electricity was made at a later stage than the actual investment [was done] into the corex plant is not true. Electricity generation from corex gases was always a key element of the project investment (this is a well known fact in India) and thus the assertion that during March 2001, the JTPCL management took the decision for the current project activity is blatantly wrong. Moreover, the first tranche (130 MW) of the project started production well before 2000 and thus that tranche is not eligible for the CDM. See the publication (which does not mention the CDM at all and is another indicator that CDM was not seriously considered!) by the project participants – Dwijendra Ghorai, Friedrich Bräuer, Helmut Freydorfer, Dieter Siuka, L'unité COREX® chez Jindal Vijayanagar Steel : une réussite sur toute la ligne, Rev. Met. Paris, N°3 (March 2001), p. 239-250; (English version, COREX operation at Jindal Steel: a success story in Millennium Steel, 2001, p. 20-25.' It has also been alleged that

there has been a price fix for the electricity being generated: one arm of the Jindal group is charging another arm of the company a higher price for the electricity it generates under the CDM than normal, so as not to make the waste heat plant financially attractive without the CDM. These and other questions about the Jindal projects' additionality and CDM norm violations were raised in the course of a Channel–4 programme in early 2007.

Windfall Profits

The extremely serious objections to the 'CDM'ness of the projects, however, went unheeded. The Jindal group went on reaping enormous profits from the projects (Dr Michaelowa said during an interview with Channel 4 that the projects can gross up to 20 million euros annually by selling CERs). One has to remember that this is an early 2007 estimate when the average price of a secondary CER (CERs issued by the UNFCCC and coming from a project that is handled by a reputed broker) was about 15 Euros, and the CDM market touched its zenith of 27 Euros/CER in July–August 2008, before the recession effects started to be visible. The two CDM Jindal projects in Karnataka have been issued 7,843,000 CERs so far (till 26 August 2009). Because of all Indian companies' typical habit of holding on to their CERs (for fetching better price at a later date), it is difficult to assume exactly how much money a particular project has earned.

Jindals admitted to have earned, till late 2007, 1.1 billion rupees (it could be much more) from selling supposedly 'reduced emissions' (1.3 million CERs) at their steel plant in Karnataka. According to company sources, this boosted other incomes, and helped the Jindal Steel Works to record their best ever quarter in terms of profit. If we consider the presently issued CER figures, the total earning from their 'profitable' clean projects can be anything between 100 and 150 billion rupees! According to another estimate, at the current market price of 15.5 Euros per CER in early 2007, the company stood to gain 109 million Euros over a 10 year period from the sale of CERs; and interestingly enough, JSW Steel is expected to 'save' on an average 0.77 million CERs per annum that can be sold in the open market, which means that the company would hold on to its CERs in wait of even bigger 'profit'!

The Jindals have 9 CDM projects in their kitty, only 3 of which have a collective potential of generating no less than 24,378,000 CERs by 2020. These are all located in the JSW area at Torongulu. One of these (the biggest, with 8,589,000 credits) has not been registered with the UNFCCC yet.

Notes

* The article has been taken from – Mate N., and S. Ghosh (2009) 'The CDM Scam: Case Study on Jindal CDM Projects in Karnataka', *Mausam*, 1-2(2-5), pp. 27-29.

14

The MSPL Wind Power CDM Project[*]

Nishant Mate and Soumitra Ghosh

Location

MSPL Ltd, part of the Baldota Group, owns three wind power projects, located at Sogi, Jogimatti, and Jajikalgudda in the districts of Bellary, Chitradurga, and Davangere, respectively, in Karnataka. Bellary, Chitradurga, and Davangere are at 300, 200, and 317 kilometres from the capital city of Bangalore.

What the PDD says

Project Overview

The MSPL, with a view of being in line with the sustainable development priorities of India, is promoting project activities to generate sizable volume of green power by tapping wind energy in the 'barren' land of Karnataka, deficit in energy and peaking power. The project proposes to generate 125.15 MW equivalent of clean electricity with efficient utilization of the available wind energy through adoption of efficient and modern technology. The project will replace energy produced through combustion of fossil fuels with equivalent volume of clean energy. Green power of 303.3 million units per annum will be fed to the KPTCL grid, a part of the southern regional grid.

The project involves three concerns of the Baldota Group – the MSPL Ltd, the RMMP Ltd, and PVS & Brothers. As per an agreement among these three, the MSPL Ltd has the ownership rights for this CDM project and is the sole transaction entity with the Executive Board of the UNFCCC (United Nations Framework Convention on Climate Change).

The 125.15 MW wind power project comprises 83 WEGs (wind-energy generators) of 1250 KW capacity, 17 WEGs of 950 KW capacity, and 7 WEGs of 750 KW capacity. The project activity has been planned and executed in two phases, with capacities of 27.65 MW in phase 1 and 97.50 MW in phase 2. The plan and WEG allocations of the two phases are described in Tables 1 and 2.

S. No.	Company	No. of WEGs	Capacity (KW)	Make
1	MSPL	7x750 KW	5250	NEG Micon
		17x950 KW	16150	NEG Micon
		5x1250 KW	6250	Suzlon

Table 1: Phase 1: Total 27650 KW

S. No	Company	No. of WEGs	Capacity (KW)	Make
1	MSPL	14x1250 KW	51250	Suzlon
2	RMMPL	31x1250 KW	38750	Suzlon
3	PVS	06x1250 KW	7500	Suzlon

Table 2: Phase 2: Total 97500 KW

Sustainable Development

The MSPL Ltd is a proactive business entity and firmly believes that effective and efficient generation of green power, coupled with responsible environmental considerations, is vital to maintain a competitive edge. This has been a guiding factor towards their initiative in the conceptualization and installation of the 125.15 MW wind power project. To be competitive in the open market economy of India, the group is developing this project as a CDM project under the UNFCCC, which would appropriately reduce the use of coal and other fossil fuels in power generation, helping in significant reduction of GHG emissions and also promoting sustainable economic growth and conservation of the environment through use of wind as a renewable resource.

The project primarily assists the State of Karnataka – and India as a whole – in stimulating and accelerating the commercialization of grid-connected renewable energy technologies. In addition, wind power projects of this magnitude, as conceptualized by this project activity, demonstrate the viability of larger grid-connected wind farms, which improve energy security, air quality, and local livelihoods, as well as assisting in the development of a sustainable domestic renewable energy industry. The specific goals of the project are as follows.

– Operationalizing sustainable development through generation of eco-friendly power

– Increasing the share of renewable energy power generation in the regional and national grid

– Bridging India's energy deficit in the business-as-usual scenario

– Providing national energy security, especially when global fossil-fuel reserves threatens the long-term sustainability of the Indian economy

– Strengthening India's rural electrification coverage

– Reducing GHG emissions compared to business-as-usual scenario

– Reducing pollutants, such as oxides of sulphur, oxides of nitrogen, particulate matters, etc., resulting from the conventional power generation industry

– Contributing towards the reduction of power shortage, especially in the state of Karnataka

– Demonstrating and helping in stimulating the growth of the wind-power industry in India

– Enhancing local employment in the vicinity of the project, which is a rural area

- Capacity-building and empowering vulnerable sections of the rural communities dwelling in the project area
- Conserving natural resources, including land, forests, minerals, water, and the ecosystems

What do the Field Studies Reveal?

Sogi Village

During a visit to the Sogi village area, R Manjunath Nayak (a civil engineer), B Hallya Nayak (a gram panchayat member of Govindpur village), and other villagers informed us that the wind mill project did not provide the village with any facility; the road built through the village is primarily for the transport purposes of the wind mill project and not meant for the villagers. The project, which is located on the nearby plateaux, usurped people's lands without paying the right price; while the market rate of one acre of land is 200,000 rupees, the company has paid a maximum amount of 80,000 rupees per acre. Worse, agricultural land lost for the construction of the road was compensated with only 5,000 to 6,000 rupees per acre. Moreover, any damage caused to the road by heavy vehicles of the company is now repaired using the gram panchayat fund. Patches of agricultural land lost to make way for tower lines for transmission of electricity were hardly compensated for – people have got a ridiculous payment of 500 to 1200 rupees. In many cases, people have not even been paid anything after losing land to the project.

The four power stations installed by the company occupy an area of 4/5 acres each, but the company has acquired 40 acres of village and forest lands. The area is mostly inhabited largely by scheduled-caste and scheduled-tribe communities, with a spattering of other communities; and for them losing land without any alternative economic options in place is a huge economic setback.

The company has not taken the necessary legal permission from the gram panchayat for setting up the project. Villagers said that the company gave assurance to provide jobs to them and thus duped them into signing some papers. However, no one has been employed in the project; a couple of local residents who were working as security guards earlier have now also been removed from their jobs.

Most youth in the village are educated; there are even some engineers and diploma-holding technicians from ITIs (industrial training institutes). But, the company shows no interests in employing them in the project, not even on contract basis. For instance, R Manjunath Nayak, a civil engineer, has long been trying to get a job in the project. But, the company management turned down his request by saying that it was against the company policy to employ local people on such type of jobs. His hope to get at least some contract works is also shattered, as all such works go to outside contractors. While the wind turbines were being installed, only 20 per-cent people from the village got some work. The rest of the workforce was brought from other areas. While the construction was on, some workers were killed in an accident. The company hushed up the case by providing some paltry compensation of 25,000 to 50,000 rupees to the victim families. All the workers who died were from local communities.

Local villagers blame the coming up of the wind-mill systems for the erratic rain fall they are experiencing of late. Due to the working of wind mills, the monsoon is changing its course, they say. The local economy is primarily dependent on agriculture, and the change in rain-fall patterns has proved disastrous for them. The noise from the wind mills has a huge deleterious impact not only on humans but on the entire biodiversity of the area. While people cannot sleep at night, cattle and the wildlife are frightened. No wild animal is seen in the forest now. Villagers were also promised free electricity supply by the project; but that too has turned out to be a false commitment.

Jogimatti Village

Village Jogimatti – a slum settlement near the wind mill project – is located in the urban area of Chitradurga city. The project itself is located in the forest area of Chitradurga. People of the village mostly work as labourers and have not lost any land to the project. The villagers, however, complained of the noise from the wind mill as a huge problem due to which they cannot even sleep at night.

Jajikalgudda Area

Upon visiting Jajikalguda, it was revealed that it is in fact a cluster of many villages dotting the hilly area – Chitegiri, Adeveli, Nichapur, Nazirnagar, Hombergatta, Deverlimmalapur, and Tipahakaguhadli. From the Chitegiri village, the company has acquired four acres of land at a price of 40,000 rupees per acre. The Forest Department has 'given' a whopping 200 acres of forest land to this wind mill project. A power station has been constructed in Nazirnagar village from where electricity generated at the wind mill is supplied to other areas. Here, only one person from the village has been employed, as a security guard. In Hombargatta village, the wind mill project took ¼ acre of land from one person and, in turn, employed him at a tower construction site saying that it was a government job and gave him only 1800 rupees per month. After a few months, he was asked to leave. The construction work has been given to a non-resident contractor, named Rajesh, who erects tower lines on villagers' agricultural lands without paying any compensation.

The CDM Hoax

The field visit clearly establishes that all the sustainable development indicators as described in the CDM guidelines have not been satisfied by the project authorities. However, it seems, the project has managed to achieve a CDM status just by producing an impressive PDD (which looks and reads very much like the NSL PDD!). None of the promises made in the PDD about the project's contribution towards sustainable development – strengthening India's rural electrification coverage; enhancing local employment in the vicinity of the project, which is a rural area; capacity building and empowerment of vulnerable sections of the rural communities dwelling in the project area; and conserving natural resources including land, forest, minerals, water, and ecosystems – has been met. Nor is there any hint of a rural development programme to show that the company is bothering at all about those promises.

The MSPL did not even bother to conduct the necessary consultation process with the local population, village panchayat, and the local elected body of representatives before initiating the project. But the PDD claims to have roped in the people in 'playing a big role in mitigating the climate crisis' by directly participating in this CDM project. In reality, the local residents are not even aware about the nature and activities of the project, let alone concepts of CDM and carbon trading. Since the project has just led to a lot of woes for the people instead of benefitting them, the local populace hate the project. However, the wind mill project is reaping huge profits both by selling electricity to the state grid and from its CDM component. On 12 February 2007, the project was issued 267 666 CERs for the verification period 22 March 2004 to 31 March 2006. Other parties in the CDM project are the United Kingdom and Northern Ireland. In monetary terms, this meant a windfall of no less than 4 million Euros (going by the average secondary CER price of 15 Euros during early 2007), provided that the project had sold all its credits at that time.

So much money...for doing what? We have seen how the apparently benign wind projects can usurp people's commons and destroy livelihoods. We have also seen how the nicely worded and sleekly laid out PDDs can be full of unabashed lies. But, what about the tall claims of emissions reduction? Wouldn't the projects have come up anyway, with or without the CDM money? Is any wind CDM project in India truly additional?

Additionality
The additionality of the wind energy CDM projects in India has always been under the scanner mainly because of the existing subsidy regime – both the state governments and the Government of India offer a range of subsidies to any renewable energy project including wind mills. Besides, there is this stipulation that the certain portion of the total electricity supplied to the grid and thereafter distributed to industrial consumers has to come from renewable sources. The UNFCCC has rejected a number of Indian wind projects on additionality grounds, including a Bajaj Auto wind project from Maharashtra. Now that the Indian government proposes to extend 1 incentives to wind farms for 10 years, the additionality of all wind CDM projects becomes doubly suspect. Perhaps the incentive move is due to the fact that wind is big business now; with the presence of corporate giants like Tata, Reliance, ONGC, and Suzlon, the government plans to extend GBI (generation-based incentive) to wind farms for a period of up to 10 years. Under this scheme, benefits equivalent to accelerated depreciation of 80 per cent at NPV (net present value) will be made available to private investors every year. The move is supported by the Planning Commission of India and the MNRE (Ministry of New and Renewable Energy), in tune with the National Action Plan for Climate Change. Currently, the wind mills can enjoy 80 per cent depreciation benefit only during the installation period in the first year. However, the new move is likely to boost up wind energy production considerably as the country plans to double its installed

capacity in this segment from the current 10 500 MW as on 31 March 2009 to 20,000 MW in the next five years. The plan is to add 2000 MW every year. The government plans to attract an investment – mostly from the private sector – to the tune of 40,000 crore (400 billion) rupees in the next five years to create this additional capacity. 'There are about 3000 private investors in this sector already and the proposed new benefit is likely to attract more investments from them,' according to a senior government official.

With the forum of electricity regulators adopting the RPS (renewable purchase standards) on behalf of all states, private investors are likely to get assured returns for the excess power they generate. Under the RPS, states will have to commit to buy a certain per cent of their electricity needs from renewable resources. The key states in the wind sector are Tamil Nadu (with an installed capacity of 4300 MW), Gujarat (1560 MW), Maharashtra (940 MW), Karnataka (1327 MW), and Rajasthan (738 MW).That the wind projects in themselves are extremely lucrative financial propositions is proved by the fact that a company's stock prices soar as soon as it announces a wind energy programme. For instance, the shares of Gujarat NRE Coke soared 8 per cent on its windmill expansion plan in 2 December 2007; and the stocks of Suzlon, the biggest wind operator in the country, showed a consistent 3 upswing throughout the first quarter of 2009.

Notes

* The article has been taken from – Mate N., and S. Ghosh (2009) 'The CDM Scam: Wind Power Projects in Karnataka', *Mausam*, 1-2(2-5), pp. 30-34.

15

The Deogad Hydroelectric CDM Project[*]

Nishant Mate and Hadida Yasmin

Location

The DHP (Deogad Hydroelectric Project) – owned by the Gadre Marine Export – is located in village Ghonsari in Kankawali taluka of the Konkan region of Sindhudurg district in the state of Maharashtra. The dam site is approachable from Phonda village on the Kolhapur–Ratnagiri road (state highway no. 49), which is 18 km from the Mumbai–Goa national highway.

What the PDD says

Project Overview

The Gadre Marine Export is to generate power from the irrigation releases of the Deogad dam, utilizing the variable head. The intake structure consists of the trash racks, the stop log gates, the air vent, and the steel penstock in the body of the dam. The powerhouse is to be located at the surface downstream. The tail water will be guided through the tail canal into the steel penstock. The 2.2-m-diameter steel penstock is designed to carry a peak discharge of 10 cubic metres per seconds. On the downstream of the dam, a 'Y' piece is to be provided to this irrigation-cum-power-outlet in order to let out water directly into the river for irrigation whenever the powerhouse is to be closed. A butterfly valve in the powerhouse is provided for controlling discharge to the turbine. No additional storage or forebay, etc., is contemplated.

Sustainable Development

The main objective of the DHP is to produce clean electrical energy in a sustainable manner, optimizing the utilization of water – a renewable resource. The electricity generated by the project activity will replace the electricity produced by thermal power plants that utilize fossil fuels in the grid. In the wake of power shortage and the ever-increasing demand for electricity in Maharashtra, implementation of the proposed project, with an installed capacity of 1.5 MW, contributes to help meeting the demand. The Designated National Authority for the CDM in India, which is under the MoEF (Ministry of Environment and Forests), has stipulated indicators for sustainable development in the interim approval guidelines for Indian CDM projects. Each of these indicators has been studied in the context of the project activity to ensure that the project contributes to sustainable development.

Socio-economic Well-Being
- The proposed project activity leads to alleviation of poverty by establishing direct and indirect benefits through employment generation and improved economic activities by strengthening of the deficit grid of the state electricity utility. This includes improvement of electricity quality, frequency, and availability.
- The construction work will generate employment for the local population. There will also be various kinds of mechanical works on the site, generating employment opportunities on a regular and permanent basis. The transportation of various project components to the project site will also create work opportunities, thereby adding to the income of the local population.
- There will also be various kinds of mechanical works on the site, generating employment opportunities for the local populace on regular and permanent basis.
- The project will create indirect employment opportunities for 50-100 unskilled workers for a period of two years (during construction), which would not happen in the absence of the project. In addition, the project creates direct permanent employment for about 35 persons for the operation of the project.
- By promoting the decentralization of economic power, the project contributes in bringing economic sustainability around the plant site.
- The project activity also leads to the diversification of the national energy supply, which is dominated by conventional fuel-based generating units.

Environmental Well-Being
- The hydroelectric project has no negative environmental impacts because it relies on existing irrigation releases and it does not involve any tree felling or submersion, etc. Furthermore, adequate provisions are made for plantation and greeneries, making the area more environment-friendly.
- The project utilizes hydro energy for generating electricity replacing polluting fossil-fuel-based power plants, thus contributing to reduction in specific emissions, including GHG emissions. Use of hydro energy – which is a renewable resource – to generate electricity contributes to resource conservation. Thus the project causes no negative impact on the surrounding environment, leading to environmental well-being.
- As hydro power projects produce no end products in the form of solid waste, such as ash, the problem of solid waste disposal encountered by most other sources of power is eliminated naturally.

What do the Field Studies Reveal?

A visit to the Ghonsari village, however, reveals something strikingly contrary to what the PDD claims. Villager Sakshat Prakash Parker told us that work on the so-called irrigation dam started 15 years ago and many villagers have lost their land and houses to this project. The population of the Ghonsari village is about

3000, and the primary source of livelihood of the villagers is agriculture. Out of the 12 *wadas* (specific community based clusters) in the village, two have completely gone under water. People who have lost their land and houses were supposed to get the resettlement money in two phases.

In 2005, when the second phase of receiving compensation was due, people started protesting against the construction of the dam. Nevertheless, construction work was carried on by the contractor Nobel India Construction of Jaipur. The local people consider the Gadre Marine Export to be the sub-contractors for electricity generation, and because the irrigation project was not completed they now think that this project is only for electricity generation, which will not be for their use. Moreover, because people from the village got no employment during the construction work of either project (let alone any other benefit they are entitled to under the CDM norms, and which the PDD so loudly proclaims), they are angry with everything concerned with the irrigation/hydropower project. 'The contractor got all the workers from outside and we got nothing,' said the villagers in unison.

Sulentin Karlu Raise, another villager, who has lost half acre of agriculture land to the dam project, corroborated the fact that the company had indeed cheated the people on the assurance that local people would get employment during the dam construction and other project activities.

Jeron Baren, a clerk in the local gram panchayat informed that the company did not hold any public meeting with the villagers before commencing the project. 'They only sent a letter to the gram panchayat just to inform that it was going to construct the dam and the hydroelectric project,' said Jeron, 'It did not seek any NOC (no-objection certificate) from the gram panchayat.'. The village experiences power-cuts for eight hours a day.

The CDM Hoax

The DHP has is a registered CDM project and the company in its PDD has announced scores of development programmes including poverty alleviation and environmental well-being.

But, we found that there has been no such project activity including the construction work that ensures direct or indirect employment generation or alleviation of poverty in this interior rural area. The PDD had, in fact, promised employment to at least 50 to 100 unskilled workers during construction of the dam and more employment during construction of the power generation facility. But, the company did not keep any promises made in the PDD. According to the PDD, at least 35 local villagers were to be recruited on permanent basis in the project; the company did appoint none.

Further, the company has not involved the local people in the transportation process, thereby depriving them of a possible income opportunity after acquiring their land. The claim of the company as regards environmental well-being in the PDD that it would promote plantation and rejuvenate green areas has also turned out to be mere promises, as no such activity was visible in the area. Any CDM project has to ensure participation of the local communities. In case of this project, the local people have little or no idea about the project activities, let alone its clean environment processes or the concepts of 'carbon trading' and 'carbon credit'. Only after the work started people came to know that this project was to generate electricity.

While the company made no efforts to make the people aware about the CDM, carbon trading, or carbon credit through which the project would reap fat profits (the project claims to reduce 37,000 tons of CO_2 equivalent, meaning an equal amount of carbon credit) its PDD trumpets that the local people are playing a big role in addressing the climate crisis and in the reduction of carbon emissions!

Notes

* The article has been taken from – Mate N., and H. Yasmin (2009) 'The CDM Scam: Case Studies on CDM Projects in Maharashtra', *Mausam*, 1-2(2-5), pp. 35-40.

16

Offsetting Lives and Livelihoods: Atmospheric Brown Cloud and the Targeting of Asia's Rural Poor*

Soumya Dutta

What is the 'Atmospheric Brown Cloud'?

In the last 7-8 years or so, in the middle of the intense engagement on the climate crisis/justice debate, and the intricate wheeling-dealings between transnational entities like the UN, nation-states, large corporations, NGOs, international funders and multi-lateral agencies, a new spin has been added. This is the issue of the Atmospheric (earlier called Asian!) Brown Cloud (ABC), containing, along with other aerosols/particulates, the Black Carbon (BC) aerosols, which are now 'discovered' to be having such a large impact on climate and glaciers that it is 'considered' almost as important as the climate changing global carbon dioxide emissions from the burning of fossil fuels (coal, petroleum products like petrol, diesel, kerosene, natural gas, etc). And the UNEP (United Nations Environment Programme) study that projected this has one Prof. Ramanathan from the well known Cripps Institute of Oceanography, as lead author/researcher. Many of these studies were coming out of data from a large scale international experiment conducted from 1996 to 1999, called INDOEX (Indian Ocean Experiment). The year 2002 saw this issue come into international focus, whereas 2008 saw the 'study' focusing on Asian sources and impacts, and consequent 'plans' to tackle this by the 'carbon bazaar'.

In short, the studies 'discovered' that the large number of poor families in Asia burning wood and other biomass for their daily cooking etc, are emitting large quantities of dark soot or black carbon particles – along with other aerosols coming from fly ash (large scale coal burning in Thermal Power Plants and industries), sulphates (these are generally from petroleum or coal burning) as well as natural particles like dust and fine salt – which are forming a sort of Brown Cloud in the atmosphere. This aerosol laden cloud is being driven by the North-East monsoon winds – from the Indian landmass to the Bay of Bengal, Arabian sea and the Indian Ocean. Thus, its concentration over these seas is highest from December to March – with a November to April presence.

Projected Impacts of the ABC

This dark coloured 'cloud' – sometimes as thick as three kilometers at its largest atmospheric depth – is said to be absorbing a large amount of incoming solar heat, thus increasing the temperature of the lower atmosphere of the earth, disturbing normal tropical air circulations. By obstructing some solar heat from reaching the surface, it is also reducing the warming of the sea surface significantly, when present over ocean surfaces – reducing evaporation. This is supposed to be causing a significant drop in monsoon rainfall over both South-West Asia, and India (though the effect on Indian summer monsoon is not well marked by this modelling study). It is also said to be reducing the winter time average temperature over the Indian land mass, thereby masking the impact of global warming – during winter time at least.

Another harmful impact of the black carbon was found to be the increased rate of melting of Himalayan glaciers, as these dark coloured BC particles travels to the Himalayas and settles down on the 'white' glaciers, thus decreasing their 'albedo' or reflectivity. This increases the heat absorption by the darkened glacial surface and consequently, its melting rate. The black carbon or soot is also a major health threat for the poor families who use solid fuels or other biomass as fuel in their 'improper' chulhas/cooking stoves.

Selective Targeting

It is also recognized that the 'smoke' from the large numbers of fossil fuel powered vehicles – particularly those with older and less efficient engines, as well as other industries burning fossil fuels are also contributing to a large extent to this problem of creating a heat absorbing cloud. To the surprise of many, the 'smoke and noise' about the problem though has been focused on the millions of poor family kitchens. As the Himalayas are well within the two large and emission increasing nations – China and India, the problem of rapid melting of Himalayan glaciers are supposed to be accelerated by such black carbon in the northern regions of India and southern parts of China.

Along with the Western nations, the Asian ruling elite and their corporate controlled media is now highlighting the urgency of tackling this ABC induced 'climate problems' (along with global warming induced climate change), demanding that this heat trapping Atmospheric Brown Cloud (ABC) be reduced – largely by addressing the 'problem' of biomass burning by the large numbers of poorer families in Asia. So, now the industrialized world, who are supposed to be no major contributor to this 'dangerous' ABC – has another stick to beat the poorer developing countries in Asia, after the collapse of the 'cattle produced methane' beating stick. They are also playing to create further space for their 'way of life' of individual cars, big houses, lit up shopping malls, mass production of uniform goods, frequent flying etc – by occupying more atmospheric pollution space to be vacated by Asia's biomass burning poor. The ABC and black carbon issue has become another pawn in the global climate-political chess game.

GHG driven Climate Change, Black Carbon and Issues of Equity and Sustainability

The worldwide engagement on the global climate change issues cluster and the resulting international negotiations at various levels have been focusing on the so-called greenhouse gases (GHG) mostly, and with very good reasons. The industrialized countries have extracted and used/burned an overwhelmingly large proportion of the total cumulative fossil fuels over the last couple of centuries. The 'waste products' of burning fossil fuels – mainly carbon dioxide (CO_2) – has been dumped by these nations into the global common atmosphere, without any care or concern of the implications.

As the total atmospheric CO_2 dumping each year (about 30 GT in the year 2006 – just from burning fossil fuels) is far above the CO_2 recycling capacity (about 14 to 16 GT) of the earth's atmosphere-hydrosphere-biosphere-lithosphere systems, the 'excess' dumped quantity of CO_2 has been building up in the atmosphere and have pushed up the CO_2 concentration in it from around 280 ppmv (parts per million by volume) during early industrial periods to the present value of around 388 ppmv. Over the last couple of centuries, the 'developed' industrialized countries have contributed to nearly 90% of this CO_2 build up. This increased concentration of CO_2 in the atmosphere is absorbing an increasing amount of the low energy and long-wave heat radiation going out from the earth's surface, thus increasing the temperature of the lower atmosphere and the surface. This has disturbed many balanced climate systems on the earth – which are critically dependent on surface (both land and water) temperatures and temperature differences, causing unpredicted variations in the climate and inflicting untold miseries to natural resource and cycle dependent poorer populations.

Logically, all effort should go towards reducing the emission of these GHGs, bringing them down to the levels which the earth's carbon cycle can handle. The most direct and logical implication of this should have been a drastic reduction in the consumption of fossil fuels by those countries and societies (both in the Global North and South) who are the over-consumers. And the Kyoto protocol and its subsequent processes were expected to address that, putting pressure on the over-consuming industrialized nations to cut fossil fuel energy use by large margins.

Any just action would also require/demand that the world's majority of forced under-consumers – most of whom are denied even subsistence level energy and other consumables – be provided with enough to have a dignified life. Without establishing a 'reasonable equity' of access to all the global commons (including the common atmosphere), there is very little chance of a sustainable solution to the problem of chaotically changing climate system, as the resource deprived will rightfully attempt to access whatever they can get hold of, irrespective of its contribution to climate crisis or otherwise. For them, it is a question of 'survival comes first', unlike the 'lifestyle demands' of the richer countries/societies. And a reality check of this is the actual average CO_2e (carbon dioxide equivalent) emissions figures of India ~ 1.4 tons/person/year,

USA ~ 19 tons/ person/year, EU ~ 9 tons/person/year. And if you take up the 'black carbon emitting' poor of South Asia, the approx. 70% of the poorest here, who are the target of this motivated ABC/BC campaign? Their average CO2e emission is less than 0.6 tons/person/year. So why is the focus of this twisted, ABC induced, climate crying on these people? There are hidden agendas behind this.

In refusing to carry out any significant reduction in their forcible over-consumption and the resultant excessive waste dumping onto the global commons (like the oceans, the atmosphere etc) – the total GHG emissions by the Annex I countries have increased by nearly 20% from their 1990 levels, instead of coming down, as envisioned in the Kyoto treaty – the capitalist-industrial societies have consistently tried to pass the burden of emission reduction to the already under-consuming societies – across nations as well as within nations. This has been done through market based mechanisms like carbon emissions trading and the Clean Development Mechanism (CDM). Now, one more potent weapon for emissions trading space has been discovered in the shape of the ABC, and the multilateral agencies along with the Western research institutes gleefully accepted the new found opportunity.

Here, it is necessary to clarify that overwhelmingly large part of the solid fuels that South Asia's (and other similar society's) poor are forced to use for cooking, are not fossil fuels, as they are part of the 'active carbon cycle' of the earth and thus do not add to the net GHG emissions. This holds true to the extent that the wood or other biomass used by these poor families do not exceed the regeneration of these biomass resources within comparable time periods. The black carbon aerosols emitted by these fuels would surely contribute to the ABC and its climate impacts, but in a less significant way than assumed by Ramanathan's and others studies, as I will explain/argue in a later part of this chapter.

The apparently 'clean' electricity used for cooking by the majority of families in the Global North, and by the upper classes in the Global South as well – is not so clean after all, as most of this electricity comes from either the dirtiest GHG emitter – coal (about 70% in India and Australia, 80% in China, over 50% in USA), or equally damaging dam based hydro-electricity (which also causes large emissions of the potent GHG methane by submerging huge forest biomasses, along with adding net CO2 emission by destroying this large forest carbon sink), or the dangerously polluting nuclear fission energy (which, apart from the millennia long radioactive contamination threat, also has a large carbon footprint – contrary to common perception – from their large embedded energy of construction, maintenance and decommissioning).

What are the Projected Facts About Atmospheric Brown Cloud?

The Intergovernmental Panel on Climate Change (IPCC) as well as well known climate scientists, such as Hansen, Jacobson and Ramanathan's UNEP study – all have tried to quantify the amount of global warming potential – called radiative forcing – of this Atmospheric Brown Cloud. There are wide variations in these calculations (from 0.2 watts/sq.mtr to over 1 w/sq.mtr – or a factor of five!), with the UNEP study leading the pack, 'showing' the highest amount of warming contribution of ABC, by taking into account the effect of glacial snow melting by the black carbon settling down on Himalayan glaciers. No wonder that the Western and Asian media have selectively picked up this highest estimate, and the supposed 'contribution' of the poor families.

The Himalayas on the Tibetan side (trans-Himalaya) are said to have warmed by more than one degree Celsius over the last century or longer – significantly higher than northern Indian plains or the global average, and the glaciers over there are supposed to be melting even faster. There are said to be 'hotspots' of this brown cloud, one of the prominent being identified as 'Indo-Asian-Pacific Plume', where one can 'see' the brown cloud rising up, even in the satellite images. In the so-called 'HinduKush-Tibetan-Himalaya' region, the warming by the black carbon factor is now being said to be almost equal to the warming by the additional carbon dioxide in the atmosphere.

As an additional impact, the health cost to the poor families using these biomass stoves is also shown to be a serious concern (which, incidentally, is supported by independent and unrelated studies), adding strength to the ABC argument.

Some Concerns, Doubts and Contradictions

There are several questions, concerns and doubts – scientific, social and economic – about these projections and calculations, their selective nature included. Let us take up the issue of the sources of, and proportionate contributors to, this ABC. No doubt that a majority of Asia's poor households burn a large quantity of wood and other biomass every day, for their cooking needs as they cannot afford to buy commercial fuel or have no access to such fuels. Nearly 70% of India's households use some kind of solid cooking fuel, and a large part of them – being extremely poor (about 70% of Indians live on little more than half-a-dollar per person per day for all their consumption expenses – as per a Government of India commissioned report. What polluting fuels can they buy and burn any way?), use collected wood and biomass. The percentage of poor households in Bangladesh, Nepal or Pakistan using solid fuels would be similar or higher. The finer parts of the soot from these are expected to rise up somewhat into the lower atmosphere by the heat created updraft, as well as by natural wind.

BUT, similar would be the case – only in a much larger scale – from the large numbers of forest fires in California, Australia, Russia, Mongolia, southern Europe etc (and also in Asian countries including Indonesia, China and India). Several such forest fires in California (USA) and Australia burn hundreds of square kilometers each, thus sending up black carbon/dark soot from many millions of whole trees – year after year. And as a burning forest fire would produce huge amounts of heat in a large area, the updrafts created are bound to be incomparably larger and stronger, carrying much larger amounts of black carbon ever higher into the atmosphere than is possible for an equivalent amount of wood/biomass burned in a few million distributed kitchen fires over a year. These would then be prone to 'transportation' to all around the globe. The most recent 'station fire' in Acton in California itself burned up nearly 44,000 hectares of forests! How many complete trees would that be?

Also consider the fact that a very large number of American and European families still burn large quantities of wood for their heating needs in winters. An average poor family kitchen uses about 7-9kg of wood in a day's cooking, whereas an average heating family will use around 30-40kg in a winter day in USA or northern Europe. Why do none of the Black Carbon/ABC studies point these out? Obviously, there is a hidden agenda here.

Over and above this, most kitchen fires are lit – naturally within partly closed kitchens (which causes large health damages to the poor women and their young children though, partly because of lack of ventilation – which causes dangerous SPM build ups, in some cases measured to be over 1000 micrograms per cubic metre of air inside the kitchen), reducing the amount of heat/smoke that leaks out and rises into the atmosphere. With climate change drying out forests in many places, forest fires are increasing almost everywhere, putting evermore amounts of smoke/BC into the atmosphere. In contradiction, the number of poor households using biomass and other solid fuels is coming down as a result of some increased access to other fuels and some better designs of wood/biomass stoves.

Thus, it is open to serious questioning as to how much of the visible smoke comes from kitchen fires, and what percentage comes from forest fires, industrial activities of fossil fuel burning, and from fossil fuelled automobiles. The quantity of fossil fuels being burnt by the increasing industrial-capitalist segment of South Asian societies (in the darkened footsteps of the 'modern' Western industrial-capitalist society) – mainly in their cars, airplanes and in mass production facilities – are rising fast in many countries of the region. The fuel consumption of the poor is not rising fast.

It is not a secret that the amount of coal and petroleum products being burned in the richer countries is far higher than that being burned in the South Asian nations. Thus, it is clear that these countries of the Global North are contributing far larger amounts of aerosol emissions (in addition to the GHG emissions), and are far larger contributors to the Atmospheric Brown Cloud and Black Carbons.

And if there are rising clouds visible from space, caused by the biomass burning Asian poor, there are much larger dark soot clouds which also are seen from space: those caused by large scale forest fires in many industrialized countries.

With all their 'advanced' technology and economic resources, it should be far easier for these governments to control and even eliminate most of these fires. Just one such large fire in California two years ago burned down nearly 1800 sq. km of forests, sending as much dark soot to form brown clouds as maybe a hundred million poor families would do during their daily meal cooking. These huge forest fires are in no way 'equivalent' to the compulsions of daily cooking by the Asian poor families. Neither are the rich people's lifestyle demands of large amounts of electricity, big cars, etc.

There are large numbers of industries and coal fired thermal power plants in northern India (as well as in other countries in the region near the Himalayas – with the enhanced glacial melting concern in focus) burning millions of tons of coal and oil and emitting hundreds of thousands of tons of dark suspended particulates into the atmosphere every year. Just in Delhi-NCR (National Capital Region, including peripheral towns like Gurgaon, Noida, Faridabad, Ghaziabad etc), there are over 16 lakh (1.6 million) cars (and about 4.4 million motorbikes to add) consuming many million tons of petrol/diesel, and emitting lakhs (one lakh is a hundred thousand) of tons of dark soot each year. Don't these travel to the Himalayas and get deposited on the glaciers? Or only the soot from the poor family kitchens has that bad habit?

Out of the 145,000+ MW of installed electricity generation capacity in India, about 95,000 MW is from coal fired thermal power plants. These consume over 350 million tons coal each year, generating over 100 million tons of ash (Indian power coal often contains 35% of minerals – which generates the ash), a significant part of which flies off to the atmosphere as fine dark particulates. Does anyone show any concern about how these tens of millions of tons of fine dark particles contribute to the formation of the Atmospheric Brown Cloud, or adds to the warming of the northern Indian mountain region, or how much they contribute to the Himalayan glacier melting? Probably not – because the electricity thus generated is consumed mostly by the influential upper and middle classes in India. The 15 lakh+ cars in Delhi-NCR must be contributing a much larger dark soot component than the tens millions of poor family kitchens, but no State wants to confront or challenge the dominating ruling classes and their consumption and emissions. This is the reason why the corporate controlled media also selectively picked up the black carbon from the poor family's kitchen as a big problem, hiding the fine particulate emissions from all these sources of the modern industrial-capitalist icons – which might have a much larger contribution.

Another carefully hidden fact is that the atmospheric life of black carbon generated from kitchen fires ranges from a few days (during periods of high humidity/precipitation) to weeks at most, while the fine dark particles generated by power plants and car industries stay much longer; the atmospheric life of carbon dioxide is measured in centuries. Which should we concentrate on, for greater impact reduction? With most countries, including the giant economy of China, following in the same emission intensive industrial-capitalist pathway (China is reported to be completing about 50 new coal fired power plants of 1000 MW capacity each year, as well as being the fastest growing car market – and the Tibetan Himalayan glaciers will also get black carbon from these Chinese emissions), contribution by the 'kitchen fire' of the poor households pales into insignificance.

The Hidden Agenda Behind Targeting Asia's Poor – Forest Control, Offsets and Carbon Bazaar

As the industrial-capitalist system contributes far more to the ABC than poor families' kitchen fires, why is attention being focused on Asia's poor? It is estimated that about 70% of Indian households still use solid fuels (including wood, coal, briquettes, dry leaves and grass, cow dung cakes, etc) as cooking fuel, as these are the only 'affordable' and somewhat accessible options. The 'great dream projection' of the Indian State – of providing 'clean LPG' to every family – is now clear to be impossible and undesirable, considering that the inevitable peak oil and gas situation is drawing nearer. Also, from the perspective of the people, why would one want to dispossess the rural and forest dwelling poor from their self-accessible fuel source, forcing them into the cooking fuel market?

But this is precisely the game plan: to bring this vast number (over 170 million households in India alone) of partly fuel self-dependent families into the commercial fuel and stove market, opening more profit avenues for the large corporations. If they remain self-dependent in terms of their cooking fuels, how do these capitalists 'grow' their fuel business? How does the GDP grow – to enable more coal fired plants, more steel plants, more cars?

Once you are successful in isolating the forest/forest-fringe dwellers from sustainable fuel and other supplies from the forests, these forest areas become available for closer government and corporate controls, including carbon trading and offset schemes – think REDD (see Lang in this volume).

Why not create a big new emission trading/offset arena in the shape of the replacement of Asian poor's kitchen fuel? It makes sense. The Western/Northern economies can keep belching CO_2 and 'offset' their polluting emissions by buying cheap credits from millions of 'improved' kitchen fires of South Asia, without having to reduce their own emissions in any way. Doesn't matter if the total emissions keep increasing, upsetting crucial climate systems of the earth. Doesn't matter if hundred million or more poor families are forced into buying unaffordable fuel, forgoing some critically needed food in the bargain. Doesn't matter if a few tens of millions of poor people's lives are 'offset'. Doesn't matter if a few million livelihoods, dependent on the collection, sorting and selling of biomass fuel, get lost in the process. All these will probably be considered 'collateral damage', a la George W. Bush and the Iraq war.

There is no doubt that the families – particularly the women and younger children staying close to their mothers – are badly affected by the high levels of particulate pollution created by the burning of these biomass fuels in pollution belching stoves and poorly ventilated kitchens. One fairly large study in Orissa (an eastern province in India) showed that the average suspended particulate (SPM) pollution inside such kitchens were as high as 1200-1500 micrograms per cubic metre of air, which is frighteningly higher than the prescribed safe levels for long hours of exposure. Thus, the need to provide these families with something much less polluting and health impacting is obvious. The problem arises with the commercialization and market linkage of this need. Once this is tied up with carbon trading type market mechanisms, the vulnerable poor families will be subjected to all kinds of exploitations and forced dispossession by the powerful trading entities and their 'skilful' technical consultants, operators, verifiers and certifiers coterie.

The State has the responsibility of providing its poor with reasonable levels of means of life, and it cannot be allowed to wriggle out of this by tying up with commercial/trading interests for any of these. Easy access to cooking fuel, which do not slowly kill you by poisoning the air you breath, should be part of the fundamental rights of every citizen of this planet, and there is no place for trading this right. What we need to get clear is what interests are behind these periodic 'scientific discoveries' and targeted media campaigns. We need to analyze and challenge these malicious attempts of further privatization of essentially common resources.

Notes

* The author has written a similar article for *Mausam*. Dutta, S. (2009) 'The ABC of How to Torment Asia's Rural Poor', *Mausam*, 1-2(2-5): 23-26.

III

CRITIQUES

Having presented the case studies, Part III offers a broader critique of carbon markets. What the seven chapters collected in this part of the book show is that carbon markets are not merely mechanisms to combat climate change. Instead, they must be seen in relation to the historical development of capitalism. In fact, carbon markets can be seen as the expansion of the capitalist project to new spheres which so far have escaped commodification. This is then where quite an existentialist question needs to be asked: can we trust capitalist markets to deal with such a grave and global problem as climate change, given that capitalist production and consumption regimes have created the problem in the first place? The authors of Part III argue that there is now overwhelming evidence that carbon markets will not help us mitigate climate change by commodifying the atmosphere, which should be seen as a common good shared by humanity. Instead, carbon markets will lead to more exploitation, inequalities and perverse speculations and financial bubbles of the kind the world has seen explode in 2008. Therefore, carbon markets should be seen as a dangerous diversion from the need to drastically change lifestyles and economic, social and political structures that will help to free ourselves from the world's addiction to fossil fuels.

17

Regulation as Corruption in the Carbon Offset Markets*

Larry Lohmann

Introduction

When a particular commodity market cannot be regulated, the attempt to regulate it can do no more than create an illusion of regulatability. Deflected into a *cul de sac*, official action to correct abuses sustains the underlying problems, or makes them worse. Regulatory acts become a danger to society. Governance becomes a part of corruption. All this happens regardless of the good intentions of regulators or anti-corruption fighters.

This chapter argues that the carbon offset market is an example of such an unregulatable market, and that attempts to regulate it will only entrench its status as a locus of international corruption and exploitation. But to set the scene, it may be useful to begin with the example of another such market that has been much in the news since 2007: the market in complex new financial derivatives that lies at the root of the recent global economic crash.

These derivatives were unregulatable. Instead of reducing or spreading risk, they amplified it and hid it.[1] Because the risk measurement models used by both companies and regulators gave the illusion that everything was under control, they made things worse. 'Giving someone the wrong map is worse than giving them no map at all,' the options trader and risk expert Nassim Nicholas Taleb pointed out.[2] US and UK officials, clinging to the dogma that regulation could handle any surprises thrown up by the explosive financial innovations of the 1990s and 2000s (or that the innovations regulated themselves), refused to consider the possibility that certain kinds of product, and certain kinds of market, were simply too dangerous to be allowed to exist. As the market for the opaque new financial products became larger and larger, so did the scope for abuses, cheats and corruption.[3]

The capture of finance policy by the private sector had a lot to do with the refusal to face up to the unregulatability of the new market. Former derivatives traders keen to stoke the booming markets, such as Robert Rubin from Citigroup and Hank Paulson from Goldman Sachs, occupied some of the highest positions in the US government. (Only ex-Wall Street executives, the reasoning went, could understand the vastly complicated world of finance well enough to govern it.) Private companies' own mathematical models were seen as a reasonable basis for regulation at both national and international levels. Orthodox economists in positions of regulatory responsibility such as successive US Federal Reserve Chairmen Alan Greenspan and Ben Bernanke were trained in ways that gave them the same faith in the inherent manageability of the new derivatives markets. Such long-entrenched forms of 'legal corruption'[4] were difficult for ordinary people either to speak against or to counter. There was little space for participating in policy or for questioning the doctrines that everything could be regulated and that 'learning by doing' would provide the answers to all problems.

A similar analysis applies to the carbon offset markets. Carbon offsets are inherently unregulatable, for unalterable scientific and logical reasons. Instead of reducing climate risk, they increase it and conceal it, along the way reinforcing environmental and social abuses of multiple kinds.[5] No one is sure how to measure them or indeed exactly what they are.[6] Partly for these reasons, offset projects have encountered persistent implementation problems, many of them documented in this book. Hundreds of projects and millions of credits are accused of being fraudulent, scams for shoring up business as usual – or worse. Scandal after scandal regarding the offset market is splashed across the front pages of newspapers. As former proponents desert the cause of carbon markets[7] and a growing crowd of prominent climate scientists and economists join the chorus of criticism,[8] the larger carbon markets of which carbon offsets are an integral part are poised on the edge of breakdown.[9]

Yet the illusion endures that carbon offset markets could someday be redeemed through reform, regulation or certification. Improved methodologies, it is said, might allow carbon credits to be calculated accurately. Greater oversight could stop fraud. Gaming could be prohibited. Land grabs could be curbed. Best-practice standards and certificates could transform the trade. A transition to renewable energy could be effected. Improving local capacity could safeguard local interests and democratize the process. With proper reforms and better regulation, carbon offsets could someday switch from being a climate danger to being a climate benefit and their generally deleterious social effects ameliorated. 'Let's not throw out the baby with the bath water,' has been the constant refrain of beleaguered carbon market proponents. 'Instead, let's practice "learning by doing" and maybe eventually the problems will become manageable.'

This illusion has practical effects. Under the 'air cover' of the claim that it is regulatable, an unregulatable offset market is taking over more and more territory at a time when it should be forced to retreat in an orderly and decorous fashion. As carbon offsets invade first the EU Emissions Trading Scheme, Australian and Japanese trading programmes, and now the incipient US carbon market, with its billions of tons of potential demand, the idea that offsets can be regulated has become a major threat to dealing effectively with climate change as well as a cause of social strife.

The illusion of offset regulatability is sustained partly because climate policy has been captured on both national and international levels by an elite alliance comprising big business, commodities traders, financial firms, neoclassical economic theorists and an influential group of professionalized, middle-class environmentalists. All are bent on seeing offset trading expand rather than be abolished.[10] Invented and developed by derivatives traders as well as economic theorists of the Chicago School and elsewhere, carbon trading has dominated global climate policy ever since being forced into the Kyoto Protocol in 1997 by the US delegation led by then Vice President Al Gore, who himself became a big carbon market player.[11] For more than a decade, governments, international agencies and private corporations alike have invested enormous resources in building up infrastructure for offset markets. The largest buyers of Kyoto Protocol Clean Development Mechanism (CDM) offset credits today are speculators on Wall Street and in the City of London and other financial districts,[12] some of which have poured millions of dollars into lobbying for a US offset market from which they also hope to benefit.[13] CDM offset regulators tend to be either offset buyers and sellers or former or current executives in private-sector carbon businesses, all of whom have a vested interest in seeing the trade expand as well as privileged access to information useful in navigating and promoting it.

In elaborating on these themes, this chapter will suggest responses to the problem of corruption in the carbon markets that look beyond 'technical fixes' that attempt to regulate malpractice and administrative abuse. Because the problems of carbon markets go much deeper than is ordinarily understood, it will argue, they demand meticulous and thoroughgoing attention to structural issues of power, knowledge and democracy.

Carbon Market Corruption: The Conventional Understanding

Beware the Carbon Offsetting Cowboys, warns the Financial Times.[14] 'Irregular Carbon Credits Cause Upheaval in the Government of Papua New Guinea,' reports The Economist.[15] 'Pollution Credits Let Dumps Double Dip', reveals the Wall Street Journal.[16] 'The Great Carbon Credit Con: Why are We Paying the Third World to Poison its Environment?' asks the Daily Mail.[17] 'Secretive U.N. Board Awards Lucrative Credits with Few Rules Barring Conflicts,' according to ClimateWire.[18] 'UN Suspends Top CDM Project Verifier over Lax Audit Allegations,' reports Business Green.[19] 'Europol Expects More Arrests in Carbon Fraud Probe,' notes Reuters.[20]

As such headlines attest, uncovering carbon market scandals is by now a minor journalistic industry. The prospective supply of further shocking stories, moreover, is limitless. Dirty installations ranging from industrial pig farms in Mexico to polluting sponge iron works in India are availing themselves of revenues from the trade, with hundreds of enterprises – including most of the 763 Chinese hydroelectric projects applying or planning to apply for carbon credits[21] – eager to take advantage of an opportunity to get a bit of extra free money for conducting business as usual. According to Peter Younger of Interpol, 'in future, if you are running a factory and you desperately need credits to offset your emissions, there will be someone who can make that happen for you. Absolutely, organized crime will be involved.'[22]

Countering such scandal stories with reassurances that regulation can solve the problems has also become a profitable industry, providing employment to hundreds of technicians, bureaucrats, academics and political figures. The CDM needs 'not something new, but rather a change of culture and professional working practices,' legal scholar Ray Purdy complacently assures his readers: 'more permanent and temporary staff … clear professional service standards … better knowledge-bases and methods of communication.' Moreover,

> to allow more transparent oversight and avoid real or perceived conflicts of interest, the [CDM] Executive Board needs to recognize the governance requirements of accountability and clearly distinguish between supervisory and executive roles.[23]

Other observers blandly recycle boilerplate about 'due process safeguards',[24] 'enhanced dispute resolution',[25] 'capacity building,' an 'internal review mechanism'[26] and improvements in 'domestic CDM structures.'[27] To quote Al Gore in recent testimony before the US Congress, 'I think there is general agreement that in Copenhagen significant reforms of the CDM, uh, Collective Development Mechanism, uh, Cooperative Development Mechanism, have to be implemented.'[28]

The understanding of corruption and regulation that enables and limits this discussion is narrow. The stories that most journalists and academics tell about corruption in the carbon markets tend to be traditional ones of con artistry, abuse of public office for private gain, and payment of bribes to government officials, as well as, occasionally, a somewhat broader narrative featuring more general abuses of power and wealth that undermine democratic governance and the cause of social justice. Although it has been out of fashion for some time, there are signs, too, that the customary story of conflict of interest may soon be revived as a framework for understanding corruption in carbon trading.

For many journalists and academics, such stories have the great virtue of being familiar and easy to tell and understand. They identify bad guys who are getting away with murder. For many technicians, bureaucrats and politicians, these stories are attractive because they imply that there is a familiar job for them to do: catch the bad guys and formulate and enforce rules that will prevent more bad guys from being tempted into abuses. In these narratives, the problems plaguing carbon markets are due to relative lawlessness, lack of technical standards and incomplete enforcement – problems well within the capability of the prospective heroes of the stories to handle.

On the surface, there is a great deal to be said for these narratives. Many examples spring to mind. However, probe a little deeper and complexities emerge that suggest a less comforting story. What follows will explore both the usefulness and the limitations of three stories that are often told about corruption and regulation in carbon markets, along the way assembling materials for a more politically and scientifically informed narrative.

Corruption as Confidence Trickery?

Everyone who participates in or studies the carbon offset market knows that it is a haven for con artists. Businesses and even international financial institutions[29] understand that, as long as they provide clever enough documentation, carbon offsets can become a source of extra funding for ventures they are engaged in that have nothing to do with climate change mitigation: even gas pipelines,[30] fossil fuel-fired generating plants,[31] coal mines[32] and oil wells.[33] An investigation of projects in India by a carbon offset market proponent found that a third were simply business as usual.[34] By the UN's own rules, most hydropower projects in the Kyoto offset pipeline arguably should not be allowed to produce carbon credits at all.[35] According to one prominent carbon banker, project proponents 'tell their financial backers that the projects are going to make lots of money' at the same time they claim to regulators 'that they wouldn't be financially viable' without carbon finance.[36] Carbon consultants often freely fabricate information required on official forms,[37] and the more convoluted offset accounting methodologies become, the more opportunities for fraud emerge. An investigation of Nigerian carbon offsets devised by Western oil companies and carbon consultant firms, for example, found that it was nearly impossible to determine whether the gas that the companies claimed will be diverted from flaring to productive use will not in fact come from dedicated gas extraction operations, whose production is not flared.[38] Businessman Marc Stuart of the carbon offset trading firm EcoSecurities admits that new schemes for generating carbon credits out of forest conservation involve such a 'brutal potential for gaming' that 'getting it wrong means that scam artists will get unimaginably rich while emissions don't change a bit.'[39]

Is regulation capable of defusing such dangers? Can reform address the relevant problems? Is it possible to 'get offsets right', as Stuart suggests it is? There are several powerful reasons for answering 'no' to all of these questions. The abuses of power and wealth that constitute carbon market corruption do not derive merely from the misdeeds of individual carbon consultants and profiteers, but inhere in market architecture itself. They are an integral technical component of commodity formation. While individual consultants can and do make use of this market architecture for the gain of their clients and themselves, it is the architecture itself that performs the central abuses. Accordingly, what are conventionally classed as scams or frauds are an inevitable feature of carbon offset markets, not something that could be eliminated by regulation targeting the specific businesses or state agencies involved. Because the underlying problem is not, essentially, a matter of poor implementation or individual malefactors, it can only be eliminated by eliminating the offset market itself.

One central difficulty is that for every offset project, carbon consultants must identify a unique storyline describing a hypothetical world without the project, and then assign a number to the greenhouse gas emissions associated with that world. They then must show that the project makes carbon savings 'additional' to those of this baseline world. By subtracting the emissions of the project world from those of the baseline world, they derive the number of carbon credits that the project can sell. Carbon accountants, that is, must present the counterfactual without-project scenario not as indeterminate and dependent on political choice but as measurable, singular, determinate and a matter for economic and technical prediction. This assumption, as Kevin Anderson, Director of the UK's Tyndall Centre for Climate Change Research, observes, is a 'meaningless concept in a complex system.' As Anderson explains, the counterfactual 'baseline' against which the purported emissions savings of a carbon offset project must be measured must be calculated over 100 years to correspond with the approximate residence time of carbon dioxide in the atmosphere. For example, a wind farm in India may claim to be generating carbon credits because it is saving, over a century, fossil fuels over and above what would have been saved without the project. However,

> the wind turbines will give access to electricity that gives access to a television that gives access to adverts that sell small scooters, and then some entrepreneur sets up a small petrol depot for the small scooters, and another entrepreneur buys some wagons instead of using oxen, and the whole thing builds up over the next 20 or 30 years. ... If you can imagine Marconi and the Wright brothers getting together to discuss whether in 2009, EasyJet and the internet would be facilitating each other through internet booking, that's the level of ... certainty you'd have to have over that period. You cannot have that. Society is inherently complex.[40]

There will thus be no general scientific consensus about the number of credits, if any, generated by a particular carbon project. Even the question whether a project goes beyond business as usual in saving carbon, as carbon trader Mark C. Trexler and colleagues noted years ago, has 'no technically 'correct' answer'[41]; as the US General Accounting Office concluded in 2008, 'it is impossible to know with certainty whether any given offset is additional.'[42]

It follows that it is also impossible to know with certainty whether any given offset is non-additional. Hence it is a misdiagnosis of the recurring scandals in carbon offset markets to say that they are due to consultants claiming falsely that non-additional projects are additional. The problem goes deeper. Scientifically speaking, there is no such thing as 'additionality' or 'non-additionality', and thus no standard that either market participants or regulators could use either to clarify the accounting rules or to prevent scamming.[43] If it is impossible to distinguish between fraudulent and non-fraudulent offset calculations, regulators' power to enforce climate benefit becomes illusory.[44] They have no choice but to fall back on aesthetic, political or pseudo-scientific criteria in deciding whether to wave projects through. As Lambert Schneider of Germany's Öko-Institut notes, 'If you are a good storyteller you get your project approved. If you are not a good storyteller you don't get your project through.'[45] The problem, in other words, is not that the tools for regulating the offset market need further development or that they are not being used correctly. The problem is that no such tools exist.

But if the offset markets cannot be regulated, then proceeding as if they could be will inevitably encourage both unscrupulous manufacturers of carbon credits and the Northern fossil fuel polluters who are only too happy to buy them without inquiring too closely into their validity. The central 'abuse of public office for private gain' in the carbon offset trade does not stem from individual corporations getting special treatment from individual public officials in return for bribes. It derives, rather, from the way that public officials across the world acquiesce in the use of fake mathematics and science to benefit a fossil fuel-dependent corporate structure as a whole at the expense of public welfare. It is less the antics of market players than the attempt to construct an unfeasible market that is corrupt, and corrupting.

Carbon offset accounting's need to isolate a unique storyline describing a hypothetical world without an offset project leads also to a second abuse of power and wealth inherent in the trade. Offset accounting frames the political question of what would have happened without carbon projects as matter of technical prediction in a deterministic system, while at the same time framing project proponents as free decision-makers whose carbon initiatives 'make a difference'. Carbon offset mathematics dictates that, in any given situation, 'no other world is possible' as an alternative to business as usual except that created by corporations wealthy enough to be in a position to sponsor carbon offsets. This suppression of unknowns built into offset mathematics entails suppression of climate alternatives pursued by the less powerful and wealthy. Among the first observers to call attention to this built-in bias were social activists from Minas Gerais, Brazil campaigning against the attempt of a local charcoal and pig iron company, Plantar, to get carbon credits for the environmentally-destructive eucalyptus plantations it had established on occupied land. The activists categorized the company's argument that without carbon credits it would have to switch from eucalyptus charcoal to more-polluting coal as an energy source as a 'sinister strategy ... comparable to loggers demanding money, otherwise they will cut down trees':

> What we really need are investments in clean energies that at the same time contribute to the cultural, social and economic well-being of local populations.[46]

For the activists, carbon accounting's suppression of knowledge of the plurality of choices amounted to an abuse of power blocking popular pathways to an alternative future.

Carbon offset accounting methodology also drives corrupt activity in another, more indirect way, through yet another of its intrinsic features: its promiscuous drive to establish that different technologies in different places are somehow climatically 'the same'. In its push for liquidity, the carbon offset market incentivizes thousands of technical experts to undertake a relentless search for far-fetched equivalences among the most distant activities. On one day, carbon consultants may devise calculations that make diverting Nigerian methane from flaring to productive use 'the same as' shutting down a Nebraska coal-fired power plant. On the next, they will come up with techniques that render the annexation of forested land in the Democratic Republic of Congo 'the same as' making efficiency improvements in Spain's housing stock. Rather than seeking ways to effect a structural shift away from fossil fuels in Northern countries, that is, offset market actors are driven toward constructing ever more fanciful equations for shifting climate burdens onto the South in the name of increased liquidity and cost-effectiveness. In political economy terms, the proliferation of such equations reflects a use of expertise and money to take advantage of a multitude of local resources and local political weaknesses across an expanding global field that is ever more difficult to police. Market expansion, far from being a solution to the market's problems, thus not only increases the ecological debt of the North to the South, but is also a recipe for growing obscurity, evasions and cheats of all kinds, greatly advantaging centralized market actors while weakening the possibility of local oversight. As Willem Buiter of the London School of Economics notes, the fact that offset accounting requires

> the impossible verification of how much carbon dioxide equivalent would have been emitted in some counterfactual alternative universe. ... makes one shout out: impossible! Fraud! Bribery! Corruption! Wasteful diversion of resources into pointless attempts at verification! And indeed this is what is happening before our eyes. Enterprises get paid for not cutting down trees and for installing filters and scrubbers they would have installed in any case. The new Verification of the Carbon Counterfactual industry is growing in leaps and bounds. The amounts of money involved are vast and the opportunities for graft, bribery and corruption limitless. The offset proposal has birthed a monster.[47]

Such a 'vastly complicated apparatus,' agrees Clive Crook of the Financial Times, is by its nature a 'playground for special interests.'[48]

Corruption as Erosion of the Rule of Law by Money and Influence?

The carbon markets abound in stories of offset developers finding ways of evading the law through bribery or abuses of influence. Officials allied to offset developers may receive land concessions that communities are denied.[49] Faulty project documents are routinely approved by government departments.[50] As Interpol observes, moreover, bribery and intimidation are certain to be ingredients of the growing forest carbon offset market;[51] recently, a nephew of Papua New Guinea's Prime Minister was accused of pressuring villagers to sign away their land for carbon deals despite there being no carbon trade laws in place.[52]

The conventional response to such stories – including that of many environmental NGOs – is to repeat the mantra that regulation is capable of saving the alleged 'real potential' of offset markets from the menace of corruption.[53] Such responses again overlook the extent to which the erosion of the rule of law is part of the design of carbon trading, not an incidental feature that can be remedied by applications of 'good governance'. For an illustration of the point it is useful once again to turn to the Niger Delta.

There, for 50 years, energy companies have been burning off the great bulk of the methane they find in underground oil reservoirs. Although methane is a valuable fuel, it is cheaper for Shell, Chevron and other firms simply to flare it on site than to use it in power plants or reinject it underground. As a result, local people are subjected to continuous noise, light and heat, acid rain, retarded crop yields, corroded roofs, and respiratory and skin diseases. Although flaring is prohibited by law in Nigeria, oil companies have so far contented themselves with paying penalties for non-compliance. In this context, one focus of local and international environmental activism is simply to insist on the rule of law. The Clean Development Mechanism, however, takes breaches of the law in Nigeria as the 'baseline' for carbon accounting. The Italian oil corporation Eni-Agip, for example, plans to buy some 1.5 million tons per year of cheap carbon dioxide equivalent pollution rights from a project at an oil-gas installation at Kwale that was registered with the UN in November 2006.[54] The core of the credit calculation is that

> whilst the Nigerian Federal High Court recently judged that gas flaring is illegal, it is difficult to envisage a situation where wholesale changes in practice in venting or flaring, or cessation of oil production in order to eliminate flaring will be forthcoming in the near term.[55]

Accordingly, the project creates an incentive for the Nigerian authorities to replace legal sanctions with prices and the rule of law with markets for environmental services. In such cases, carbon trading undermines any attempt to tackle the underlying reasons why environmental law is flouted in Nigeria by ensuring that money is made by treating violations as a given.

In many other host countries as well, the Kyoto offset market is creating incentives for emissions-related environmental laws not to be enforced or promulgated, since the greater the 'baseline' emissions, the greater the payoffs that can be derived from carbon projects.[56] These incentives are explicitly spelled out in UN policy. In August 2007, for instance, the CDM Executive Board published forms for the submission of applications for a new type of carbon project called programmatic CDM or 'programmes of activities' (PoA). A PoA, it stated, could be additional and thus acceptable as CDM even if a law already existed that mandated the measures that the PoA would bring about, if that law was not being 'enforced as envisaged but rather depend[ed] on the CDM to enforce it', or if the PoA would 'lead to a greater level of enforcement of the existing mandatory policy/regulation than would otherwise be the case'.[57] Here as elsewhere, corruption – interpreted as the erosion of the rule of law by financial interest – is a structural principle of carbon offset trading. Regulation curbing corruption would have to outlaw offset trading itself.

Corruption as Conflict of Interest?

Everyone working in carbon offsets is aware of the conflicts of interest that pervade the trade. These conflicts are present at all levels, but particularly afflict the carbon markets' regulatory systems. For example, Lex de Jonge, head of the carbon offset purchase programme of the Dutch government, is the chair of the Board of the Clean Development Mechanism (CDM), the UN offset market's regulatory body.[58] Other members of the board have meanwhile been accused of being 'very active in defending projects that come from their country or that are hosted in their country, or where some companies have a particular interest.'[59] Barclays Capital, a major speculator in the carbon markets, boasts openly that 'two of our team are members of the Executive Board.'[60] In addition, like credit ratings firms in the financial markets, private sector carbon auditors approved by UN regulators have a strong interest in gaining future contracts from the companies that hire them; unsurprisingly, they wave through an overwhelming majority of projects under review.[61] Meanwhile, banks that own equity stakes in carbon offset projects, or are 'going long' on carbon credits, may also be carbon brokers or sector analysts, 'creating a temptation to bid up carbon prices to increase the value of their own carbon assets.'[62] For example, Goldman Sachs owns a stake in BlueSource, a carbon offset developer, while JPMorganChase has acquired Climate Care, another offset specialist, and is to buy carbon offset aggregator EcoSecurities for US$204 million.[63]

Within the insular, tightly-knit professional climate mitigation community, moreover, experts are constantly passing through revolving doors between private carbon trading consultancies, government, the UN, the World Bank, environmental organizations, official panels, trade associations and energy corporations. For example, Martin Enderlin, a CDM board member from 2001 to 2005, is now director of government and regulatory affairs at EcoSecurities.[64] As one principal of a carbon asset management firm who is also a member of the UN's CDM methodology panel noted at an industry meeting in London in October 2008, 'I helped set the rules; now my firm plays by those rules.'[65]

Revolving doors host a flow of traffic to and from many other zones of the carbon market as well. James Cameron, an environmental lawyer who helped negotiate the Kyoto Protocol, now benefits from the market he helped create in his position as Vice Chairman of Climate Change Capital, a boutique merchant bank that recruited as staff members Kate Hampton, former climate chief at Friends of the Earth, and Jon Sohn, formerly of World Resources Institute. Hampton was then seconded by Climate Change Capital to the UK's Department for Environment, Food and Rural Affairs (DEFRA) as a senior policy adviser during the UK's G8 summit (which focused on climate change) and EU Presidency. Climate Change Capital's Vice President for Carbon Finance, Paul Bodnar, took charge of climate change finance at the US State Department in 2009. Henry Derwent, a former director of international climate change at Britain's DEFRA, who was responsible for domestic and European climate change policies, is now president and chief executive of the International Emissions Trading Association, the industry alliance. Sir Nicholas Stern, author of the British government's Stern Report on Climate Change, has meanwhile championed the initiative of his private firm, IDEACarbon, to set up a carbon credit ratings agency – which many observers are likely to see as subject to the same type of conflict of interest that earlier afflicted Moody's and other credit ratings agencies that depended for their income on the companies whose products they were rating.[66] In the unregulated 'voluntary' markets for carbon credits, conflict of interest is also deeply entrenched. Laurent Segalen, formerly a carbon trading manager at the failed Lehman Brothers investment bank, expressed a wide consensus when he affirmed that 'traders should be the ones designing and determining the standards.'[67] The secretariat of the UK's All-Parliamentary Committee on Climate Change, which proposes regulatory policy for the voluntary carbon offset market, is housed at The Carbon Neutral Company, whose business depends on such regulation.

Is it possible to get rid of such pervasive conflict of interest through regulation? No, because conflict of interest is inherent in offset market structure. First, the fact that supply and demand in this trade, as well as the nature of the commodity itself, are dependent on decisions made by small elites within governments, all of whom, whether buyers or sellers, are interested mainly in creating as many carbon credits as possible, means that there is little incentive on any side to inquire too closely into whether the manufacture of those credits is good for the climate or not. While buyers of blue jeans care about whether they will wear out or not, acting as a check on the temptation of manufacturers to cut corners, buyers of carbon credits care only about whether regulators will accept them in lieu of local compliance.[68] And while most markets have regulators whose careers depend on checking to see whether the goods on sale are what they say they are, regulators in the carbon offset market, as often as not, are buyers or sellers themselves, whose interests lie elsewhere. 'I don't see us as police,' the chair of the CDM Executive Board confirmed in 2007.[69] European Commission coordinator for carbon markets and energy policy Peter Zapfel, a disciple of US economist-advocates of pollution trading and an instrumental figure in convincing European bureaucrats and governments to commit themselves to carbon trading,[70] meanwhile has openly urged 'cross-fertilization between regulators and regulated.'[71] Nor could environmental impact assessments (EIAs) compensate for the lack of market incentives working in favour of climatic stability, even if carbon project EIAs were tasked with assessing climate impacts, which they are not. Throughout the world, conflicts of interest are also an inherent part of the EIA process, since consultants contracted to perform EIAs are typically paid by project developers themselves as a part of regular and accepted practice.

Second, the trade in carbon commodities, like that in advanced credit derivatives, is both so complicated and so lucrative that the experts best qualified to regulate it are almost certain to have vested interests, whether they are involved in making money out of it directly, in advising interested governmental parties to it, or in designing it. As early as 2000, top Intergovernmental Panel on Climate Change scientist John Houghton admitted it was impossible to staff his scientific panel on forestry offset accounting without recruiting experts with financial interests in selling carbon credits.[72] Today, when the largest buyers of carbon credits are financial-sector speculators bent on creating complex new instruments with them, including Goldman Sachs, Morgan Stanley, Barclays Capital, Deutsche Bank, Rabobank, BNP Paribas Fortis, Sumitomo, Kommunalkredit, Cantor Fitzgerald, Credit Suisse and Merrill Lynch, meaningful regulatory oversight has become even less likely. Any more general public understanding of the tricks of the trade, meanwhile, is virtually ruled out at the start by the complicated nature of the commodities on offer. The recent temporary suspension of the accreditation of the leading verifier of CDM credits, the Norwegian firm Det Norske Veritas,[73] on the comparatively trivial ground that a company employee had signed off on five projects without surveying them, unwittingly reveals the impossibility of regulators' coming to terms with the central issues involved, much less engaging in meaningful action. So does the ineffectual UN reaction to rumblings about corruption on the CDM Executive Board – which has been to admit that determining whether members are subject to conflict of interest is left to 'their own individual discretion' and that they need do nothing more than state under oath that they have 'no financial interest in any aspect of the Clean Development Mechanism.'[74]

Conclusion

Preliminary reactions to corruption and abuse in the carbon offset trade – scandal stories in the news media, a few arrests or suspensions, calls for better regulation – have served a useful purpose in that they have been a first indicator of fundamental problems in market structure. But this first reflex response needs now to be supplemented with analysis of what underpins the scandals: by themselves, knee-jerk calls for 'reform' and 'regulation' are likely in the end to function only to deepen the roots of social exploitation and climate danger.

A first step is to understand that the principal problems of corruption in carbon markets are not located in the transgressions of individual firms, government officials or rogue traders seen as acts of corruption such as fraud or bribery. That is, the essential problems are not 'carbon cowboys' or 'bad apples.' Rather, they are to be found in the architecture of the markets themselves, which have been the creation of economists, traders, policy wonks, ministers, UN officials, NGOs, scientists and other experts as well as of the corporate sector. As argued above, the contradictions built into the markets – unverifiability of carbon credits, mutually-reinforcing relationships between carbon commodity production and erosion of checks and balances and the rule of law, systematic bias entrenching the power of fossil fuel-dependent corporations at the expense of public interest, and so forth – cannot be resolved by regulation any more than they can be addressed by 'learning by doing'. To continue to claim that carbon offset markets can be regulated is to legitimize continued corruption and to undermine popular struggles against it, as well as to harm the causes of climate action and climate justice.

By the same token, because the problems are systemic rather than criminal in a conventional sense, to call for the suspension, arrest, prosecution or shaming of the US and European economists, officials, policymakers and experts who have created carbon offset products or promoted their official acceptance is neither appropriate nor necessary. Despite the responsibility of such elites for entrenching inherently corrupt and damaging trading systems in national and international law, the correctible problem lies in the existence of those systems itself, not in their inventors and advocates; in any case, presumably, no clear legal basis exists for claims of causality or intent to defraud. No more useful purpose would be served by pursuing the officials and experts responsible than by attempting to prosecute the individuals responsible for the development and spread of certain hazardous chemicals or financial instruments such as collateralized debt obligations.

It should be sufficient, instead, for society to take the perfectly conventional, well-worn and easily implementable self-protective path of simply abolishing the trade in question, just as it has banned, or could ban, the manufacture or trade of certain chemicals, weapons or financial derivatives. Any reasonably thorough investigation into the corruption built into the carbon offset markets shows that they require not purification, but elimination. Once the systemic problem is tackled, petty or individual corruption will no longer be an issue: if illegal offset trading aimed at easing compliance with government-mandated emissions limits were carried out at all, it would have to be carried out in public. Doing away with this trade would be a simple, adult and effective approach to preventing a type of corruption which is threatening not only ordinary landholders, workers and victims of pollution but also human flourishing and survival itself.[75]

Notes

* This article was first published in Reddy, T. and A. Ferrial (forthcoming) *Climate Change and the Governance of Carbon Trading: A Critical African Review*, ISS Corruption and Governance Programme Publication.

1 Lohmann, L. (2009) 'Regulatory Challenges for Financial and Carbon Markets', *Carbon & Climate Law Review*, 3(2): 161-71 and Lohmann, L. (2009) 'When Markets are Poison: Learning about Climate Policy from the Financial Crisis', *Corner House Briefing Paper No. 40*, September 2009. www.thecornerhouse.org.uk/subject/climate.

2 Taleb, N. (2009) Preface to Triana, P. *Lecturing Birds on Flying: Can Mathematical Theories Destroy the Financial Markets?*, Wiley.

3 Lohmann, L. (2009) 'When Markets are Poison: Learning about Climate Policy from the Financial Crisis'.

4 Standing, A. (2008) 'Corruption and Industrial Fishing in Africa', *Anti-Corruption Resource Centre*, Bergen. p. 9, for an interesting discussion of legal corruption.

5 Lohmann, L. (ed.) (2006) *Carbon Trading: A Critical Conversation on Climate, Privatization and Power*. Uppsala: Dag Hammarskjold Foundation; Lohmann, L. (2008) 'Carbon Trading, Climate Justice and the Production of Ignorance: Ten Examples', *Development*, 51(3): 359–365.

6 Lohmann, L. (2005) 'Marketing and Making Carbon Dumps: Commodification, Calculation and Counterfactuals in Climate Change Mitigation', *Science as Culture*, 14 (3): 203-235.

7 Webb T. and T. Macalister (2009). 'Carbon Trade Wrong, says Lord Browne', *The Guardian*, 8 March. Even the academic economists who first mooted the idea of pollution trading in the 1960s are sceptical about the effectiveness of today's carbon markets.

8 These now include James Hansen, Jeffrey Sachs, Joseph Stiglitz, William Nordhaus, Kevin Anderson and Gregory Mankiw.

9 The hedge fund Pure Capital, for instance, sees a 30 per cent chance of carbon market collapse. See Lawrence Fletcher, 'Hedge Fund Firm Pure Capital Targets Carbon, Food,' Reuters, 18 June 2009.

10 State or regulatory capture occurs when private firms gain undue influence in the shaping of regulation and other policies that affect their own interests. For example, corporations may contribute to a political party's election fund in return for lower environmental standards, or treasury ministries may be staffed by financiers or traders who plan to return to the private sector after promulgating policies that benefit their old firms or harm their competitors. State capture is as prevalent in the North as in the South, and tends to be exacerbated by economic liberalization. State capture is particularly prevalent in carbon markets, since its very product is created by government action and, as financial analyst John Kay explains, 'when a market is created through political action, business will seek to influence market design for commercial

advantage' ('Why the Key to Carbon Trading is to Keep it Simple,', *Financial Times*, 9 May 2006).

11 Lohmann, L. (2009) 'When Markets are Poison: Learning about Climate Policy from the Financial Crisis'.

12 United Nations Environment Programme Risoe Centre on Energy, Climate and Sustainable Development, *CDM Pipeline*, http://www.cdmpipeline.org/.

13 Taibbi, M. (2009) 'The Great American Bubble Machine', *Rolling Stone*, Issue 1082-1083.

14 Harvey, F. (2007) 'Beware the Carbon Offsetting Cowboys', *Financial Times*, 26 April.

15 The Economist (2009) 'Money Grows on Trees,' 6 June.

16 Ball, J. (2008) 'Pollution Credits Let Dumps Double Dip: Landfills Find New Revenue in Trading System Meant to Curb Greenhouse Emissions', *Wall Street Journal*, 20 October 2008.

17 Ghouri, N. (2009) 'The Great Carbon Credit Con', *Daily Mail*, 1 June.

18 Gronewold, N. (2009) 'Secretive UN Board Awards Lucrative Credits with Few Rules Barring Conflicts,' *Climate Wire*, 4 July.

19 Young, T. (2008) 'UN Suspends Top CDM Project Verifier', *Business Green*, 1 December 2008, http://www.businessgreen.com/business-green/news/2231682/un-slaps-cdm-verifier.

20 Chestney, N. and M. Szabo (2009) 'Europol Expects More Arrests in Carbon Fraud Probe', *Reuters*, 20 August 2009.

21 Haya, B. (2007) 'Failed Mechanism: How the CDM is Subsidizing Hydro Developers and Harming the Kyoto Protocol', http://www.internationalrivers.org/files/Failed_Mechanism_3.pdf.

22 Creagh, S. (2009) 'Forest CO_2 Scheme Will Draw Organised Crime: Interpol', *Reuters*, 1 June.

23 Purdy, R. (2009) 'Governance Reform of the the Clean Development Mechanism after Poznan', *Carbon & Climate Law Review*, 3(1): 5-15.

24 Von Unger, M. and C. Streck (2009) 'An Appellate Body for the Clean Development Mechanism: A Due Process Requirement', *Carbon & Climate Law Review*, 3(1): 31-44.

25 Millar, I. and M. Wilder (2009) 'Enhanced Governance and Dispute Resolution for the CDM', *Carbon & Climate Law Review*, 3(1): 45-57.

26 Jacur, F. (2009) 'Paving the Road to Legitimacy for CDM Institutions and Procedures: Learning from Other Experiences in International Environmental Governance', *Carbon & Climate Law Review*, 3(1): 69-78.

27 van der Gaast, W., and K. Begg (2009) 'Enhancing the Role of the CDM in Accelerating Low-Carbon Technology Transfers to Developing Countries', *Carbon & Climate Law Review*, 3(1): 58-68.

28 International Rivers Network, 'What's in a Name? Corker Mentions Our CDM Work in Congress,' http://www.internationalrivers.org/en/node/3817.

29 Lohmann, L. (ed.) (2006) *Carbon Trading: A Critical Conversation on Climate, Privatization and Power*, p. 147.

30 Lohmann, L. (ed.) (2006) *Carbon Trading: A Critical Conversation on Climate, Privatization and Power*, pp. 292-94.

31 Brahic, C. (2009) '"Green" Funding for Coal Power Plants Criticised', *New Scientist* 2697.

32 See, for example, United Nations Framework Convention on Climate Change, 'Yangquan Coal Mine Methane (CMM) Utilization for Power Generation Project', Shanxi Province, China, http://cdm.unfccc.int/Projects/DB/TUEV-SUED1169658303.93.

33 Gardner, T. (2007) 'Blue Source To Capture Kansas CO_2, Up Oil Output', *Reuters*, 22 August, http://www.planetark.com/dailynewsstory.cfm/newsid/43843/story.htm.

34 Channel 4 (UK) (2007) 'Dispatches: The Great Carbon Smokescreen'.

35 Haya, B. (2007) 'Failed Mechanism: How the CDM is Subsidizing Hydro Developers and Harming the Kyoto Protocol'.

36 Financial Times, 16 February 2005.

37 Point Carbon (2005) 'Consulting Firms Deny Wrongdoing in Drafting Indian PDDs', 11 November, http://www. pointcarbon.com.

38 Osuoka, I. (2009) 'Paying the Polluter? The Relegation of Local Community Concerns in "Carbon Credit" Proposals of Oil Corporations in Nigeria'.

39 'REDD – The Basis of a "Carbon Federal Reserve"?', CleanTech Blog, http://www.cleantechblog.com/2009/05/redd-basis-of-carbon-federal-reserve.html.

40 Kevin Anderson, testimony before the UK Parliamentary Environmental Audit Committee, 23 June 2009, http://www.parliamentlive.tv/Main/Player.aspx? meetingId=4388.

41 Trexler, M., D. Broekhoff, and L. Kosloff (2006) 'A Statistically Driven Approach to Offset-Based GHG Additionality Determinations: What Can We Learn?', *Sustainable Development and Policy Journal*, 6: 30.

42 United States General Accounting Office (2008) 'International Climate Change Programs: Lessons Learned from the European Union's Emissions Trading Scheme and the Kyoto Protocol's Clean Development Mechanism', GAO Report GAO-09-151, p. 39.

43 Perhaps partly for this reason, it has been repeatedly proposed that the additionality requirement be eliminated. However, all other proposals for defining what an offset is have proved no less problematic. For example, proposals for 'sectoral' or policy-based CDM again leave judgements about whether carbon credits are climatically effective up to officials with vested market interests, with insufficient or nonexistent checks and balances.

44 All regulation currently proposed for carbon markets assumes incorrectly that the distinction between fraud and non-fraud can be made and enforced. Under the Kyoto Protocol, this assumption forms the basis of the work of the Clean Development Mechanism Executive Board. In the US, it is the unexamined assumption of, for example, the Emissions Allowance Market Transparency Act (S. 2423) proposed by Senator Dianne Feinstein, the Waxman-Markey Act, and the Climate Market Auction Trust and Trade Emissions Reduction System (HR 6316) introduced by Congressman Lloyd Doggett.

45 Schneider, L (2007) Presentation at conference on Review of the EU ETS, Brussels, 15 June.

46 FASE et al., 'Open Letter to Executives and Investors in the Prototype Carbon Fund', Espirito Santo, Brazil, 23 May 2003; A. P. L. Suptitz et al., 'Open Letter to the Clean Development Mechanism Executive Board', Minas Gerais, Brazil, 7 June 2004. Recent moves by the World Bank and other UN agencies to open up native forests to carbon accounting are similarly viewed as providing an opening for governments to threaten to destroy their forests if they are not granted carbon credits. See, e.g., *World Rainforest Movement Bulletin*, December 2008, www.wrm.org.uy.

47 Willem Buiter, 'Carbon Offsets: Open House for Waste, Fraud and Corruption', http://blogs.ft.com/maverecon/2007/ 07/carbon-offsets-html/.

48 Crook, C. (2009) 'Obama is Choosing to be Weak', *Financial Times*, 8 June 2009.

49 Lohmann, L. (ed.) (2006) *Carbon Trading: A Critical Conversation on Climate, Privatization and Power*, p. 243.

50 Lohmann, L. (ed.) (2006) *Carbon Trading: A Critical Conversation on Climate, Privatization and Power*, p. 271.

51 Creagh, 'Forest CO$_2$ Scheme Will Draw Organised Crime: Interpol'.

52 Gridneff, I. (2009) 'PNG PM's Nephew "Pushing Carbon Deals"', *The Age*, 3 July 2009, http://news.theage.com.au/ breaking-news-world/pngs-pm-nephew-pushing-carbon-deals-20090703-d7g8.html.

53 See, for example, the presentations of Patrick Alley of Global Witness and colleagues at the Bonn climate negotiations, 3 June 2009, http://unfccc2.meta-fusion.com/kongresse/090601_SB30_Bonn/templ/ply_page.php?id_kongresssession=1757&player_mode =isdn_real; http://www.redd-monitor.org/2009/06/05/forests-corruption-and-cars-why-redd-has-to-be-about-more-than-carbon/.

54 United Nations Environment Programme Risoe Centre, *CDM Pipeline.*

55 Det Norske Veritas (DNV), 'Clean Development Mechanism Project Design Document Form for Recovery of Associated Gas that Would Otherwise be Flared at Kwale Oil-Gas Processing Plant, Nigeria', 2004, http://www.dnv.com/focus/climate_change/upload/final%20pdd-nigeria%20ver.21%20%2023_12_2005.pdf.

56 Lohmann, L. (ed.) (2006) *Carbon Trading: A Critical Conversation on Climate, Privatization and Power*, pp. 148, 292.

57 Figueres, C. (2007) 'The CDM and Sustainable Development', *Environmental Finance*, December, pp. S50–S51.

58 Point Carbon (2008) 'CDM Market in Good Shape: Official', *2* April.

59 Gronewold, 'Secretive UN Board'.

60 Leeds, C. (2008) 'Carbon Markets and Carbon Trading: Greener and More Profitable', presentation, 13 June.

61 Ball, 'Up In Smoke'.

62 Chan, M. (2009) 'Subprime Carbon? Rethinking the World's Largest New Derivatives Market' Friends of the Earth, http://www.foe.org/subprime-carbon-testimony.

63 Szabo, M. and P. Sandle (2009) 'JPMorgan To Buy EcoSccurities for $204 million', *Reuters*, 16 September.

64 Gronewold, 'Secretive UN Board'.

65 Notes from 'Carbon Finance 2008', *Environmental Finance* Conference, 8-9 October.

66 Harvey, F. (2008) 'Carbon Credit Ratings Agency is Launched', *Financial Times*, 25 June.

67 Reklev, S. (2007) 'Cowboys or Cavalry?' *Trading Carbon*, December, pp. 27–28. Similarly, the International Emissions Trading Association has argued in a letter to US Senators Dianne Feinstein and Olympia Snowe, who had introduced a carbon market governance bill, that '[t]he market itself recognizes the importance of integrity and exerts discipline on participants … Trading companies set their own trading limits to guard against excessive speculation. The market itself punishes firms that exceed responsible limits by downgrading credit ratings, lowering lines of credit or barring individuals or firms from trading' (IETA letter to Sens. Feinstein and Snowe, 4 March 2008, http://www.ieta.org/ieta/www/pages/getfile.php?docID=2938).

68 Driesen, D. (2003) 'Markets are Not Magic', *The Environmental Forum*, November/December, pp. 18–27, p. 22.

69 Nicholls, S. (2007) 'Interview with Hans-Juergen Stehr,' *Environmental Finance*, December, p. S42.

70 Braun, M. (2009) 'The Evolution of Emissions Trading in the European Union – the Role of Policy Networks, Knowledge and Policy Entrepreneurs', *Accounting, Organizations and Society*, 34 (3-4).

71 Notes from 'Carbon Finance 2008', *Environmental Finance* Conference, 8-9 October 2008.

72 Lohmann, L. (2001) 'Democracy or Carbocracy? Intellectual Corruption and the Future of the Climate Debate', *Corner House Briefing Paper No. 24* October 2001, http://www.thecornerhouse.org.uk/subject/climate.

73 Young, 'UN Suspends Top CDM Project Verifier'.

74 Gronewold, 'Secretive UN Board'.

75 The argument of this article that corruption is inherent in the carbon offset market and is only furthered by regulatory efforts also applies to the second component of carbon markets, cap and trade. For example, the climatic efficacy and 'climatic equivalence' of emissions cuts undertaken at different places, times, and using different technologies cannot be verified under cap and trade any more than can the climatic efficacy of offsets, making it impossible to distinguish between abuse and non-abuse. Similarly, bribery is a structural feature of all existing cap and trade systems in the form of the 'grandfathering' of allowances, regardless of

the legal or illegal conduct of individual allowance grantees. However, such topics lie beyond the scope of this chapter. Some of them are taken up in, for example, Lohmann, L. (2009) 'Regulatory Challenges for Financial and Carbon Markets', *Carbon & Climate Law Review*, 3(2): 161-71; Lohmann, L. (2006) *Carbon Trading: A Critical Conversation on Climate, Privatization and Power*; Lohmann, L. (2009) 'When Markets are Poison: Learning about Climate Policy from the Financial Crisis', *Corner House Briefing Paper No. 40*, www.thecornerhouse.org.uk/subject/climate; and Lohmann, L. (forthcoming) 'Uncertainty Markets and Carbon Markets', in *New Political Economy*. Many thanks to Trusha Reddy, Joe Zacune and Andre Standing for helpful comments on and criticisms of a draft of this chapter.

18

The Politics of the Clean Development Mechanism: Hiding Capitalism Under the Green Rug

Joanna Cabello

Introduction

Climate change is a consequence of capitalism. An on-growing extractive system entirely dependent on the use of fossil fuels as a cheap energy source that has driven unsustainable practices with socially and environmentally destructive consequences. The climate crisis also embodies the complexities of unequal distribution of impacts, historical responsibility for emissions, the right to use atmospheric capacity, as well as political, economic and social injustices. Within this context, a hegemonic world polity and ideology based on liberal or free-market environmentalism[1] started mandating how involuntarily interdependent states should deal with 'common problems' by devolving power to global market forces and non-state actors. This led to the international response in 1997, through the Kyoto Protocol, of establishing the carbon market as the only 'efficient' solution to deal with climate change.

While the Protocol binds industrialized countries to reduce their emissions to an average of 5.2 per cent compared to 1990 levels by 2012, the core deal of this agreement was held together with the creation of the so-called 'flexible mechanisms'[2] from which the 'Clean Development Mechanism' (CDM) is the only one that involves developing countries. The CDM enables investment in 'emission-saving' projects in developing countries in exchange for carbon credits that industrialized countries (also their companies or financial institutions) can use to meet their targets under the Protocol or to trade within the carbon market. Therefore, if a corporation needs to emit above its permitted level, it can buy cheap credits within the carbon market to cover this increase. The assumption is that as greenhouse gases (GHG) are emitted they will result in a contribution to the global increases of temperature, regardless of where or which is the source. However, it allows corporations and governments to buy their way out of the problem by offsetting their pollution somewhere else.

Under the Emissions Trading mechanism (also known as cap-and-trade), industrialized countries have distributed their initial allocation of credits or 'rights to pollute' to their dirtiest industries, which can be bought and sold between them as a market commodity. Conversely, it also allows trading with the 'Certified Emission Reduction' (CER) credits, acquired under CDM projects in developing countries, thereby inflating the fixed caps. By April 2009, there were over 1,500 registered CDM projects and over 4,000 projects awaiting approval.[3] European corporations are the main buyers of CERs and currently, offsets are predicted to deliver more than half of the European Union's planned reductions to 2020.[4]

This article argues then that the CDM, as a keystone of the carbon market, is a central element in the expanding agenda of capitalism in two fundamental ways. First, materially, it allows the creation of new financial markets, securing the conditions for accumulation and capital reproduction while allowing polluters to avoid making any real structural change. And second, ideologically, it searches to legitimize the ongoing commodification of nature (the atmosphere in this case) reinforcing a 'green capitalism'[5] whose legitimacy is an essential part of its own existence.

In this regard, the New Carbon Finance[6] agency shows that in spite of the global economic recession, the volume of trading in the carbon market in the first quarter of 2009 grew by 37% compared to the forth quarter of 2008. Moreover, they expect that most of the growth will come from increased liquidity in the secondary market of CDM carbon credits. Similarly, a recent analysis by Friends of the Earth[7] highlights the problems with carbon tradings's financial growth, which currently is fundamentally a derivatives market on which speculators do the majority of trades. This speculative nature – which also led to the recent financial crisis – can generate a carbon bubble and stimulate the development of subprime carbon (future contracts to deliver carbon that carry a relatively high risk of not being fulfilled), particularly with CDM credits.

The constant need for legitimacy is at the same time inherent to the carbon market's accumulation ambition, whereby the 'green' discourses have managed to disguise an economic treaty as an environmental treaty.[8] The CDM appears then like an ideal strategy for maintaining the status quo: while creating a new commodity, the right to pollute, it simultaneously establishes the apparatus which gives the illusion of having 'carbon neutral' governments, corporations, industries or life-styles without making any real reduction or structural change. Moreover, as nature is considered a form of capital, 'environmental sustainability' has also been redefined to provide the basic conditions for preserving capital as 'economically sustainable'.

The international negotiations, on the other side, have framed the climate crisis as a technical issue rather than political, marginalizing voices for alternative knowledges; and as a result, there is a perception of having governance without politics, while these unaccountable and undemocratic institutions are, on the contrary, embedded with political as well as economic interests. As Welford argues,

> the dominant corporate culture (…) believes that natural resources are there for the taking and the environmental and social problems will be resolved through growth, scientific advancement, technology transfer via private capital flows, free trade and the odd charitable hand-out.[9]

In this vein, key tasks for implementing, financing and monitoring the scheme have been 'outsourced' to the private sector, giving them the space to legitimize their own actions.

The UN Commission on Global Governance indicates that 'governance has been viewed primarily as intergovernmental relationships, but it must now be understood as also involving non-governmental organizations, citizen's movements, multinational corporations, and the global capital market'.[10] With this understanding, as the professor Sangeeta Kamat[11] analyses, power relations are seen as non-existent. Profit-seeking corporations and marginalized groups are considered equal legitimate actors and private interests are represented in the form of 'partnerships'. Consequently, during the Bali and Poznań UN climate negotiations (2007 and 2008 respectively), the corporate lobby group International Emissions Trading Association – with 172 corporate members such as BP, Chevron, ConocoPhillips, E.ON, Goldman Sachs, PetroBras, Repsol YPF, Schell, Rio Tinto or The Carbon Neutral Company[12] – was the largest represented 'Non-Governmental Organization'.[13]

In this regard, as David Harvey[14] argues, the acceleration of privatization and financialization are creating a form of accumulation in which states exercise their power to preserve property rights and other market institutions while dispossessing, in this case, those who live in and with a privatized environment. Therefore, the CDM, which masks a mechanism for land grabs, local conflicts and pollution, dispossesses local communities not by the conventional form of property rights but by the application of ownership constructs at the global level.

It's Not Only About the Climate!

The CDM was a late intervention in the Kyoto negotiations. It emerged from the Brazilian delegation proposition, accepted by the Group of 77 and China, to create a 'Clean Development Fund (CDF)' on the basis of the 'polluter pays' principle. It would apply penalties for industrialized countries that exceeded their targets in order to finance clean energy projects for mitigation (actions to avoid and reduce emissions) and adaptation (actions that deal with the impacts of climate change) in developing countries. However, during the Kyoto negotiations in 1997, the CDF was transformed into the CDM and, as the researcher from the Corner House, Larry Lohmann, stated 'fines were transformed into prices; a judicial system was transformed into a market'.[15]

Each project – including hydropower dams, efficiency improvement in coal-fired power plants, wind farms, monoculture plantations, biomass power plants, etc. – must go through a UN registration process designed to ensure 'real, measurable and verifiable' emission reductions that are 'additional' to what would have occurred without the project. This additionality characteristic is crucial, but at the same time, it is its most fundamental flaw. There is no sound way to show that a project would not have happened without the CDM. As the professor of the Öko Institute, Lambert Schneider, stated,[16] 'If you are a good storyteller you get your project approved. If you are not a good storyteller you don't get your project through'. Yet, if a project was going to happen anyway, no real offset is being made since new emissions should need new 'emission-saving' projects. However, new markets in benefit of the same private and governmental actors have been made in order to help the capitalist system entering another phase with 'green' legitimation.

The additionality requisite requires identifying one distinctive business-as-usual storyline to compare with the storyline that comprises the project. With countless 'without-project' scenarios, the selection of which one is to be used in measuring the carbon credits is a matter of political decision rather than economic or technical conjectures.[17] As the organization International Rivers[18] highlights, as of 1 October 2008, 76 per cent of all registered projects were already completed by the time they were approved as eligible to sell credits. In China for instance, more than 200 large-scale hydro plants are at the CDM validation phase even though hydro is a major component of the Chinese five-year governmental plan. Since constructions began before CDM registrations, these projects would have continued even if they were not registered as CDM projects.[19]

On the other side, the Designated Operational Entities (DOE) or the so-called 'validators', which are mainly large risk management firms, verify and validate each project's emission-reductions and removals. The CDM Executive Board accredits the DOEs so that they can be hired by project developers as external auditors for validating the project documents (assessing projects in accordance with CDM rules) and verifying the emission reductions in the field (assessing if the project is reducing emissions as claimed and according to the stipulated methodology).

This outsourcing of 'expertise' for supervising the CDM places a heavy reliance on profit-driven private actors for transparency and accurate reports. Moreover, the few registered DOEs have made the system a practical oligopoly: they are able to set prices for their services and they can collude among themselves to ensure that projects are approved in order to receive all of the proposed CERs. The CDM Executive Board itself has stated that there is a 'clear and perceived risk of collusion'[20] between the DOEs and the companies that hire them to review their offset projects due to their strong interests in having future contracts.

Consequently, as Heidi Bachram and others[21] affirm, 'as all scramble for a piece of the emissions trading pie, no equivalent level of activity is seen from credible verifiers or monitors'. This on-going marketization and privatization of climate governance has turned the negotiations into structures for legitimized accumulation – with corporate powers at the heart of it – that sustain and increase old relations and imbalances and relations of power between rich and poor, North and South, as well as the idea of maintaining continuous business-as-usual growth on a finite planet.

A Greenwash Scheme

The research organization CorpWatch defines greenwash as 'the phenomenon of socially and environmentally destructive corporations attempting to preserve and expand their markets by posing as friends of the environment and leaders in the struggle to eradicate poverty'.[22] Moreover, they affirm that it also involves 'any attempt to brainwash consumers or policy makers into believing polluting mega-corporations are the key to environmentally sound sustainable development'. In this regard, the severe ideological reductionism of the negotiations, which has transformed ecological politics into managerial strategies, is indeed constantly helping powerful actors to overcome the capitalistic intrinsic tension between accumulation and legitimation ambitions.

In this regard, the Protocol states that CDM projects *are* emission reductions, however, planting trees, fertilizing oceans, burning methane from landfills to generate electricity, or setting up wind farms cannot be verified to be climatically equivalent to reducing fossil fuel consumption.[23] Moreover, since these offset projects generate CERs that will allow emissions somewhere else, then there is no reduction happening at the global scale. On the contrary, they are creating new credits for the Emissions Trading scheme, underestimating the already inadequate caps established in the Protocol. Northern polluters can continue to pollute, and even increase pollution legitimately, with the help of the carbon market without being concerned about abatement actions.

As the New York Times highlights, 'if a company or a country is fined for spewing excessive pollutants into the air, the community conveys its judgement that the polluter has done something wrong. A fee, on the other hand, makes pollution just another cost of doing business, like wages, benefits and rent'.[24] The focus is thus no longer on reducing emissions but on trading and claiming credits. In another words, the wealthiest actors are – one more time – enabled to buy their way out.

BP and Shell, for example, have been cultivating 'progressive' corporate images and positioned themselves at the forefront of the offsets market. The opportunity to greenwash their activities in order to present themselves as environmentally responsible is legitimizing their destructive forms of production and extraction. On the other side, several offset companies offer citizens, companies and governments the illusion of being 'carbon neutral' by buying some offset credits. No change is required. As a result, the space for ecological political opposition or organized acts of resistance, mainly in Western societies, has been significantly reduced.

Similarly, the carbon market's ideological fabrication reflects the battle of interests and powers at play during the negotiations for persuading partners and possible allies towards hegemonic convictions, whereby the various actors have to deal with an involuntary ecological interdependence. Consequently, the Kyoto debate has been instrumental for re-affirming capitalistic interests in moments of global governance legitimacy crises.[25] According to Henry Bernstein,[26] this crisis has been indeed alleviated in part by the success of the free-market environmentalism in which the climate negotiations subordinated environmental purposes for economic goals.

Sustaining the Inequalities

The core objectives of the CDM are to help industrialized countries meet their commitments under the Protocol and at the same time, to promote sustainable development in developing countries. The latter was crucial for earning the support of the Group of 77 and China block. Hence, in order for a CDM project to be registered it must first be approved by the Designated National Authority (DNA), which is selected by each developing country and who's prerogative is to validate whether the project contributes to its sustainable development.

Market liberalism's compatibility with sustainable development has been constantly disproved by many.[27] In the climate case, the absence of a concrete definition in the Protocol presents an assumption that projects that are good for carbon abatement must also be good for sustainable development. Moreover, the construction of the concept with poverty is linked with the historical usage of the term 'development' and consequently, the responsibility and pressure of achieving it is being pushed towards developing countries.

For that reason, while the accounting for emission reductions is subject to a stringent international assessment, the sustainable development objective is considered unnecessary to assess at the international level and has been entirely left to the approval of the DNA at the Project Document stage (before the implementation). All the responsibility for monitoring each project's sustainability therefore depends on developing countries. Even more importantly, as capitalism depends on exploiting and intensifying global inequalities to further its own growth, the sustainable development discourse is being used as a way to legitimize this new colonialist scheme.

Consequently, there is a trade-off between the two objectives in favour of the one that has a price in the market. As Bobby Peck from the South African environmental justice organization 'GroundWork' notes,[28] 'companies that are able to avoid reducing GHG through carbon trading are also not going to be reducing the other pollution that causes harm to local communities next to these industries'. Furthermore, most large-scale renewable energy projects (such as windmills, dams and plantations) are silent in their need for big quantities of land and resources for implementation as well as in the social impacts that this conveys, such as the massive evictions of local communities, land-grabbing, migration to the cities, direct human and indigenous rights violations, repression of social movements, and many more.

Since developing countries' interest to participate in the CDM scheme essentially rests on obtaining further funds, strict sustainability requirements are then undermined in order to facilitate the entrance of new investors. In this way, the CDM is legitimizing a type of sustainability whose definition is not contested at the governance decision-making tables and whose legitimization is more important than even its attempt to accomplish it. As Cathleen Fogel mentions, 'global discourses emphasize that *standardized* carbon units can be produced through *standardized* sequestration projects in *standardized* developing countries. In order to be efficient and hence, to economically benefit from global institutions, the 'local' must accept its construction as compliant, homogenous and safe, which is to say, as absent'.[29]

This false notion of sustainability, as the activist Vandana Shiva[30] affirms, is then assigning primacy to capital, depending on capital, and substituting nature as capital. Therefore, words that were meant to speak about politics and power have become co-opted and meaningless for the service of alternative interventions and mobilizations, by framing them not only as neutral but also turning them into merely policy buzzwords.

During the 7th Session of the UN Permanent Forum on Indigenous Issues in May 2008, an Indigenous representative declared that

> The Report doc E/c.19/2008/L.2 does not take into account the proposals and concerns of the Indigenous Peoples regarding the initiative to reduce emissions from deforestation in developing countries known as REDD or the CDM or the Carbon Market (...) The adopted recommendations (...) made by the Forum experts are not the position of indigenous organizations (...) We are also concerned that the initiatives of CDM are considered examples of 'good practice'. [31]

Many local and Indigenous groups in India, Thailand, Indonesia, South Africa, Nigeria, Brazil, Peru, Paraguay, among many more around the world, besides trying to incorporate counter-hegemonic discourses around the negotiation tables, are strongly resisting the carbon market locally, specifically CDM projects.

From a developing country perspective, the economic incentives and technology transfers for Southern big polluters and governments are clear. But it must also be made clear that this represents a payment to ensure that the 'North' and wealthiest actors can continue polluting and accumulating as well as deepening the intrinsic inequalities of the world political economy. The CDM has become an instrument of foreign policy that creates new structural dependencies. We are facing a new form of colonialism whereby the expansion is not only the cooptation of resources and land but also of atmospheric capacity.

Who Received an Invitation?

The CDM project cycle heavily relies on a diverse set of actors, including governments, corporations, auditors, science boards, financial investors, international and local NGOs, local communities, etc. However, the institutionalization of an 'invited'[32] participation has paved the way for establishing a structure that imposes boundaries and excludes certain actors and views from entering the arenas in the first place and hence, obstructs critical and different discourses and epistemologies.

For this reason, when the Protocol was ratified, the accepted line of reasoning was that a market-driven mechanism is 'the only possible' alternative.

This dominant idea strengthens the ideological hegemonic stance. As the CDM Executive Board Secretary, Yvo de Boer, stated during the COP in Bali, 'market-based mechanisms need to be at the heart of things. It's the only way of achieving the goal'.[33] Therefore, this 'development' thinking is reduced to certain social actors (i.e. UN bodies) and a certain social transformation (i.e. technology transfer), while marginalizing other social actors and trivializing other alternatives for change.[34]

In this regard, it is interesting to highlight some of these actors. Industrialized governments on one side carry a convenient dual role. They, and 'their' corporations, are buyers of CERs on the market while simultaneously deciding upon the rules of this market as Parties of the decision-making process. Moreover, they channel important donations to the UNFCCC secretariat for its operation, as well as to the World Bank, UNEP and UNDP, which are institutions heavily involved in the finance and implementation of CDM projects.[35]

Similarly, the role of the World Bank in the management of carbon funds is more than controversial due to its self-assigned role as a facilitator or broker of the carbon market while making money out of its commissions on projects. Even more fundamentally, through its initial position in the market as well as in the regulatory field, the Bank is influencing CDM regulation in its own interest under a facade of political neutrality. The Group of 77 and China block have clearly stated during the negotiations that they do not consider the World Bank as the suitable institution to manage the climate funds and would prefer a body directly accountable to the UNFCCC. Furthermore, the World Bank's role turns out to be ironic since it still funds heavily polluting industries and is not willing to mainstream climate change considerations into its own energy projects or country strategies.

On the completely opposite side, the small existing 'consultative' space where local communities can give their input on the projects has other constraints embedded in the politics of participation.[36] While formally the CDM has different opportunities for public involvement, they only take place when the design of the project is already decided. The language used in most of the documents is English and their translation into local languages is not required. Moreover, most information is communicated through the Internet, which is most of the times not a culturally appropriate way to reach local communities. Consequently, local and indigenous peoples are not considered actors of their own development but on the contrary, the CDM is establishing a homogenous 'sustainable development' path which is constructed in international arenas for accomplishing specific colonialist purposes.

Conclusion: A Mechanism for Dispossession

The CDM structure – created in the name of 'mitigating global warming' while transferring 'clean' technology to the developing world in the name of 'sustainable development' – has become an instrument used to expand capitalist globalization whereby the wealthiest actors continue to accumulate by dispossessing the excluded. Is in this sense, local and social movements striving for climate justice are in essence struggling against the capitalistic model.

The CDM and the carbon market are based on the idea of economic growth within the extractive system of capitalism, and at the same time, on a climate governance that has been pursuing a regime of expanding accumulation. On one side, developing countries will have to bear the consequences of being industrialized countries' carbon dump and 'pay the bill' for not having the right to pollute. Moreover, since emissions are growing faster than ever and climate change impacts are affecting the poorest parts of the world; countries with the most to lose are being more dispossessed, intensifying long-standing exploitative and dependent relations which started in colonial times.

On the other side, local communities intervened by CDM projects in most of the cases are being dispossessed from their lands, forests, water sources and traditional ways of living. In the name of 'sustainable development' an imposed 'development' is determining their path by hegemonic and capitalistic values. The transferred large-scale 'clean' technologies, which serve powerful global interests, are undermining the traditional ways for sustaining local and indigenous peoples' livelihoods, which are an invaluable source of ecological sustainable alternatives. Moreover, the incorporation of the environment into the heart of liberal market institutions, such as the World Bank, enables a more rooted institutionalization of green capitalism. Hegemonic discourses are trying to persuade us that a green capitalist economy could achieve the miracle of sustainable development *and* continuous 'growth'. However, global policies that intensify inequalities, social injustice and accumulation by dispossession practices are false solutions.

The lobbying and political pressures for the summit at Copenhagen in December 2009, where the negotiations for the post-Kyoto agreement will be carried out, are trying to deepen the process within these market mechanisms. Proposals for other kinds of offsets have been presented, with the same underlying logic, the same profit-driven incentive, and still no real structural changes. Countries and corporations continue to seek ways to avoid their reduction obligations by deepening the process of accumulation under a green capitalism.

Conversely, the alternatives are strong and diverse: community-led renewable energy, food sovereignty, reverse over-production and over-consumption, small scale agriculture systems, respect and learn from indigenous and traditional ways of living, and many others within the scope of people-centred approaches. The need for climate justice cannot be neglected or postponed any more.

The world needs a radical change in its fundamental economic pillars. Technological solutions are limited and do not address the historical and structural problem of the ideological and material foundations of capitalism. For that reason, building alternatives to capitalism's inexorable accumulation forces is necessary for achieving no-carbon economies within a social justice framework.

Notes

1   Bernstein, S. (2005) 'Legitimacy in Global Environmental Governance', *Journal of International Law & International Relations*, 1: 139-166.

2   The 'flexible mechanisms' are Emissions Trading, Joint Implementation and the Clean Development Mechanism. Further information: Carbon Trade Watch (2009), http://www.carbontradewatch.org/index.php?option=com_content&task=view&id=89&Itemid=79.

3   UNFCCC (2009) 'Expert Meeting on Trade and Climate Change: Trade and Investment Opportunities and Challenges under the Clean Development Mechanism'. Geneva: United Nations Framework Convention on Climate Change.

4   EU: 'Towards a comprehensive climate change agreement in Copenhagen', http://eur-lex.europa.eu/LexUriServ/LexUriServ.do?uri=CELEX:52009DC0039:EN:NOT.

5   Further information on the 'green' concept: Wall, D. (1993) *Green History: A reader in Environmental Literature, Philosophy and Politics*. London: Routledge.

6   New Carbon Finance. (2009) 'Carbon Market volume up 37% in Q1 2009', http://www.newcarbonfinance.com/download.php?n=20090427_PR_Carbon_Markets_Q12009.pdf&f=fileName&t=NCF_downloads.

7   Chan, M. (2009) 'Subprime carbon? Re-thinking the world's largest new derivatives market', *Friends of the Earth*. United States.

8   Paterson, M. (1996) *Global Warming and Global Politics*. London: Routledge.

9   Welford, R. (1997) 'Introduction: What are we doing to the world?', in R. Welford (ed.), *Hijacking Environmentalism. Coroprate responses to sustainable development*. London: Earthscan Publications Limited, pp. 251.

10 Commission on Global Governance (1995) *Our Global Neighbourhood*. Oxford: Oxford University Press.

11 Kamat, S. (2004) 'The privatization of public interest: theorizing NGO discourse in a neoliberal era', *Review of International Political Economy*, 11(1): 155-176.

12 International Emissions Trading Association (IETA) http://www.ieta.org/ieta/www/pages/index.php?IdSiteTree=84.

13 World Development Movement (2007) 'Who is the biggest NGO in Bali?', www.wdm.org.uk/news/archive/2007/biggestngoinbali06122007.html.

14 Harvey, D. (2003) *The New Imperialism*. Oxford: Oxford University Press.

15 Lohmann, L. (2006) *Carbon trading: a critical conversation on climate change, privatisation and power*. Uppsala: Dag Hammarskjold Foundation.

16 Schneider L. (2007) 'Practical Experiences with the Environmental Integrity of the CDM'. Presented at the 4th Meeting of the ECCP Working Group on Emissions Trading in the review process of the EU ETS: Linking with emissions trading schemes of third countries. Brussels, 15 June.

17 Lohmann, L. (2006) *Carbon trading: a critical conversation on climate change, privatisation and power*.

18 International Rivers (2008) *Rip-offsets: The failure of the Kyoto Protocol's Clean Development Mechanism*, Berkley, USA.

19 http://www.internationalrivers.org/node/1892.

20 Ball, J. (2008) 'UN Effort to Curtail Emissions In Turmoil', *The Wall Street Journal*, http://online.wsj.com/article/SB120796372237309757.html.

21 Bachram, H., J. Bekker, L. Clayden, C. Hotz and M.A. Adam (2003) *The sky is not the limit: the emerging market in greenhouse gases*. Amsterdam: Transnational institute - TNI.

22 http://www.corpwatch.org/article.php?id=242.

23 Lohmann, L. (2006) *Carbon Trading: a critical conversation on climate change, privatization and power.*

24 Sandel, M. (1997) 'It's Immoral to Buy the Right to Pollute'. *New York Times,* p. A29. December 15th.

25 The systematic destabilization of national economies such as Mexico, South East Asia, etc; creation of bubble economies, mainly high-tech and housing markets; governance scandals such as the collapse of LTCM; more recently, the financial crisis; etc.

26 Bernstein, S. (2005) 'Legitimacy in Global Environmental Governance', *Journal of International Law & International Relations,* 1: 139-166.

27 Driesen, D. (2007) 'Sustainable Development and Market Liberalism's Shotgun Wedding' Syracuse, NY: Syracuse University, College of Law; Kysar, D. (2005). 'Sustainable Development and Private Global Governance', *Texas Law Review,* 83: 2109-2114.

28 Lohmann, L. (2006) *Carbon Trading: a critical conversation on climate change, privatization and power.*

29 Fogel, C. (2004) 'The Local, the Global, and the Kyoto Protocol', in S. Jasanoff et al. (eds), *Earthly politics: local and global in environmental governance.* Cambridge, Mass: MIT Press.

30 Shiva, V. (1992) 'Recovering the real meaning of sustainability', in D.E. Cooper and J. A. Palmer (eds.) *The Environment in Question: Ethics and Global Issues.* Routledge, London.

31 Sommer, R. (Director) (2008) Permanent Forum on Indigenous Issues. Earth Peoples (Producer).

32 Further information on 'invited spaces' for participation: Cornwall, A. (2004) 'Spaces for transformation? Reflections on issues of power and difference in participation in development', in S. Hickey et al. (eds) *Participation: From Tyranny to Transformation.* London: Zed Books, pp. 75-91.

33 Cundy, C. (2007) 'Taking the next steps', *Kyoto and the Carbon Markets* special supplement: 48.

34 Sachs, W. (1999) *Planet Dialectics: Explorations in Environment and Development.* London: Zed Books.

35 Wittneben, B. (2007) *The Clean Development Mechanism: Institutionalizing New Power Relations.* Rotterdam: Erasmus Research Institute of Management.

36 Further information: Lovbrand, E., J. Nordqvist and T. Rindefjall (2007) 'Everyone loves a winner - Expectations and realisations in the emerging CDM market', Paper presented at the conference on *Human Dimensions of Global Environmental Change.*

19

Rent Seeking and Corporate Lobbying in Climate Negotiations

Ricardo Coelho

The Political Economy of Carbon Markets

Rent seeking behavior can be defined as using resources to influence changes in legal rights to obtain an increase or avoid a decrease in income wealth.[1] The concept was devised by Gordon Tullock[2] to define the process by which industrialists engage in lobbying government officials or competing to become civil servants, in order to achieve monopoly power. Anne Krueger[3] broadened the concept to include lobbying for protectionism. In this chapter, we use the same framework for analyzing corporate lobbying inherent to carbon markets, as industries compete to get a greater share of the cake.

Economically, rent seeking results in a waste of resources with the lobbying process. Politically, it leads to corruption and undermines democratic decision-making. Yet, and in spite of this, the so-called 'international community' approved a treaty, the Kyoto Protocol, which created a heaven for rent-seekers: an artificial market for carbon emissions, complemented with offset provisions under the umbrella of the Clean Development Mechanism (CDM).[4]

When a new carbon market is born, such as the European Union Emissions Trading System (EU ETS), an enormous rent is delivered on a silver platter to the greatest polluters, corresponding to the value of emissions permits given away for free. The potential profits (or avoided losses) from trading (or using for compliance) cheap carbon credits from the CDM (the Certified Emissions Reductions, in Kyoto's jargon) add to this rent. If we further add profits from financial services offered by consulting agencies, investment banks and traders, we have more money than Uncle Scrooge had in its money bin.

As time goes by, more and more big polluters, bankers and traders, are aware of the potential for profits in an international carbon trading system. This has led companies to form partnerships with NGOs, allegedly with the objective of inducing corporate awareness to global warming mitigation, to lobby for the schemes that allow trading of carbon emissions. One example is US Climate Action Partnership (USCAP), incorporating big polluters like General Electric, General Motors, BP America, Dow Chemical, Duke Energy, Ford, Dupont and Rio Tinto and the NGOs Environmental Defense Fund (EDF), Natural Resources Defense Council (NRDC), The Nature Conservancy and World

Resources Institute (WRI). Its recent recommendations for US legislative action includes a mandatory cap-and-trade program for greenhouse gases (GHG), allowing for emissions offsetting.

Other partnerships went further, creating voluntary emissions trading schemes. In 1995, the Pilot Emissions Reductions Trading Project (PERT, renamed later as CleanAir Canada) launched an emissions trading program in the Canadian province of Ontario. The program was reformed in 1997 to include CO2 emissions and one year later the Greenhouse Gas Emissions Reduction Trading Pilot (GERT) allowed the trading of GHG emissions in Canada. Both programs were conceived with the input of NGOs, unions, industries and local and federal authorities.

The US followed this example, and in 2000, EDF joined BP, Dupont, Shell and other companies from Canada and France to set up the Partnership for Climate Action (PCA), another voluntary GHG trading system. Three years later, this initiative gave way to the Chicago Climate Exchange (CCX), formed by a group of big polluters, including BP, Ford and Dupont. CCX is the axis of international voluntary carbon trading, joining forces with other similar exchanges in Montreal (Canada) and Tianjin (China) and with European Climate Exchange (ECX), the biggest exchange operating within the EU ETS.

We can see, then, that the majority of the world's greatest polluters support and even lobby for international carbon trading. But it would be extremely naïve to think that this happens because their CEOs are deeply committed to saving the planet from runaway climate change. On the contrary, what determines corporate lobbies' tactics is rent seeking behavior.

Tactics for Corporate Lobbying

We can distinguish three tactics for corporate lobbying in climate negotiations: denial, influence and greenwashing.[5] The first one consists on creating and financing think tanks and fake grassroots groups that launch campaigns, supported by some scientists (the so-called 'skeptics') to deny global warming science. If it succeeds in creating doubts about the consequences of rising GHG concentrations, reducing the influence of the environmental movement, then it will be much more difficult to enact a strong climate agreement.

The first tactic has been followed most notably by Exxon-Mobil.[6] This company was instrumental in the creation of the Global Climate Coalition (GCC), a denialist lobby group, after the first meeting of the Intergovernmental Panel on Climate Change (IPCC), in 1989. The GCC gathered representatives from the oil, auto and mining industries and was successful in persuading governments to oppose binding targets in the Earth Summit, in 1992. In 2002 the group was dismantled, with a statement supporting George W. Bush's proposal for (ineffective) voluntary agreements with the industry.

The second one consists on lobbying governments in order to assure a more corporate-friendly attitude by national delegations in the climate negotiations. This might involve the persuasion of government officials or even the appointment of corporate-friendly officials to integrate governmental delegations. Again, Exxon-Mobil provides an example of this tactic. In 2001, this company issued a memo to the US government suggesting that Robert Watson, a radical scientist working within the IPCC, was substituted by a 'skeptic'.[7] The scientist was in fact substituted in 2002 by the Bush administration, elected with a generous financial support from Exxon.

The third tactic is more of a public relations and marketing stunt. By enacting voluntary agreements through partnerships with NGOs, governments or the UN, companies can clear their image and give the impression that they are doing their part in the effort to mitigate global warming, even if the reality is very different. This allows them to gain bargaining power in climate negotiations, namely because NGOs' opposition to their presence will be much lower. In more extreme cases, NGOs funded by corporations show more concern with polluter's profits than with the environment, leading them to praise ineffective policies like emissions trading.[8]

These three tactics aren't mutually exclusive and any company or business association can resort to more than one of them. For the corporate lobby, anything that helps in its effort to delay climate action is worth the effort. BP, for instance, was a founder of the GCC but left in 1997, reversing its denialist position. In 1999 BP set up an internal emissions trading scheme and one year later it launched its 'beyond petroleum' campaign, announcing its (fake) effort to reposition itself as a renewable energy company. In 2002, the company played a central role in the formation of an voluntary carbon trading scheme in the UK, which distributed 111 million pounds to participating companies to continue 'business-as-usual',[9] given the over-allocation of permits to pollute. After this 'success', BP lobbied for the expansion of the experiment at the EU scale.[10] Meanwhile, the aggressive 'greenwashing' campaign shielded the oil company against criticisms from environmentalists.

How an Industry-Friendly Treaty was Signed

The negotiation process for an international agreement on climate change mitigation started in the late 80s. After some meetings and the publishing of the IPCC's first assessment report in 1990, it became evident that a framework convention was necessary. To this endeavor, an International Negotiating Committee (INC) was created.

Two aspects from the negotiations in the INC deserve our attention. The first is that, from the start, both the US and the EU opposed binding targets for emissions. This lead to a convention, the United Nations Framework Convention on Climate Change (UNFCCC), approved in the Earth Summit, that merely recommended the stabilization of emissions at 1990 levels by 2000.[11]

The second is that even before the UNFCCC was approved, research on carbon trading was already being undertaken within the United Nations Conference on Trade and Development (UNCTAD). This UN agency set up a department on this instrument in 1991 and a few years later was already proposing its implementation.[12]

The famous Kyoto Protocol was approved in 1997, incorporating the trading of permits to emit GHG as a way to reduce emissions by about 5% until 2012 from 1990 levels. From the start, two major loopholes were introduced to assure the stabilization of permit prices at low levels: 'hot air' and the CDM.

The first loophole was created indirectly by allowing countries to choose 1990 as the base year for emissions reductions. Eastern European economies collapsed in the period 1990-91, so their GHG emissions dropped sharply during this period. Nevertheless, 1990 was chosen as the base year to calculate emissions reductions targets, so these countries could claim huge emissions reductions and sell excess permits. Swamping the carbon market with this 'hot air' allows Western European countries, as well as Japan, to buy cheap credits from the former USSR countries, while the latter can claim a huge rent. Also, significant emissions reductions were achieved in the UK and Germany because of a switching from coal to gas in electricity generation, for economic reasons, and deindustrialization in the former German Democratic Republic. Had the base year been set at 1992, and industries would have to comply with much more stringent targets for GHG emissions reductions.

The second loophole became known as 'the Kyoto surprise', given the way it was introduced in the negotiations. Brazil had by 1997 set forth the proposal to create a Green Development Fund, aimed to finance mitigation projects in less developed countries and financed by the fines imposed on the countries that didn't comply with required emissions reductions. At the time, developing countries, gathered in the G-77, as well as China, endorsed this proposal and opposed the idea of allowing industrialized countries to exceed their emissions quotas by financing emissions reductions projects in other countries. Nevertheless, Brazil reversed its position just one month before the negotiations in December, presenting instead a joint proposal with the US to create the CDM.

This was just one of the successes by the Clinton-Gore administration in its effort to water down Kyoto. During the next years, a confrontation emerged between the EU and the US considering restrictions on carbon trading. The EU wanted that at least half of the emissions reductions by industrialized countries were met with domestic action, instead of using permits bought from other countries, a position backed up by the least developed countries. The US, on the other hand, defended unrestricted carbon trading and offsetting as the only feasible way to comply with the emissions reductions required by Kyoto. This position had the support from Australia, New Zealand, Canada and Japan.

The US delegation, led by Al Gore, was also a key player in the tackling of opposition to the CDM by the end of 1998. The EU and the G77 were by then pointing its guns at the US' pretension to count forests and agricultural fields as carbon sinks, in order to get credits to pollute more. At the same time, the group of least developed countries, along with China, were being bought into accepting carbon trading by promises of technological transfers and financial aid. In the end, the 'Buenos Aires Plan of Action', approved in the fourth

Conference of Parties (COP), set up as a priority the design of the CDM,[13] the biggest corporate bail out in environmental treaties.

At the same time, the US was already preparing its way out of the Protocol. The Senate Resolution 98, approved by unanimity, stated that the US government shouldn't be a signatory to any protocol that didn't impose emissions reductions targets for developing countries and/or would seriously harm its economy.[14] This resolution is still valid, implying the impossibility of ratification by the Senate. Nevertheless, the US government decided to continue negotiating, hoping to gather the necessary support to shape the Protocol according to the needs of its industry.

Over the next years, negotiations continued to reflect the confrontation between the EU and the US, with the latter continuing its pressure to get more credits, unrestricted carbon trading and offsetting and the exemption of penalties for no compliance. The dispute led to the failure of negotiations at COP-6 in Den Haag, in 2000. In 2001 Bush's administration announced its abandonment of the climate negotiations and the dispute ended. But by this time EU's opposition to unrestricted carbon trading and offsetting had been overcome by the growing pressure from the industry and it ended up reversing its position. What we have now is a compromise stating that the use of flexibility mechanisms should be 'supplemental' to domestic action, a meaningless rule given that no quantitative restrictions have been set.

In 2001, COP-7, in Marrakesh, ended up defining the main features of the CDM. Given the disagreements on forestry projects, though, the rules of procedure were only completely defined two years later.

Corporate Lobbies and Corporate-Friendly NGOs Paving the Way to Environmental Injustice

To understand how the US position on carbon trading prevailed even after its withdrawal from negotiations, we must analyze the role of corporate lobbying. Not only were corporate lobbies incorporated in climate negotiations as 'stakeholders', but also its representativity clearly surpasses that of the NGOs,[15] even if we don't consider the number of lobbyists that manage to attend negotiations as government representatives. Worse, the UN has been so overwhelmed with corporate culture that it went so far as to encourage the growing participation of these lobbies in COPs.

In 1999, UNCTAD joined forces with the World Business Council for Sustainable Development (WBCSD), a corporate lobby that emerged in the Earth Summit, and created the International Emissions Trading Association (IETA). IETA is now the most powerful corporate lobby with the explicit purpose of supporting emissions trading, counting with the support of powerful companies like BP, Shell, KPMG and PriceWaterhouseCoopers. Yet, it was created by an organization aimed to protect peace, human rights and the environment.

Another international institution crucial in shaping Kyoto's Protocol according to industry needs was the World Bank (WB). In 1999, the WB established the Prototype Carbon Fund (PCF), a $180 million mutual fund, to fuel the CDM and JI markets. At this time, Kyoto had not been ratified by enough countries to guarantee its survival[16] and the rules of procedure for CDM projects had not been established, but many companies were eager to invest in the new carbon market, anticipating future profits from being the 'first movers'. The PCF was obviously a tool to force the implementation of Kyoto's flexibility mechanisms. Mainly because of this fund, in 2004, before Russia ratified Kyoto, assuring that the Protocol would be implemented, more than 120 transactions of carbon credits had already been registered.[17]

As for corporate lobbies, besides IETA, two other groups have had a central role in climate negotiations. The first is the mentioned WBCSD, which gathers CEOs from about 140 of the world's largest transnational corporations.[18] The council is a keystone in the rebranding of corporations as environmentally friendly (read greenwashing) and it had a major role in the lobbying for carbon trading.

The second is the International Chamber of Commerce (ICC), the most influential corporate lobby group. From the start, the ICC has lobbied against government regulation and for emissions trading and offsetting.[19]

These industrial lobbies were successful in their campaigns for 'cost effective' environmental policies. We can see this influence very clearly in the creation of a European carbon market, molded according to industry's needs.

A Market for Carbon is Born

The design of the EU ETS, the only carbon trading scheme implemented so far, is symptomatic of the importance of corporate lobbying in the politics. By the early 1990s, a carbon tax was seen as the basis of EU's climate policy. But this proposal never got the necessary unanimous support from the Council of Financial Ministers and was strongly opposed by industry. This lead to a stall, which ended in 2000, when the EU started discussing the creation of a region-wide carbon trading system.

To implement the ETS, the European Commission (EC) undertook a consultation with the main stakeholders. In 2000, a Green Paper on emissions trading was published and representatives from industries and environmentalists were asked to express their opinions.[20] Unsurprisingly, industries lobbied for the right to use the flexibility mechanisms of the Kyoto Protocol (CDM and JI) to comply with the emissions reductions requirements, while environmentalists defended that the use of these mechanisms should be very restricted. Also unsurprisingly, the final position from the EC largely reflected the interests from the industry, allowing for the member states to establish limits for the use of the flexibility mechanisms.

Another contentious issue between environmentalists and industrialists was the allocation of permits to emit CO_2. While Climate Action Network – Europe (CAN-E), representing big environmental NGOs, defended auctioning of permits, according to the polluter-pays principle, the largest corporate lobbies argued that they should be given for free according to historic emissions ('grandfathering'). Again, the EC's decision reflected the needs of the industry, giving them permits that they could later sell for a profit.[21] For power companies, what this means is that they can pass the costs of the permits to the consumers, raising electricity bills, and then sell them for a profit if they end up with a surplus. The scandal was not corrected in the second phase (from 2008 to 2012), and a report by emissions trading advocate World Wildlife Fund (WWF) and the consultancy agency Point Carbon estimated that windfall profits in this phase could reach 23 to 63 billion Euros.[22]

Meanwhile, emissions reductions were not induced by the ETS in its first stage. There was a substantial over-allocation of permits in Phase I, exceeding emissions in 3%, and the carbon market collapsed in 2007, with the permit price dropping to near zero. The allocation in the second stage was lower but still the permit price collapsed again recently, in the midst of the financial crisis. The contraction in production increased the number of permits held in excess by companies, while the need for liquidity made companies sell those permits. In February 2009, a ton of carbon was worth only about 8€.

As a result, the ETS has failed to provide an environmental gain. The EU-15 has committed itself to reducing its GHG emissions by 8% until 2012, from 1990 levels, a target that could easily be reached given the significant emissions reductions achieved in Germany and in the UK in the early 1990s, following a switch from coal to gas in electricity generation. But the lack of a coherent policy to phase out fossil fuels has led the EU-15 to miss the target: according to estimates by the European Environment Agency, GHG emissions reductions by 2012 will amount to a mere 3.6%, in a 'business-as-usual' scenario. This doesn't mean, however, that the EU won't comply with Kyoto. Thanks to the extensive use of the CDM, the EU not only anticipates compliance with the 2012 objective but has even announced more ambitious targets for the future.

Nevertheless, the EC is still committed to carbon trading as the cornerstone of its climate policy after Kyoto, which shows how it weighs industries' profits against climate change mitigation.

More Loopholes in the Horizon

In present climate negotiations, the hot topic for a post-Kyoto agreement is the broadening of the scope of the CDM. We can see this by the list of proposals presented recently by the Ad-Hoc Working Group on Further Commitments for Annex-I[23] Parties under the Kyoto Protocol (AWG-KP).[24] In the next climate summit at Copenhagen, in December 2009, world leaders are going to discuss these proposals.

The most important one is the issuance of carbon credits for preserving forests – the Reduced Emissions from Deforestation and Forest Degradation (REDD) proposal. This proposal was already incorporated in the Bali Action Plan, approved at COP-13. When forests are commodified, a massive land grab is to be expected in Latin American, African and Asian countries, and communities that live a low carbon life will be treated as climate criminals for wanting to occupy 'carbon sinks'. With the objective of creating yet another source of cheap carbon credits, the fate of the world's forests will be at the hands of the unpredictable financial markets.[25]

Other proposals include the generation of carbon credits by two dangerous technologies: carbon capture and storage (CCS) and nuclear energy. CCS is the basis of the 'clean coal' fantasy. By sequestering carbon emissions at coal plants and burying them underground, the energy industry hopes to continue to use fossil fuels and, simultaneously, earn carbon credits. But CCS won't be available in the next decades and it will be an uneconomic, dangerous and environmentally unfriendly technology.[26] As for nuclear energy, the unresolved issues of radioactive waste and leakages show that we are merely trading one environmental disaster for another.

Another issue that will be discussed in Copenhagen is the creation of a sectoral crediting mechanism. This would allow for a non-Annex I country to generate carbon credits if the emissions from a given sector are below the projected emissions (an absolute target) or if the carbon intensity of emissions is below the projected carbon intensity (a relative target). If the proposal is approved, we can expect the carbon market to be swamped with cheap carbon credits.

The AWG-KP is merely expressing the needs of the industry and the financial sector in its proposals. This is of no surprise when we see that its chair is Harald Dovland, a former consultant for Poyry, which is a consulting and engineering agency that profits from carbon trading.[27] Once again, the conflict of interests is evident.

A similar analysis can be made of the Council of the EU proposals for Copenhagen. Despite the failure of the ETS in providing significant emissions reductions, the EU maintains its commitment to carbon trading and proposes the creation of a OECD-wide carbon market by 2015, to be expanded in 2020 to industrialized developing countries through sectoral crediting and trading mechanisms.[28] Simultaneously, the EU is presenting its '20-20-20 target' (20% reduction in GHG emissions and 20% renewable energy by 2020) as a sign of good will but in reality the emissions reductions after 2012 will only amount to 4-5%, as Stefan Singer, director of global energy policy at WWF concludes.[29] The extensive use of the CDM will be determinant for compliance with the target, so it is no wonder that the EU supports many of the AWG-KP proposals, like REDD and sectoral crediting mechanisms.

Conversely, the substitution of the CDM by a fund that assures the transfer of clean technology and the assistance to adaptation in the least developed countries in such a way that real emissions reductions are made, the standard of living is raised and people's participation is assured is not on the table.[30] A profound reform of the CDM, to assure that non-additional offsets are purged from the system, is supposedly on the negotiating table, but in reality no industrialized country wants to lobby for a much tighter certification system that would make them resort more to domestic action to comply with Kyoto.

Certification problems occur because two conflicts of interest prevail in the CDM. The first is the conduction of certification by private companies working for the sellers of carbon credits. It is in the best interest of both parties in the process of certification that the number of rejected projects will be as low as possible, so there is an obvious incentive for cheating. On the other hand, the competition between certification companies will lead to adverse selection, as the ones that spend less time and resources with the process of certification will offer lower prices and, consequently, get a greater market share.

The second conflict of interest arises because both sellers and buyers of carbon credits are very much indifferent to quality. It doesn't matter at all that a certain project delivers no emissions reductions and even harms local people, as long as the seller is able to earn a profit and the buyer is able to use the cheap credits to comply with Kyoto. Again, there is an incentive for strategic cooperation.

To be fair, in recent years some modifications in the certification process have been introduced, raising the number of projects rejected by the CDM Executive Board. But the pressing issues regarding conflicts of interests are not seriously discussed in present climate negotiations, turned into a trade fair by the increasing power of corporate lobbying. Neither is the basic question of the impossibility of accurately defining 'additionality'. In the end, the result of climate negotiations reflects the balance of power between NGOs and corporate lobbies, with the latter having much more negotiating power.

Building a Movement Against Climate Change Profiteers

Real solutions for climate change won't come from international negotiations riddled with rent-seeking from corporate lobbies. Nor will they come from miraculous new technologies or from market-based policies. On the contrary, the problem of global warming can only be seriously addressed when its main source, the use of fossil fuels, is stopped. The full decarbonization of industrialized societies may be a condition for its survival as a civilization. Moreover, the degradation of life conditions in the global South by 'natural' catastrophes, droughts and plagues as a result of global warming, a problem caused by the industrialized North, is an unacceptable form of social injustice.

All this seems consensual, as we can find this sort of considerations in global leader's speeches. Yet, new coal plants, highways and airports continue to be built in industrialized countries, even when it implies a departure from emissions reductions targets. This incoherence can be explained by the political economy of global warming, given that the phasing-out of fossil fuels implies reducing profits from major companies, namely from the energy and transport sectors. Corporate lobbies will then oppose any regulation that would hurt them. Only by forming strong social movements can we make politicians work for the ones that elected them, instead of obeying those who finance their campaigns.

Notes

1   Hartle, D.G. (1983) 'The Theory of 'Rent Seeking': Some Reflections', *The Canadian Journal of Economics / Revue canadienne d'Economique*, 16(4): 539-554. Note that rent is defined as an economic gain not obtained from trading products or supplying labor or capital.

2   Tullock, G. (1967) 'The Welfare Costs of Tariffs, Monopolies and Theft', *Western Economic Journal*, 5: 224-232.

3   Krueger, A. O. (1974) 'The Political Economy of the Rent-Seeking Society', *The American Economic Review*, 64(3): 291-303.

4   The Joint Implementation (JI) mechanism also allows the generation of carbon credits for compliance with Kyoto's demands but, unlike the CDM, finances projects for emissions reductions located in industrialized countries. Its importance in global carbon markets is small, so we won't discuss this mechanism here.

5   Transnational Institute (2003) *The Sky is Not the Limit: The Emerging Market in Greenhouse Gases*. Carbon Trade Watch Briefing.

6   http://www.exxonsecrets.org.

7   Original memo in http://www.nrdc.org/media/docs/020403.pdf.

8   For instance, Environmental Defense Fund, a strong proponent of corporate-NGO partnerships, has lobbied for market-based policies, using the same arguments as the industry (see http://www.edf.org/page.cfm?tagID=1085). The corporate-backed NGO was a major player in the design of emissions trading schemes both in the US and in the EU.

9   As was mentioned by the Conservative Edward Leigh, on the House of Commons Select Committee on Public Accounts' session on the UK Emissions Trading Scheme. http://www.publications.parliament.uk/pa/cm200304/cmselect/cmpubacc/604/ 604.pdf

10 Corporate Europe Observatory (2009)' BP - Extracting Influence at the Heart of the EU', http://www.corporateeurope.org/climate-and-energy/content/2009/01/bp-extracting-influence-eu.

11 Halpern, S. (1992) *United Nations Conference on Environment and Development: Process and documentation. Providence: Academic Council for the United Nations System*, http://www.ciesin.org/docs/008-585/unced-ch1.html#PC-climate.

12 Transnational Institute (2003) *The Sky is Not the Limit: The Emerging Market in Greenhouse Gases*.

13 See Decision 7 in unfccc.int/resource/docs/cop4/16a01.pdf.

14 http://www.senate.gov/legislative/LIS/roll_call_lists/roll_call_vote_cfm.cfm? congress=105&session=1&vote=00205.

15 A list of observers from the 'civil society', i.e., NGO and industry representatives, is available at the UNFCCC site, http://maindb.unfccc.int/public/ngo.pl?mode= wim&search=A.

16 The Kyoto Protocol would not have been implemented if it wasn't ratified by countries representing 55% of 1990 GHG emissions.

17 A description of key issues with the PCF and other investment funds operating in carbon markets and managed by the WB, see Institute for Policy Studies (2008) 'World Bank: Climate Profiteer', http://www.ips-dc.org/getfile.php?id=181.

18 Transnational Institute (2003) *The Sky is Not the Limit: The Emerging Market in Greenhouse Gases*.

19 ICC (1997) 'Statement by the International Chamber of Commerce, the world business organization, at the conclusion of the Third Conference of the Parties to the Framework Convention on Climate Change', http://www.netcase.net/collection4/folder165/id367/printpage.html?newsxsl=&articlexsl=.

20 CEC (2000).

21 Directives 2003/87/EC and 2004/101/EC. For a greater detail on rent-seeking in the EU ETS design, see Ricardo Coelho (2008) 'Rent Seeking and Capture in the EU ETS', http://sites.google.com/ricardosequeiroscoelho.

22 WWF and Point Carbon (2008) 'EU ETS Phase II – The potential and scale of windfall profits in the power sector', at http://assets.panda.org/downloads/point_carbon_wwf_windfall_profits_mar08_final_report.pdf.

23 Annex I Parties, in the Kyoto Protocol, are the industrialized countries, subject to emissions targets.

24 http://unfccc.int/resource/docs/2009/awg7/eng/l02.pdf.

25 To learn more about REDD and alternative, fund-based proposals, see for instance World Rainforest Movement (2008) 'From REDD to HEDD', http://www.wrm.org.uy/publications/briefings/From_REDD_to_HEDD.pdf.

26 For more details see Greenpeace (2008) *False Hope: Why carbon capture and storange won't save the climate*, in http://www.greenpeace.org/raw/content/usa/press-center/reports4/false-hope-why-carbon-capture.pdf.

27 Bullard, N. (2008) 'Who is Harald?', *New Internationalist*, http://www.newint.org/columns/currents/2008/06/01/climate-negotiations/.

28 Council of the European Union (2009) 'Council Conclusions on the further development of the EU position on a comprehensive post-2012 climate agreement', http://ec.europa.eu/environment/climat/future_action.htm.

29 Interview for EurActiv.com, in www.euractiv.com/en/climate-change/wwf-singer-eu-cheating-world-climate-change/article-181121.

30 This was proposed by International Rivers, for instance. See http://internationalrivers.org/node/3498.

20

Forests, Carbon Markets and Hot Air: Why the Carbon Stored in Forests Should not be Traded

Chris Lang

Introduction

Reduced emissions from deforestation and forest degradation (REDD) is, in theory at least, a simple idea. Governments, companies, forest owners, local communities or indigenous peoples in the South should be rewarded for keeping their forests instead of cutting them down. The devil, as always, is in the details. Marc Stuart of EcoSecurities describes REDD as

> the most mind twistingly complex endeavor in the carbon game. The fact is that REDD involves scientific uncertainties, technical challenges, heterogeneous non-contiguous asset classes, multi-decade performance guarantees, local land tenure issues, brutal potential for gaming and the fact that getting it wrong means that scam artists will get unimaginably rich while emissions don't change a bit.[1]

None of this prevents Stuart from supporting the trade in carbon stored in forests. This is perhaps not surprising since as a founder of one of the biggest carbon consulting firms in the world, EcoSecurities, he has made his fortune from carbon trading.

What is REDD?

The idea of making payments to discourage deforestation and forest degradation was discussed in the negotiations leading to the 1997 Kyoto Protocol, but it was ultimately rejected in part at least because of the problems that Marc Stuart describes. REDD developed from a proposal in 2005 by a group of countries calling themselves the Coalition of Rainforest Nations (more on them later). Two years later, the proposal was taken up at the Conference of the Parties to the UN Framework Convention on Climate Change in Bali (COP-13). An agreement on REDD is planned to be made at COP-15 which will take place in Copenhagen in December 2009.

The 'Bali Action Plan' Calls for:

> Policy approaches and positive incentives on issues relating to reducing emissions from deforestation and forest degradation in developing countries; and the role

of conservation, sustainable management of forests and enhancement of forest carbon stocks in developing countries.[2]

The above paragraph (paragraph 1b(iii)) is referred to as 'REDD-plus'. It is worth reading closely. 'REDD-plus' includes activities with potentially extremely serious implications for indigenous people, local communities and forests:

- 'conservation' sounds good, but the history of the establishment of national parks includes large scale evictions and loss of rights for indigenous peoples and local communities;[3]

- 'sustainable' management of forests' could include subsidies to commercial logging operations in old-growth forests, indigenous peoples' territory or in villagers' community forests;

- 'enhancement of forest carbon stocks' could result in conversion of land (including forests) to industrial tree plantations, with serious implications for biodiversity, forests and local communities.[4]

In order to prevent abuses under REDD, we would hope, as an absolute minimum, to see that the UN is ensuring that international human rights instruments are reaffirmed in any agreement on REDD. Particularly important are the UN Declaration of the Rights of Indigenous Peoples and the concept of Free Prior Informed Consent. Unfortunately, the UN climate change negotiations are going in the opposite direction. In December 2008, at COP-14 in Poznań, the US, Canada, New Zealand and Australia opposed any reference to Indigenous Peoples' rights in the negotiating text and the draft text was duly weakened.[5]

While there has not yet been any agreement on how REDD is to be financed, a look at some of the main actors involved suggests that there is a serious danger that it will be financed through carbon trading. The role of the World Bank is of particularly concern, given its fondness for carbon trading.

The World Bank's main mechanism for promoting REDD is a new scheme, launched in Bali in 2007: the Forest Carbon Partnership Facility (FCPF). Under this scheme, the Bank is working with tropical countries to help them achieve 'readiness' for REDD. When the World Bank launched the FCPF, Benoit Bosquet, a senoir natural resources management specialist at the World Bank, said 'The facility's ultimate goal is to jump-start a forest carbon market that tips the economic balance in favour of conserving forests.'[6] As Marcus Colchester of the Forest Peoples Programme points out, the speed with which the FCPF is going ahead risks undermining REDD. In particular, the 'importance of securing rights [for indigenous peoples and local communities] has been played down and [the World Bank's] safeguards process allowed to drift.'[7]

What's Wrong with Trading Forest Carbon?

The problem with trading the carbon stored in forests is that we need to reduce greenhouse emissions *and* stop deforestation. We cannot trade off one against the other.

Trading the carbon stored in forests would mean that one ton of emissions reduced through avoided deforestation or forest degradation would allow emissions in the North to increase by one ton. Offsetting emissions in the North against carbon credits generated through REDD does not by definition, reduce greenhouse gas emissions. This is the model of the clean development mechanism. A recent report by the University of Zurich, Öko-Institut, Perspectives GmbH and Point Carbon explains the problem succinctly:

> A continuation of the CDM as a pure offset mechanism would not directly contribute to the achievement of this goal [of limiting warming to 2°C], since the emission reductions generated under this mechanism in developing countries allow for higher emissions in industrialized countries.[8]

Another problem is that carbon markets cannot send long term investment signals. During 2008, the global financial and economic crisis led to a slight reduction of emissions of greenhouse gases in the EU. But when New Carbon Finance released a report announcing the fall in emissions in mid-February 2009, the Financial Times described the report as a 'blow to the [carbon] market'.[9] The Financial Times explained that

> Falling emissions spell a lower carbon price because fewer permits will be needed by the heavy industries, such as power generation and steel-making, covered by the scheme.

This is a serious flaw in the carbon market. If the price of carbon goes down due to the economic slowdown, incentives for serious re-investment disappear. As soon as the economy starts to recover, all the old machinery is simply started up again.

On 1 April 2009, thousands of people set up camp in the City of London, outside the European Climate Exchange to protest against carbon trading in Europe. They had good reason to do so, given the record of the EU Emissions Trading Scheme. Kevin Smith of the NGO Carbon Trade Watch sums up the problems:

> Phase 1 of the scheme gave away the right to pollute for free. Bingo! The biggest polluters then made billions in windfall profits. Phase 2 and in the wake of market meltdown, the price of carbon is again at rock-bottom. The EU scheme is providing all manner of opportunities to pollute and make money, which is why companies from e.on to BP to BAA are all supporters. As a mechanism to reduce emissions it has been an out and out failure.[10]

Innovative Financial Mechanisms or the New Sub-Prime?

The problems with trading forest carbon are not limited to the fact that it will not address runaway climate change. Forest carbon would be one part of the global carbon markets. In 2007, Chris Leeds, then-head of emissions trading at Merrill Lynch told the New York Times that carbon could become 'one of the fasting-growing markets ever, with volumes comparable to credit derivatives inside of a decade.'[11] But the similarities between carbon trading and derivatives trading are not limited to predictions of the size of the carbon trade. There are close parallels between the way the carbon markets are developing and the way the markets in derivatives and futures developed, until they crashed spectacularly in 2008. Yet proponents of financing REDD through trading forest carbon talk about 'innovative financial instruments' as if the current global financial crisis had never happened.

For example, at a side event at the climate conference in Poznań in December 2008, Ben Vitale of Conservation International spoke positively about the role innovative financing could play in financing forest conservation. During the questions after his presentation, I asked Vitale to say something about the current global financial crisis and in particular to say something about the financial innovations that led to the financial collapse and the billion dollar bailouts. I noted that the carbon market will be extremely complex, not transparent and that it seems ironic to be talking about 'innovative financing' at this particular moment in history.

Vitale declined to answer my questions, commenting only that with this financial crisis perhaps it makes sense to have a more stable fund but, he added, the fund would have to be very, very large and it would have to grow over time. He made no mention of the bailout of the banks, the complexity of the carbon market, or what would happen if the carbon market fails.[12]

Derivatives Trading and Carbon Markets

Larry Lohmann of the UK-based research and advocacy organization The Corner House has been investigating the failure of carbon markets to address climate change for several years. In 2006, he edited a book titled *Carbon Trading: A Critical Conversation on Climate Change, Privatization and Power.*[13] In a summary of a memorandum submitted to a UK Select Committee on carbon trading, Lohmann points out the dangers of carbon trading:

> Carbon markets are characterised by a type of speculative derivatives trading and need to be evaluated as such in the light of the current financial crisis. Like financial derivatives markets, carbon markets are legitimated by (spurious or overblown) claims of efficiency but undermined by their tendency to exacerbate a crisis. Carbon markets are plagued by difficulties of asset valuation parallel to those that have contributed to the financial crisis and are themselves prone to a similar crash. Carbon markets are also characterized by inherent problems of conflicts of interest, regulatory capture and unregulatability familiar from recent analyses of the financial crisis.[14]

Even people very closely involved in the carbon markets admit that there are similarities with trade in derivatives. 'I guess in many ways it's akin to subprime,' said Marc Stuart of EcoSecurities after the value of the company's shares crashed in 2008. 'You keep layering on crap until you say, "We can't do this anymore"'.[15]

Lohmann lists some of the institutions dealing in derivatives that are involved in carbon markets, including Goldman Sachs, Deutsche Bank, Morgan Stanley, Barclays Capital, Fortis, Rabobank, BNP Paribas, Sumitomo, Kommunalkredit, Lehman Brothers, Merrill Lynch and Cantor Fitzgerald. He points out that 'The stupendous complexity of new financial instruments such as collateralized debt obligations is in some ways matched by that of carbon trading, with its reams of additionality calculations, diversity of carbon credits, daunting monitoring and legal requirements and crowd of acronyms.'[16]

Several of the same people who were involved in creating financial derivatives markets are also involved in creating carbon markets. The founder and chairman of the Chicago Climate Exchange is Richard Sandor, who in the 1970s was one of the leading developers of derivatives and futures markets.[17] The Chicago Climate Exchange offers a futures contract based on emissions allowances under a US cap and trade scheme – before such a scheme even exists.

Richard Sandor is, predictably, in favour of trading forest carbon. 'The clock is moving. They are slashing and burning and cutting the forests of the world. It may be a quarter of global warming and we can get the rate to two per cent simply by inventing a preservation credit and making that forest have value in other ways. Who loses when we do that?' he said in an interview with *The New Yorker* last year.[18]

Sandor appears to have little sympathy for local communities and even less knowledge of the complexities of tropical forest politics. The obvious answer to Sandor's question is that indigenous peoples and forest dependent communities are likely lose when someone in the USA makes their forests more valuable to outsiders. Land grabs are the almost inevitable consequence of increasing the value of forests. Of the many attempts to stop or slow deforestation, the few successful projects have been those that work closely with local communities and actively support Indigenous Peoples' and local communities' rights.

'Fleecing Landowners and Indigenous People'

Recent events in Papua New Guinea illustrate some of the problems with Sandor's simplistic approach to forests. The Office of Climate Change appears to have issued at least 40 REDD 'credits', each denoting one million tons of carbon, according to investigations carried out by a journalist with *The Economist* magazine.[19] One of the REDD carbon 'credits' relates to the Kamula Doso REDD project. As the Eco-Forestry Forum, a PNG NGO, points out 'In November 2008, the Office of Climate change issued a certificate granting the rights to 1 million tons of carbon from Kamula Doso to a company called Nupan Trading limited. This certificate was issued despite PNG having no laws that allow trading in carbon rights and the Office of Climate Change not having obtained the informed consent of landowners.' The head of the Office of Climate Change, Theo Yasause, denies any wrongdoing and says that the sample credits were created merely 'to see what it looked like'.[20] In June 2009, Yasause was suspended while an internal investigation of the Office of Climate Change is carried out.[21]

Meanwhile, conmen are travelling from village to village offering fake carbon trading deals and promising huge returns. Villagers hand over about US$500, for 'registration as a shareholder' in a carbon trading company. They receive a receipt and the conman leavers, never to be seen in the village again. Natasha Loder, a journalist with *The Economist*, comments,

> What is striking about the invention of an avoided forest carbon market is the extent to which it is quickly spawning a variety of imaginative ways of fleecing landowners and indigenous people in the rush for green gold.[22]

Can Carbon Markets be Regulated?

It is not just out in the bush in Papua New Guinea that trading in forest carbon is unregulated. Carbon markets are riddled with conflicts of interest and revolving doors between public and private institutions as well as between regulators and traders.

Carbon markets were effectively created in 1997, when Al Gore led the USA's climate negotiators in destroying the Kyoto Protocol, by allowing rich nations to buy their emissions cuts from other countries.[23] Of course Gore's film 'An Inconvenient Truth' has done a great deal to convince large numbers of people that climate change is real and the film has also helped expose the folly of climate denial. But when Gore jets around the world for his extraordinarily well-paid speaking appointments[24] he does not mention the inconvenient truth that he helped to create carbon markets and is now profiting from them. In 2004, Al Gore co-founded Generation Investment Management, together with David Blood, former chief executive of Goldman Sachs Asset Management.[25] In 2008, Gore's firm bought a 9.6 per cent share in Camco International Ltd,[26] a Jersey-based company which holds one of the world's largest carbon credit portfolios. In February 2009, Generation Investment Management increased its share in Camco International to 13.74 per cent, making it the largest shareholder in the company. The following month Generation Investment Management increased its share further to 18.94 per cent.[27]

Ken Newcombe is another key player in the development of the trade in forest carbon. As a recent article in the trade magazine *Point Carbon* notes, 'Ken Newcombe has been involved in carbon markets since their inception.'[28] From 1990 to 1996, Newcombe was chief of the global environment division at the World Bank. He led the Bank's involvement in Forest Market Transformation Initiative,[29] which the Bank describes as a 'strategic coalition of conservation NGOs, private sector forest industry leaders, researchers, development practitioners, and financiers, including the World Bank, [that] is working to develop innovative approaches to the adoption of more environmentally friendly forest management and marketing practices in the remaining forest frontiers.'[30] After that, he led the creation of the Bank's Prototype Carbon Fund.

During a press conference at the Carbon Expo Trade Fair in Cologne in 2004, Newcombe explained the purpose of the Bank's involvement in carbon markets: 'The World Bank is reducing the risk for private investors.'[31] The following year Newcombe left the Bank's carbon finance unit, by which time the Bank was managing carbon funds with a total value of US$1 billion. Newcombe became senior manager and advisor of the G8 investment framework initiative at the World Bank.

The next year he was on the move again, this time to Climate Change Capital, the largest private sector carbon fund in the world. James Cameron, Vice-Chairman at Climate Change Capital said of Newcombe, 'He has a fantastic network and knows about World Bank projects that we can now invest in.'[32] Cameron is an environmental lawyer who helped negotiate the Kyoto Protocol. From Climate Change Capital, Newcombe headed the carbon desk at Goldman Sachs in New York before launching his own company, C-Quest Capital, to profit further from the carbon markets. 'We see the voluntary market as a risk hedge strategy,' Newcombe explained to *Point Carbon*. 'We are getting our foot in the door in assets we think might be good for compliance in the future', he added. He sees international offsets as an 'inevitable part of any US scheme'.[33]

Newcombe divides people involved in carbon markets into two groups. 'There are those who see it as a way to make money, and see it as the next wave as high risk, high reward businesses', he told *Point Carbon*. 'Others are wanting to make good money, but by doing good in the process. I like to think I build teams who are the latter camp'.[34]

Newcombe is a director emeritus of Forest Trends, an organization that developed from the World Bank's Forest Market Transformation Initiative.[35] Michael Jenkins, the president and CEO of Forest Trends also came from the World Bank, where he held a joint appointment as a senior forestry advisor. Given the World Bank's disastrous record in the forests of the global South,[36] Jenkins might be considered a strange choice to head up an NGO. But Forest Trends is no ordinary NGO. Its board includes representatives from Mitsubishi International, ABN Amro, Sveaskog, The Nature Conservancy, Greenpeace Russia, Rainforest Action Network and Generation Investment Management.[37] One of the board members, David Brand, is head of New Forests, 'an investment management and advisory services firm specializing in forestry and land-based environmental markets, such as timber, carbon, biodiversity and water'.[38] (For more information about New Forests' activities in Uganda, see the chapter by Ricardo Carrere in this book.) Generation Investment Management is one of the four shareholders in New Forests and David Blood, the co-founder of Generation Investment Management, sits on New Forests' board.[39] Forest Trends publishes Ecosystem Marketplace and helped create the Katoomba Group, the Business and Biodiversity Offsets Program, SpeciesBanking.com, ForestCarbonPortal.com and the Chesapeake Fund, all of which promote market 'solutions' to environmental problems. In April 2009, at the tenth anniversary of Forest Trends, Al Gore said 'Forest Trends has become widely-regarded as the most comprehensive advocate and resource for anyone who wants to understand and help to further develop markets for ecosystem services'.[40]

Larry Lohmann notes further conflicts of interest in carbon markets.[41] Barclays Capital is a major investor in the carbon markets and at the same time boasts that 'One of our team is a member of the Methodology Panel to the UNFCCC CDM Executive Board'.[42] Lex de Jonge is simultaneously head of the carbon offset purchase programme of the Dutch government and vice chair of the Clean Development Mechanism Executive Board. Harald Dovland headed Norway's climate negotiations team for 12 years. He is vice chair of the Ad Hoc Working Group on Further Commitments for Annex I Parties under the Kyoto Protocol.[43] Dovland states that what is needed now is 'acceptance of long-term goals on a high political level, further development of markets, and innovative financing solutions'. But at the same time, Dovland is an advisor to Econ Pöyry, a company which profits from carbon trading.[44]

The magazine *Point Carbon* claims to be a 'provider of independent news, analysis and consulting services', but as the Financial Times recently noted Point Carbon is in fact 'part-owned by financial and industrial interests'. (Point Carbon is owned by Oak Investments, JP Morgan, J-Power, Mizuho, Schibsted and the employees.)[45]

Caisse des Dépôts is one of the organizations that is pushing to include forests in carbon markets, through reports such as 'Reducing Emissions from Deforestation and Degradation: What Contribution from Carbon Markets?'.[46] But as well as producing reports promoting expanded carbon markets, Caisse des Dépôts profits from the trade in carbon. It is a 40 per cent shareholder in Paris-based BlueNext, Europe's main spot EU Allowances (EUAs) exchange. In February 2009, BlueNext was earning over 2 million euros a week on transactions of EUAs.[47]

Kevin Conrad and the Coalition for Rainforest Nations

This discussion of carbon markets and forests would be incomplete without looking at the role of Kevin Conrad, ambassador and special envoy for the environment and climate change for Papua New Guinea. In December 2007, at the UN climate negotiations in Bali, Conrad told the US delegation, 'if for some reason you're not willing to lead, leave it to the rest of us. Please get out of the way'.[48]

To his credit, Conrad remains critical of the US. 'President Barack Obama's current proposal to reduce US emissions to 1990 levels by 2020 and 80 per cent below by 2050 is grossly insufficient in the near term and simply pushes true responsibility on to future US presidents', he wrote in April 2009. 'Why should the greatest emitter in history be granted 12 extra years simply to get to the starting line accepted by other industrialized countries? Is this leadership or laggardship?'[49]

Conrad is executive director of the Coalition for Rainforest Nations (CfRN), a group of tropical countries which tabled the first proposal for REDD at the UNFCCC COP11 in Montreal, in 2005. CfRN has since grown from 11 countries led by Papua New Guinea and Costa Rica, to 40 countries.[50] Conrad and CfRN promote trading of the carbon stored in forests: 'The Rainforest Coalition seeks to incorporate certified emissions offsets related to deforestation (in addition to afforestation and reforestation) within global carbon emissions markets by revising the Marrakech Accords, amending the Kyoto Protocol, or developing a linked 'optional protocol' under the UNFCCC'.[51]

Conrad is not a forester, nor does he appear to have any experience in managing or protecting forests. His qualifications are business qualifications, most recently a degree from the Columbia Business School. For the final project of his Executive M.B.A. Conrad looked at whether the money from carbon credits could equal the revenue from logging in Papua New Guinea. His supervisor for this project was Professor Geoffrey Heal, Head of Columbia Business School. When the project was completed, Conrad and Heal persuaded Papua New Guinea's prime minister, Michael Somare, to start the Coalition for Rainforest Nations.[52] In January 2005, Somare called for the formation of the Coalition for Rainforest Nations at the World Leaders Forum held at Columbia University.[53] In May 2005, Somare was back at Columbia University for the Global Roundtable on Climate Change, once again calling for the Coalition for Rainforest Nations:

> I have called for the formation of a 'Coalition for Rainforest Nations.' To support that call, my government has held discussions at the United Nations with representatives from Peru, Congo, Costa Rica, Dominican Republic, Mozambique, Tanzania and Zambia – who, together with us, would constitute the largest expanse of rainforest globally under such an issue-specific coalition.[54]

Speaking in Parliament a month later, Somare referred to the Global Roundtable on Climate Change as a 'Carbon Trading Seminar [that] I addressed at the Columbia University'.[55] Perhaps not surprisingly, the secretariat of the CfRN is in Columbia University.[56]

Geoffrey Heal, the co-founder of the CfRN, is Garrett Professor of Public Policy and Corporate Responsibility and Professor of Economics and Finance at Columbia University's Graduate School of Business, and Professor of Public and International Affairs at the School of International and Public Affairs.[57] He is also on the board of the Union of Concerned Scientists and was a Director of Petromin Holdings PNG Ltd,[58] a state-owned oil, gas and mineral company. Kevin Conrad was also hired as an advisor to Petromin.[59]

Some of Conrad's business deals are controversial in Papua New Guinea. A recent article in the Australian newspaper, *The Age*, comments that Conrad 'has been linked to a string of failed business dealings in Papua New Guinea.' In 2007, Peter O'Neill, then-opposition leader in PNG accused Conrad (among other things) of 'involvement in a failed housing scheme in the 1990s for the Public Officers Superannuation Fund where 17 million kina ($A8million) was paid but not one single house was built.'[60] In an interview with Australian Associated Press, Conrad said 'If you look at PNG every businessman has failed about as often as they have succeeded and the reason is because the government has had too much control'.[61]

Taking the Pressure of Polluters and Subsidizing Logging

In his speech at Columbia University, Michael Somare said 'Let me be clear, our intentions are NOT to take the pressure off the fossil-fuel emission reductions necessary within industrialized nations'.[62] But on its website, the CfRN states that it aims to 'Slow deforestation internationally through the Clean Development Mechanism (CDM) and other international investments in forest conservation'.[63] Trading the carbon stored in forests inevitably takes the pressure off to reduce fossil-fuel emissions in the North. One example of this is the American Clean Energy and Security Act (ACESA), about which Kevin Conrad is so critical. One of the reasons that the US can get away with such a low target is because of the offsets loophole. A critique by International Rivers and Rainforest Action Network points out that the Act is 'is seriously weakened by its heavy reliance on offsets to substitute for actual emissions cuts by large polluters'. Payal Parekh of International Rivers explains that 'If polluters indeed use the maximum allowable number of offset credits, domestic emissions in 2012 would increase by 38% rather than decrease by 3%, the reduction that the cap sets. Emissions would not dip below 2005 levels until 2026, 17 years from today'.[64]

The legislation is further weakened by the inclusion of 'sustainable forest management'. As International Rivers and Rainforest Action Network explain:

> ACESA envisions offset credits for 'sustainable forestry practices,' a widely abused term that is too often a cover for expanded industrial logging into primary tropical rainforests. Unless forest degradation is included, even heavily logging a forest, which would result in large emissions, could still generate offset 'credits' because full deforestation was avoided.[65]

The Coalition for Rainforest Nations' is also interested in developing 'Sustainable Forest Markets'. Under this initiative, CfRN's website explains that

> In cooperation with the International Timber Organization (ITTO), the Rainforest Coalition will facilitate certification of sustainable logging, develop disincentives to illegal logging and support the establishment of businesses within developing countries that can process lumber locally to the standards of, and in partnership with, end users in industrialized markets.[66]

What this means in reality became more clear in May 2009, at the World Business Summit on Climate Change. Business leaders from around the world flew to Amsterdam to discuss how they could profit from climate change. (The website of the Copenhagen Climate Council, which helped organize the event has the headline, 'Turning risks into opportunities'[67]). For industry, REDD 'presents ample opportunity for the private sector to engage all along a €50-100 billion value chain'.[68] A report produced by the ClimateWorks Foundation for the Summit explains which companies might benefits from REDD: 'Companies in forest management, pulp and paper, or construction could build new businesses around carbon abatement'.[69] In its presentation at the Summit, Project Catalyst, which brings together 'climate negotiators, senior government officials...and business executives emphasized 'the size of the prize for business'.[70]

The assumption underlying sustainable forest management is that by logging less destructively, more trees will be left standing and therefore less carbon will be released to the atmosphere. Here we enter the territory that Dan Welsh, a journalist with *Ethical Consumer* magazine describes so well: 'Offsets are an imaginary commodity, created by deducting what you hope happens from what you guess would have happened'.[71] A recent report by Global Witness, 'Vested Interests – industrial logging and carbon in tropical forests', documents how what the logging industry hopes will happen (or at least says it hopes will happen) in any case releases large amounts of carbon to the atmosphere. Reduced impact logging 'kills 5-10 non-target trees for every target tree cut, and releases between 10 and 80 tons of carbon per hectare'. Logging also makes forests more vulnerable to further deforestation and to fire. 'During the El Niño events in the late 1990s, 60% of logged forests in Indonesian Borneo went up in smoke compared with 6% of primary forest', Global Witness notes.[72]

Campaigns Against Trading in Forest Carbon

Several NGOs and networks are campaigning to expose the problems with trading the carbon stored in forests, including FERN, Friends of the Earth, Indigenous Environmental Network, the Durban Group, World Rainforest Movement, Rainforest Action Network, Global Witness, The Wilderness Society, Greenpeace and the Rainforest Foundation. By creating a huge number of carbon credits, the trade would allow business as usual to continue in the North. In an interview with *The Guardian* in November 2008, Joseph Zacune of Friends of the Earth explains that 'there is genuine risk that all of these kinds of proposals would provide a get out of jail free card to rich nations. It would allow them to buy their way out of emissions reductions. It would create the climate regime's biggest ever loophole and would remove any environmental integrity from a post 2012 deal'.[73]

A Greenpeace report released in March 2009 makes a related point: 'Including forest protection measures in carbon markets would crash the price of carbon by up to 75 percent and derail global efforts to tackle global warming'. The report, which was produced by a New Zealand-based economic modelling group called KEA3, found that including REDD credits in carbon markets would reduce investments in clean technologies worldwide, causing a 'lock in' effect, leaving high-carbon technologies such as coal-fired power stations in place for many years to come. In addition, the report points out that 'significant questions of permanence, leakage, and additionality have been raised about potential REDD credits; as well as the ability of countries to accurately measure, monitor, and report on such emissions'.[74]

Academics such as Alain Karsenty of CIRAD (the Paris-based International Centre for Cooperation on Agroforestry Research and Development) also point out the dangers of trading in forest carbon. In a paper published last year in the *International Forestry Review*,[75] Karsenty comments on the uncertainties involved in establishing the impact of REDD measures, which would 'essentially force experts to disentangle an embedded array of factors, isolating what can be the net impact of policies and measures effectively taken by the authorities to tackle deforestation (i.e. stringent law enforcement, removal of agricultural subsidies, etc.) and external factors such as (involuntary) changes in market prices for agricultural commodities, drought episodes causing forest fires (as well as abnormally high rainfalls)'.

Karsenty concludes his paper with the following statement, 'Markets instruments are very effective tools for achieving specific goals, such as improving efficiency of economic agents, but they will probably be unable to change the socio-political context underlying tropical deforestation'.

Conclusion: Forest Carbon should not be Traded

The carbon stored in forests should not be traded. There are several important reasons why not, which I've covered in this chapter. To summarize:

- First, we need to reduce greenhouse gas emissions and stop deforestation. We cannot trade off one against the other.
- Second, carbon markets do not send long term investment signals. A volatile carbon price might be great for investors willing to bet on the future price of carbon. If the carbon price drops during a recession industry is given little incentive to invest in the major changes required. When the economy recovers, the old machinery is restarted.
- Third, there are close parallels between the market in derivatives and the market in carbon. Proponents of the trade in forest carbon talk about 'innovative financial instruments', in spite of the current financial crisis.
- Fourth, carbon markets are riddled with conflicts of interest. This may not be illegal, but it certainly makes the sector very difficult (or impossible) to regulate.
- Fifth, trading the carbon in forests is bringing calls for 'sustainable forest management', from institutions such as the International Timber Trade Organization that have supported destructive forestry operations for decades. Logging of primary forests (including so-called 'reduced impact logging') would release huge amounts of carbon to the atmosphere. Offsetting the carbon supposedly stored in forests subjected to 'reduced impact logging' would allow emissions to continue in the North, would lead to forest degradation and destruction on a large scale and would provide an enormous subsidy to the timber industry.
- Sixth, trading the carbon stored in forests would create a loophole for the North, allowing industry to write cheques rather than reduce emissions at home.
- Seventh, forests are not just sticks of carbon waiting for an economist to value them correctly so that they will not be cut down. They are home to millions of people. Defending the rights of indigenous peoples, forest dwelling communities and local communities is crucial to preserving tropical forests.

The UN climate negotiations are getting more and more complex, while governments' proposed emissions reduction targets are less and less likely to address runaway climate change. George Monbiot has developed a simple test to show whether governments are genuinely commitment to stopping the climate crisis: 'whether they are prepared to impose a limit on the use of the [fossil fuel] reserves already discovered, and a permanent moratorium on prospecting for new reserves. Otherwise it's all hot air'.[76] Governments proposing to trade the carbon stored in forests fail Monbiot's test because trading REDD credits allows the continued burning of fossil fuels. Hot air, then.

Notes

1   Stuart, M. (2009) 'REDD – The Basis of a 'Carbon Federal Reserve'?', *Cleantech Blog*, 8 May, http://bit.ly/11K627.

2   'Report of the Conference of the Parties on its thirteenth session, held in Bali from 3 to 15 December 2007 Addendum Part Two: Action taken by the Conference of the Parties at its thirteenth session', FCCC/CP/2007/6/Add.1*, 14 March 2008, http://bit.ly/4rh21.

3   *See*, for example, 'Protected Areas: Protected against whom?', Oilwatch and World Rainforest Movement, January 2004, http://bit.ly/V53db.

4   The UN definition of 'forests' does not differentiate between an old-growth rainforest and a monoculture industrial tree plantation. See REDD-Monitor for a discussion about the problems with the UN's definition, http://bit.ly/bO9Yw.

5   Tauli Corpuz, V. (2008) 'International Human Rights Day 2008: A sad day for Indigenous Peoples', statement by the Chair of the UN Permanent Forum on Indigenous Issues, 10 December, http://bit.ly/wcMay.

6   'Forest Carbon Partnership Facility Takes Aim at Deforestation', World Bank press release, 11 December 2007, http://bit.ly/4deUxV.

7   Colchester, M. (2009) 'Safeguarding Rights in the FCPF', presentation at the *Forests, Governance and Climate Change* meeting organised by the Rights and Resources Initiative and Chatham House at the Royal Society, London, 8 July, http://bit.ly/jEKw4.

8   Butzengeiger-Geyer, S., P. Castro, R. Harthan, D. Hayashi, S. Healy, K. Magnus, Maribu, A. Michaelowa, Y. Okubo, L. Schneider, and I. Storrø, (2009) 'Options for utilizing the CDM for global emission reductions', *Report to the German Federal Environment Agency* , 4 June, p. 3.

9   Harvey, F. (2009) 'EU carbon prices plummet as emissions continue to fall', *Financial Times*, 17 February, http://bit.ly/F6w8E.

10 Kevin Smith (2009) 'The Climate Camp vs the Carbon Market', *Transnational Institute*, 26 February, http://bit.ly/xo2cZ.

11 Kanter, J. (2007) 'In London's Financial World, Carbon Trading Is the New Big Thing', *New York Times*, 6 July, http://bit.ly/BWDlm.

12 Lang, C. (2008) 'Day one in Poznan: UN doesn't discuss REDD, Conservation International does', *REDD-Monitor*, 1 December, http://bit.ly/14yp68.

13 Lohmann, L. (2006) *Carbon Trading: A Critical Conversation on Climate Change, Privatisation and Power.* Uppsala: Dag Hammarskjold Foundation.

14 'Memorandum submitted by The Corner House' to the UK Parliament select committee on the role of carbon markets in preventing dangerous climate change, 1 March 2009, http://bit.ly/jbNTP.

15 Ball, J. (2008) 'Up In Smoke: Two Carbon-Market Millionaires Take a Hit as UN Clamps Down – EcoSecurities Sees Shares Slide 70 Per Cent', *Wall Street Journal*, 14 April, http://bit.ly/4GjTe.

16 'Memorandum submitted by The Corner House' to the UK Parliament select committee on the role of carbon markets in preventing dangerous climate change, 1 March 2009, http://bit.ly/jbNTP.

17 Goodell, J. (2006) 'Capital Pollution Solution?', *New York Times*, 30 July, http://bit.ly/6Aak8.

18 Specter, M. (2008) 'Big foot', *The New Yorker*, 25 February, http://bit.ly/LXUOR.

19 'Money grows on trees', *The Economist*, 6 June 2009, http://bit.ly/13fEgR. I wrote a summary of research carried out by two journalists about carbon trading in Papua New Guinea here: Lang, C. (2009) 'PNG update: Yasause suspended, dodgy carbon credits and carbon ripoffs', *REDD-Monitor*, 2 July 2009, http://bit.ly/13Rcuh.

20 Gridneff, I. (2009) '"Sample" documents blamed for PNG carbon deals', *Australian Associated Press*, 15 June, http://bit.ly/bIMvE.

21 Gridneff I. (2009) 'PNG climate office director suspended', *9 News*, 1 July, http://bit.ly/pvCEw.

22 Loder, N. (2009) 'The carbon rip-off', *Overmatter blog*, 11 June.

23 Monbiot, G. (2007) 'Hurrah! We're Going Backwards!', *The Guardian*, 17 December, http://bit.ly/tjcXJ.

24 According to an article in *Klima* magazine, Gore pockets US$300,000 for each of public speeches. Plus expenses. 'Wie sich der Klima-Guru mit alten Vorträgen immer aufs Neue die eigenen Taschen füllt, *Klima Magazin*, Nr. 01, 11 January 2009, http://bit.ly/fMb6b.

25 Tucker, S. (2004) 'Blood and Gore launch firm with a difference', *Financial Times*, 8 November, http://bit.ly/1asTpb.

26 'Green firm Camco gets Blood and Gore investment', *Reuters*, 4 June 2009, http://bit.ly/YiMsU.

27 'Key Developments For Camco International Limited', *Reuters*, accessed 20 June 2009, http://bit.ly/100bSI.

28 Twidale, S. (2009) 'Ken Newcombe', *Trading Carbon*, 3(3): 16.

29 Twidale, S. (2009) 'Ken Newcombe', p. 16.

30 'Major World Bank Programs: NGO participation', *World Bank Annual Report 1997*, http://bit.ly/OHjVg.

31 Lang, C. (2004) 'The carbon spin doctors: How the World Bank explains emissions trading to journalists', *World Rainforest Movement Bulletin*, 84, July, http://bit.ly/IZoAN.

32 Desai, P. (2006) 'Carbon Emissions Market Comes of Age', *Reuters*, 31 March, http://bit.ly/jdSKY.

33 Twidale, S. (2009) 'Ken Newcombe', p. 16.

34 Twidale, S. (2009) 'Ken Newcombe', p. 16.

35 'Forest Trends' History', Forest Trends website, http://bit.ly/u9QRS.

36 *See*, for example, 'Broken Promises: How World Bank Group policies and practice fail to protect forests and forest peoples rights', World Rainforest Movement, 2005, http://bit.ly/NCCsv.

37 'Board Members', Forest Trends website, accessed 20 June 2009, http://bit.ly/ZAlbV.

38 New Forests website, http://bit.ly/2SRmhc.

39 'Board Members', New Forests website, http://bit.ly/11OnR7.

40 Zwick, S. (2009) 'Environmentalists, Financiers Commemorate Decade of Forest Trends', *Ecosystem Marketplace*, 22 May, http://bit.ly/3SSdnw.

41 Lohmann, L. (2008) 'Climate Crisis: Social Science Crisis', a chapter for *Der Klimawandel: Sozialwissenschaftliche Perspektiven*, VS-Verlag, http://bit.ly/Tulpe.

42 'Emissions Trading', Barclays Capital website, http://bit.ly/dV9Nk.

43 'AWG-KP 8 Officers', UNFCCC website, http://bit.ly/NXc1o.

44 Bullard, N. (2008) 'Who is Harald?', *New Internationalist*, 412, June. http://bit.ly/3XQktN. *See* also, Lang, C. (2008) 'Pöyry: The economic hit men of the pulp industry', in *Plantations, Poverty and Power: Europe's role in the expansion of the pulp industry in the South*, World Rainforest Movement, December 2008, http://bit.ly/kOFPM.

45 Harvey, F. (2009) 'Carbon trading poised to decline', *Financial Times*, 24 February, http://bit.ly/28gUl. I made this comment in a draft article I wrote for Point Carbon's

magazine *Trading Carbon*. Perhaps not surprisingly the comment was not published in *Trading Carbon*. 'Your comments on Point Carbon may be valid in a wider context, but were irrelevant in a feature on why REDD shouldn't be in the carbon market. Those criticisms could be made of any number of companies in a wide range of fields', the editor of the magazine, Robin Lancaster explained (in an email dated 1 April 2009). Lancaster also deleted all reference to Al Gore, Ken Newcome and Richard Sandor from my article on the grounds that in the eyes of the law they haven't done anything wrong. I agree. I'm not accusing them of having broken any laws. Lancaster added that 'all of the edits and decisions on the piece were made by me as editor as is the case with every word of editorial in the magazine.' The same issue of *Trading Carbon* included a profile of Ken Newcombe written by Susanne Twidale.

46 Bellassen, V., R. Crassous, L. Dietzsch, and S. Schwartzman (2008) 'Reducing emissions from deforestation and degradation: what contribution from carbon markets?', *Mission Climat of Caisse des Dépôts*, Climate Report No 14, September.

47 Szabo, M. (2009) 'Carbon exchanges cashing in amid EU slowdown', *Reuters*, 17 February, http://bit.ly/xq6FF.

48 'PNG's Kevin Conrad in Bali: US, Get out of the Way!', CNN – YouTube, http://bit.ly/kLLHC.

49 Conrad, K. (2009) 'Moving to the environmental age', *Trading Carbon*, 3(3): 24.

50 The Coalition for Rainforest Nations' website lists the following countries as participants: Bangladesh, Belize, Central African Republic, Cameroon, Congo, Colombia, Costa Rica, DR Congo, Dominican Republic, Ecuador, Equatorial Guinea, El Salvador, Fiji, Gabon, Ghana, Guatemala, Guyana, Honduras, Indonesia, Kenya, Lesotho, Liberia, Madagascar, Malaysia, Nicaragua, Nigeria, Pakistan, Panama, Papua New Guinea, Paraguay, Peru, Samoa, Sierra Leone, Solomon Islands, Suriname, Thailand, Uruguay, Uganda,Vanuatu and Viet Nam, http://bit.ly/wXT9a.

51 'Initiatives: Carbon Emissions', Coalition for Rainforest Nations website, http://bit.ly/2jb6g.

52 Sessions, E. (2005) 'Using the Tools of Business to Inform Environmental Policy', *The Record*, Columbia University, 31(6), 28 November, http://bit.ly/guaIu.

53 'Highlighted Events: World Leaders Forum, Columbia University (Jan. 15, 2005)', Coalition for Rainforest Nations website, http://bit.ly/4CmSol.

54 'Climate Science: What Do We Know?', Global Roundtable on Climate Change Spring 2005, The Earth Institute at Columbia University, http://bit.ly/D4EwY. Somare's powerpoint presentation, *Rainforests and Climate Change*, is available here: http://bit.ly/9QM8T, and his speech is available here: http://bit.ly/ROxN9.

55 'Answers to questions without notice – from Member for Markham, Hon. Andrew Baing', Parliament House, Monday, 20 June 2005, http://bit.ly/lrUXo.

56 'About the Coalition', Coalition for Rainforest Nations website. http://bit.ly/18uG9b.

57 'Geoffrey Heal', Coalition for Rainforest Nations website, http://bit.ly/FycP7.

58 The Union of Concerned Scientists website states that Heal is a Director of Petromin: http://bit.ly/lSvr2. He is not included in a list of Directors on the Petromin website: http://bit.ly/eBUQF. However, he is listed as a Director on an undated 'Information Brochure for the Media' on Petromin's website: http://bit.ly/OdCtc. Heal's Curriculum Vitae, dated February 2008 states 'Director, Petromin Holdings PNG Ltd. (The national oil, gas and mineral company of Papua New Guinea) 2007 on' http://bit.ly/BHZ8c.

59 'MPs in uproar over Petromin', Post Courier, 1 March 2007, http://bit.ly/3il8fQ.

60 Gridnef, I. (2009) 'Climate hero under fire in PNG', *The Age*, 8 May, http://bit.ly/5Spmi.

61 Gridnef, I. (2009) 'Climate hero under fire in PNG'.

62 Statement by Sir Michael T. Somare, GCMG KSt.J CH, Prime Minister of Papua New Guinea, Global Roundtable on Climate Change, Columbia University, New York, 12 May 2005, http://bit.ly/ROxN9.

63 'Context: Recognizing Forests' Role in Climate Change', Coalition for Rainforest Nations website, http://bit.ly/zBtLB.

64 Parekh, P. (2009) 'Waxman-Markey Bill: No Cuts until 2026!', International Rivers website, 15 April, http://bit.ly/YuOTO.

65 'Analysis of the Waxman-Markey Draft', International Rivers and Rainforest Action Network, 15 April 2009, http://bit.ly/18VGzX.

66 'Initiatives: Sustainable Forest Markets', Coalition for Rainforest Nations website, http://bit.ly/LuQlo.

67 Copenhagen Climate Council website, http://bit.ly/O9Lg9.

68 'The Business Case for a Strong Global Deal', prepared for the World Business Summit on Climate Change by the ClimateWorks Foundation, 2009, http://bit.ly/1nZgwV.

69 'The Business Case for a Strong Global Deal'.

70 Reyes, O. (2009) 'Carbon trading and cash values on forests cannot curb carbon emissions', *The Guardian*, 28 May, http://bit.ly/UTDRl.

71 'Debate – Carbon Offsetting', Together Works, 11 June 2007, http://bit.ly/FlE3V.

72 'Vested interests - Industrial logging and carbon in tropical forests', Global Witness, 4 June 2009, http://bit.ly/1fxKAJ.

73 'Deforestation: 'Genuine risk of biggest ever loophole', *The Guardian* website, 25 November 2008, http://bit.ly/18Rk4U.

74 'REDD and the effort to limit global warming to 2°C: Implications for including REDD credits in the international carbon market', prepared for Greenpeace by KEA3, 30 March 2009, http://bit.ly/edtz4.

75 Karsenty, A. (2008) 'The architecture of proposed REDD schemes after Bali: facing critical choices', *International Forestry Review*, 10(3), http://bit.ly/TfQD7.

76 Monbiot, G. (2009) 'How Much Should We Leave in the Ground?', *The Guardian*, 6 May, http://bit.ly/uNBMY.

21

Hegemony and Climate Justice: A Critical Analysis*

Vito De Lucia

Introduction

Justice has over the years become a key factor to consider in any climate negotiation. Usually framed in terms of distribution of the burdens of climate mitigation and adaptation,[1] its foundational principle in the current climate regime is that of the principle of common but differentiated responsibility.

In the last 15 years, the ethical challenge of climate change has been sharpened by the increasing severity of both projected and occurring impacts of climatic changes, and by the increasing awareness that the distribution of the damaging impacts is inversely proportional to the causative sources of the climate forcing gases.

At the Bali Climate Conference held in December 2007 a new negotiation platform[2] set the stage for a two-year negotiation efforts aimed at landing a post-Kyoto climate agreement in 2009, at the Copenhagen Conference, the hope reinvigorated as the USA re-joined the negotiation. The 2008 Poznań Conference however, has been by most accounts a disappointment, leaving the task of preparing a draft negotiating text wholly to the preparatory meetings leading up to the Copenhagen Conference.

In the meantime, two 'events' can be assigned particular significance: the election of Barak Obama as the new President of the USA, and the emergence – in Bali and then in Poznań – of the Climate Justice Now! Coalition.[3] The former gives hope of a new, progressive and climate-friendly policy orientation of the American administration, breaking decisively with the previous Bush approach, and finally putting climate mitigation on the policy agenda, both domestically and, most importantly, as a matter of foreign policy. The second 'event' represents, on the other hand, a radical, grassroots movement whose aim is to expose 'false solutions' to the climate crisis and the green washing of the climate regime,[4] and which, most importantly, insists that justice is a fundamental issue to be addressed in any climate negotiation.

The language, the discourse of justice is however embraced so widely that the question must be asked: what is climate justice? The discourse of justice brings together politics, business and civil society, a convergence which is in this article postulated to occur through a dialectical debate which oscillates between extreme neoliberal, market radicalism and a liberal, embracing, cosmopolitan articulation of equity and climate justice. In this context, justice becomes an ideological tool aimed at winning the consent of a wide social base, necessary for the renegotiation and (re)solidification of consent and hegemony, aimed at the reconfiguration of capitalism under conditions of ecological/climate crisis.

This article aims at making visible the assumptions underlying the framing of climate justice within what we will call the United Nations Climate Regime[5] (UNCR) and civil society at large. This task of deconstruction is important to the extent that climate justice is necessary for the survival of all the species including the human one. As such this article intends to be a contribution within current debates.

Mainstream Articulations of Climate Justice: (Re-)Distribution, Compensation, (right to) Development

Let us review some central contribution towards the shaping of climate justice, from academia, politics, and civil society. Already Henry Shue[6] had begun framing the question of climate justice in terms of distribution of costs and benefits. The UNFCCC and the Kyoto Protocol also centered the question of ethics on responsibility and distribution of costs and benefits.[7] Distribution is likewise the main focus of major equitable frameworks such as Contraction and Convergence[8] and Greenhouse Development Rights.[9] While the former – endorsed widely – promotes a per-capita egalitarian approach supported by technology and emissions trade flows, the latter's main contribution is the explicit incorporation of the right to development into the equation. The Buenos Aires Declaration on the Ethical Dimension of Climate Change (BADEDCC), a major attempt a laying out a comprehensive articulation of ethics in relation to climate change, proposed as key ethical questions responsibility for damages, distribution of harm and benefits, allocation of emissions budgets across countries, economic costs,[10] technology.

Kofi Annan's Global Humanitarian Forum (GHF) launched recently a Global Alliance for Climate Justice. The main priority areas are identified in financial transfers and transfer of sustainable technologies. Which is to say, (re-) distribution and (right to) development.

And if justice is framed in terms of (re-)distribution, it follows as a precondition the necessity to measure emissions and sinks, costs and benefits, by way of mapping a whole series of acts, meanings, sentiments, species and ecosystems onto monetary expression11. This regardless of how difficult or arbitrary the mapping turns out to be. Money (and monetary/monetized instrumentality) becomes the fundamental rationalizing ground for action.

This presupposition of commensurability, moreover, leads to two important consequences. Firstly, market exchanges become universally possible, hence carbon trading (in all its forms) as the 'flagship' climate policy of UNCR. Secondly, substitutions can take place. Substitution of man-made capital for nature (and 'natural capital'); of (equitable) monetary compensation for lack/loss of access to local ecological resources and means of subsistence;[12] of monoculture carbon plantations for native forests;[13] of development for livelihood, as in the construction of 'need' and 'rights', and its counterpart, dependency.[14]

Development in particular deserves examination at some length. By mapping the world over a rich-poor continuum, measured in a very specific, culturally narrow way, development becomes an obligation rooted in the universal/izing discourse of justice and human rights, which must be 'distributed' equally.

This discourse is built on the assumptions of industrial progress, which can raise the standard of living in the 'underdeveloped areas.'[15] In the context of climate change, (this) development is framed as the only adaptive path for countries, populations and communities which will feel the brunt of climatic changes. At the same time, development *causes* climate change, as economic growth – the underpinning engine of development – *grows* GHGs emissions. Development is thusly inextricably linked to technology: technology can make development climate-friendly and low-carbon.

Technology however has a distinct 'ideological footprint', as it is inevitably linked with specific power/knowledge configurations: its underlying social power relations will re-produce themselves inevitably with their diffusion, and will re-produce as well their main features of metabolizing 'time and space',[16] inevitably implicating both extraction and accumulation of ecological and social value.[17] In this light, 'transfer of technology' is not a neutral operation. Lohmann reinforces this point when he suggests that the practices of development necessarily 'ignore, displace, supplant or even eradicate knowledge possessed by their "target populations"'. [18] This displaced, supplanted knowledge often represents key social strategies for addressing the local effects of climate change.

As distribution – of costs, benefits and emissions rights – underlies mainstream articulations of climate justice, it follows that it's the market which can best and most efficiently operationalize this distributional justice, enabling finance and technology to rescue the heating planet in a just and equitable way, and the circulation of emissions rights to their 'highest and best use' through the emerging global carbon market.

Climate, Hegemony and Justice: A (neo-)Gramscian reading

Gramsci and (neo-)Gramscian critical theory offers a useful lens to interpret the UNCR and civil society's conceptions of climate justice.

Gramsci[19] articulated his idea of hegemony as the supremacy of a social group predicated on both coercion and consent. As it is the element of consent which for Gramsci allows a hegemonic social group to endure, social control must be built on intellectual and moral leadership. Hegemony is thus 'endlessly reinforced in schools, churches, institutions, scholarly exchanges, museums and popular culture.'[20] The spontaneous consent of 'the great masses' is for Gramsci historically caused through the workings of intellectuals and civil society. A key element sustaining hegemony is the concept of historic bloc, which represents the relationship established by the dominant social forces with antagonistic ones. This relationship produces an integration of different class interests, engendering a convergence of economic and political objectives, a convergence which is also, importantly, intellectual and moral.

There are three mechanisms required to establish a ruling world view: universalization, naturalization and rationalization. The first represents the projection of a historically situated and local project as universal. This can take the form of political alliances or cultural dissemination. Naturalization entails a process of reification of a given situation, abstracting it from its historical contingency, and containing any social demands necessarily within this constructed social ontology. Rationalization, finally, refers to the supporting role of an intellectual class, which produces and re-produces knowledge in order to maintain hegemony: 'theory is always for someone and for some purpose.'[21] Gramsci calls this intellectual class 'organic intellectuals'.

Organic Intellectuals and the Mythos of Science[22]

The term *mythos*, whose original meaning is 'utterance', indicates both statement and story, eluding the distinction between objective and discursive knowledge. One of the function of *mythos* is that of narratives, storing and sanctioning rituals with functional purpose. Pantheistic religions, through their associating each 'god' (a river, a mountain, a field, agriculture etc.) to specific rituals governing behavior (both social and individual) represent embedded norms of social conduct and of technical behaviour.[23] *Mythos* is also, importantly, embodied in a specific place and culture, coordinating and containing social action through its patterns of *nomos* and *ethos*. While *nomos* means either custom, convention or (positive) law, *ethos* can be rendered as the 'disposition, character, or fundamental values peculiar to a specific person, people, culture, or movement.'[24] Their etymological roots clearly indicate the relation between habit/values and place. By articulating possibilities and proprieties of socio-technical configurations, these conventions mold the natures and the cultures in particular ways. Different *ethe*[25] map to different sets of dispositions and values, and through their localization they also, significantly, express different relational engagements with particular natures/ecologies and different patterns of *nomos*, which is to say, customs and norms. Traditionally localized *ethe* function mostly through ecological exchanges (with nature) rather than economic exchanges (with markets), which determines a necessary harmonization of cultural and ecological times and rhythms, in order to 'guarantee an uninterrupted flow of goods, materials, and energy from ecosystems.'[26]

The emergence of the specific *mythos* of modern science has determined a separation of *nomos* from *ethos*. This has rather important effects, which underlie and sustain the claims for objective and universal/izing value: by separating the two elements, modern science has obscured the significance of locality and particularity, while providing the intellectual mechanism to support universal laws. However, modern science is *one* historically given mode of knowledge, which is particular to a specific culture, and which incorporates *within it* a similarly contingent set of assumptions and values.

Let us take as an example the precautionary threshold that emerges from the reports of the IPCC. The establishment of an acceptable threshold is controversial. It is the fruit of a compromise which considers already 'committed' warming, mitigation possibilities under current circumstances, mathematical modelling through which projections of future change are tested and predicted, economic trade-offs and analyses of costs and benefits.

Donald Brown[27] reminds us that the composition of the IPCC is very much skewed towards Western scientists and the Western knowledge system. The IPCC's work is based on the input of 'a narrow elite', expression of those societies which will be impacted the least by climate change and of an epistemic community largely comprised of scientists and technicians from North. Brown continues by remarking how '[t]he voices of the sufferers – people living in climate change hot spots, indigenous nations, children, disenfranchised – are not included in the assessment reports and seldom reviewed for inclusion in the work of the IPCC', and '[d]ecision on final synthesis reports, including line-by-line review of text, is made by government representatives'. The knowledge of the victims is discounted, displaced and delegitimized through the processes of production of ignorance Lohmann[28] refers to. 'The reports' concludes Brown 'are vetted by a narrow group of experts trained and privileged by larger structures of globalization'.

'Organic intellectuals' contribute significantly to the production and re-production of a specific power-knowledge configuration instrumental to the hegemonic project. It is indeed through the IPCC reports that the mitigation debate is framed in terms of technology, carbon markets and efficiency.[29] The pre-analytic vision or world view of most IPCC contributors and authors is such that those policy recommendations are seen as inevitable. Rationalization thusly leads to naturalization.

The 'Climate Ethics Consensus' and the Negotiation of Hegemony: Convergence of UNCR and Civil Society

Robert Cox distinguishes in this respect between problem-solving – which 'takes the world [...with its] prevailing social and political relations and [...] institutions [...] as the given framework for action' – and a critical, counter-hegemonic approach, which calls into question these institutions and social and power relations, aiming at decentering that very framework of action.[30]

The debate over current climate policy is by and large shaped and conducted on a problem-solving agenda,[31] with carbon markets as the centerpiece, as we shall see through a cursory review of the climate policy approach of 6 major civil society organizations (CSOs): they legitimate the UNCR's paradigm, although dialectically posing demands on it.

- *WWF International* is one of three core founders[32] of the Gold Standard, a private 'certification' whose goal is to guarantee the environmental integrity of carbon offsets available for purchase in both the CDM and the private carbon markets.[33]

- *Environmental Defense Fund (EDF)* is a major supporter of current US cap-and-trade legislation:[34] 'Our top priority is to pass national legislation that caps global warming pollution and creates a flexible emissions trading market'.[35]

- *National Resources Defense Council* also sees the carbon market as a solution, both in its domestic cap-and-trade policy form, and in offsets mechanisms.[36]

- *Climate Action Network* (CAN), a worldwide network of more than 450 NGOs, works 'to promote government and individual action to limit human-induced climate change to ecologically sustainable levels,'[37] by protecting the 'atmosphere while allowing for sustainable and equitable development worldwide'. CAN – which is a major 'sparring partner' of UNCR – endorses a three track approach.[38] The Kyoto Track: capping and pricing carbon to provide incentives to market agents. The Greening Track: markets/price incentives can funnel financial flows towards low-carbon technological development. The Adaptation Track: a corrective measure of re-distributional finance for adaptation.

- *Framtiden i våre hender*[39] is a major Norwegian Environment and Development organization. It is actively engaged in the promotion of carbon offsets, through the management of a localized version of 'My Climate'.[40]

- *Greenpeace International,* in its very recent submission[41] to the UNFCCC on the matter of the role of the CDM in the second commitment period of the Kyoto Protocol, demands sharply more strict rules to ensure the CDM's environmental integrity. However, Greenpeace still appreciate CDM's value in a future international climate regime. Greenpeace is also a supporting member of the Gold Standard.

While there are some – even significant – differences, all reviewed CSOs can be placed along the same continuum, mirroring UNCR's policy platform. The economics of pollution are the center of the UNCR's policy approach. The Polluter Pays principle links the legitimacy of the release of waste streams (or pollution damages) to one or another form of payment for the use of the waste recycling capacity of the local/global environment, following a Coasian, (carbon) market logic.[42] The 2001 and 2007 reports of IPCC's Working Group III offer ample evidence on the matter, and so does the UNCR. In the Kyoto Protocol the key policy instruments – the so-called flexibility mechanisms – are three market-based instruments: Emissions Trading (ET); Joint Implementation (JI); Clean Development Mechanism (CDM).

Against this background, civil society, while making demands on States (politics) with the aim of advancing a progressive vision of climate protection and social justice, use nonetheless the categories and methods of UNCR (economics), championing a 'more and better' approach: more stringent emissions caps, more energy efficiency, better designed carbon markets, more financing and investment in adaptation and sustainable development etc.

The integration of UNCR and civil society into what can be described as a 'Climate Ethics Consensus' (CEC) takes place through a dialectic that constitute and reproduce the hegemonic historic bloc: within it, consent is created, shaped, negotiated and maintained through the acceptance and internalization of a set of values and world-views that reinforce established power relations. While UNCR reflects the internationalization – and trasnationalization – of the (aligned) material interests of the various domestic hegemonic groups, civil society is the vehicle through which the ethics and the conception of justice promoted by the hegemonic group(s) is legitimized. In this manner, the world-view of the hegemonic social group is reinforced and validated, while protest and 'resistance' is transformed and incorporated within the hegemonic discourse, serving to further its internalization and naturalization by the 'great masses'.

Civil society becomes then a *key part* of the forming historic bloc of a 'sustainable capitalism', by either 'endlessly reinforcing' hegemony, or through the process of *trasformismo*. *Trasformismo* is for Gramsci a 'strategy of assimilating or domesticating potentially dangerous ideas',[43] and the groups and organization which promotes them. This process works towards the integration and incorporation of those ideas and groups, drawing them within the paradigm of the dominant social group.[44]

Trasformismo through Justice

Gramsci viewed justice, legitimacy and moral credibility as necessarily integrated. The UNCR, the organic intellectuals and civil society, when aligned in the CEC, constitute and legitimize that climate governance whose practices, Paterson argues, 'should be understood as a pursuit of' a coherence between accumulation and legitimacy within the context of 'an ecological regime of accumulation thus forestalling more radical critiques arguing that capitalism and sustainability are inimical'.[45]

In this respect justice becomes a tool of hegemonic groups towards the coordination of dispersed values into an ideological 'whole' supportive of their position of dominance. To this purpose, the perception of justice, and the dialectical processes whereby the 'great masses' demand (and obtain) 'more' justice,[46] are to be maintained within specific boundaries. This task is accomplished through civil society,[47] and its participation in the shaping of the UNCR. To further illustrate, an article that appeared in The Guardian[48] reported how, during the climate meeting in Accra in 2008, justice groups protested against forest carbon trading, because forest credits schemes could undermine the world price of carbon, damaging the effectiveness of the market. The arguments distinguished then between a just and an unjust market, where just maps to efficient: including forests in the carbon market could 'crash the price of carbon' and reduction of pollution in rich countries would become uneconomic: prices and markets as THE solution to climate change. Greenpeace reiterates this point in one of its submissions to the UNFCCC:[49]

> Inclusion of LULUCF and REDD activities in mechanisms generating offset credits...has the potential to Flood the carbon market with cheap credits, which in turn have the potential to significantly lower the global price of carbon and thus undermine ambitious emissions reduction targets for industrialized countries by reducing the incentive to invest in low carbon infrastructure.

CAN makes a similar distinction:

> If emissions trading is designed well it could help us substantially reduce our greenhouse pollution. If it is designed badly it could be an elaborate way to disguise a lack of action and transfer wealth to polluters.[50]

Civil society becomes thus both the 'object and the medium' of the hegemonic struggle.[51] The outcome of these dialectical processes is dynamically captured by 'soft' declarations and political statements. It then 'trickles' slowly and in diluted form into the hard rules of the UNCR. This gives a sense of participation to environmental and social movements and organizations. Their demands are watered down and re-oriented so that discontent is absorbed and kept within the framework of action,[52] providing the hegemonic social group with a mechanism to manage the demands of 'dissent' and to 'transform' potential resistance: by adhering to 'some' of the demands in some diluted form, it draws these groups within its bloc. Once integrated and transformed, civil society can become an engine of hegemony. At the same time this same process isolates the more radical antagonizing elements of potential counter-hegemony, by framing their existence outside of 'common sense'.

The emphasis on (distributional) justice and the right to development as key elements of any post-Kyoto agreement has then this effect of transforming dissenting sections of public opinion and developing countries into 'supporters' of the global capitalist vision and ideology of the dominant social group. Justice turns then into a fundamental space of ideological negotiation, where hegemony is nurtured, articulated and universal/ized. Its articulation is founded on the instrumental role that justice is to play as regards cementing the historic bloc, and its function of 'coordination of the interests of other groups with those of the leading class or fraction in the process of securing their participation in [their] social vision.'[53]

Conclusion: Towards a Climate Justice as Equity

To conclude, some remarks on future directions towards the delineation of an alternative (counter-hegemonic) climate justice. Any even provisional research agenda in this respect will need to address three crucial points. Firstly, a historical and comparative narrative of justice. This is instrumental to highlight the transition from oral to literate justice, which also maps to a transition from a local, customary and necessarily plural justice(s) towards a legalized, universal/izable singular one. The ideal, fixed, abstract, universal conception of justice, was shaped by the fixity and 'removed' properties of writing, which have historically facilitated analytical thinking, the 'objectification' of human knowledge, and its being eradicated – disembedded – from the flow of human experiences.[54]

In the fluid oral world by contrast, what one must relate to is not a singular justice, but rather justices, in the plural. Justices do not reflect an a priori set of principles, but are processes whose aim is that of conserving existing mores or restoring the propriety of the relationships within the community.[55]

The second point to address is the conceptual transition from a Universal Ethics[56] (back) to local *ethe*/equities. This can be accomplished partly by rejoining *nomos* and *ethos,* and will help towards (re-)aligning justice with people's natures and cultures. In this sense, it may be useful to mend the rip – between justice and equity: in Plato[57] equity becomes the mitigating element of 'people's justice', and justice begins to assume a technical, legal, top-down character. Equity has however opposed the universal nature of the Law with the historicity of actual facts, thusly operating to constrain the disembedding trajectory of law, re-locating, quite literally, justice in its contingent historicity.[58]

Finally, a third point to address is how Universal/izing Justice is a social construction. The Greek philosophers known as the sophists already expressed a critical, 'constructivist' view of justice: relative, contingent and linked to prevailing interests. Protagoras' famous saying that man is the measure of all things[59] is a deep commitment to criticizing any instances of 'a view from nowhere', any essentialist, universal/izing, transcendental, objective configurations of Man, virtues, values and reality. An anticipation of Sandel's critique of liberal justice and of the 'unencumbered man'[60] which it presupposes.

This relation of justice to contingency and circumstance[61] (both of place, time and culture) is inextricably linked to equity, so that the severance of the link between to two has operated functionally towards allowing the disembedding of justice from, ultimately, people. At the same time Justice enters the realm of Law. As such, universal and 'juridified', it can become a 'mode of hegemony', an instrument of the dominant social group, either towards forming/re-forming an historic bloc, or towards the maintenance of its stability.

We have seen how both *nomos* and *ethos,* given their fundamental linkages with place, and by way of articulating possibilities and proprieties of socio-technical configurations, encompass *both* natures and cultures in particular ways. Latour[62] reinforces this point, by submitting that '[f]or each Society there exists a corresponding state of Nature'.[63] Moreover, traditional/indigenous societies/*ethe* possess a 'unified' vision of nature-cultures: 'it is the impossibility of changing the social order without modifying the natural order – and vice versa – that has obliged the premoderns[64] to exercise the greatest prudence.'[65] It is significant to note how the situatedness of *ethos* within a specific 'socio-ecological place' provides solid grounding for an embodiment of justice which applies to the whole nature-culture. Further, Toledo shows the clear and solid linkage between cultural and biological diversity, centered around indigenous nature-cultures. Yet these local, indigenous knowledges and the related articulations of localized nature-cultural justice(s) are threatened by the globalizing Justice of hegemonic UNCR: universal/izing Ethics (Justice) threatens local *ethe.*

Climate change, while global as a scientific and economic problem of accumulation,[66] is local in many of his socio-ecological effects. This global-local cleavage allows 'global benefits' of GHGs reduction – and of 'climate and development financing'[67] – to stem from 'substitutions'. Projects such as large dams, or large carbon-absorbing plantations of alien fast-growing tree species[68] substitute/compensate for the destruction of *local* livelihoods consequence of land expropriation and displacements.[69] So long as they generate credits expendable in the carbon markets.

Indeed climate policy, being predominantly abstract and global, destroys time and again local justice(s). Recognition that extraction and accumulation draw resources – both human and natural, both time and space – towards the core, leaving peripheral, marginal places in 'poverty', with a loss of geographical, cultural, economic, social and political diversity, becomes then crucial. (Re-)distributing (some of) the benefits of this time/space appropriation only increases the gap between Justice and justice(s). When monetary distribution/compensation predominates, peripheral places are drawn within the global capitalist flows, losing the ability to cope independently. This incorporation of nature-cultures within global markets and the global circuit of capital is a natural consequence of the 'enforcement' of universal/izing justice, which penetrates through financial flows, the right to development, and its inevitable technological dimension – with the implications illustrated above. The UNCR provides the legal and technical means, particularly through carbon markets. Justice to the ethical justification.

This global/izing dimension of UNCR's climate policy enhances and furthers that process of socio-ecological disembedding identified by Polanyi (2001) as one of the crucial elements of the 'great transformation' spawned by the rise of industrial capitalism. Caroline Merchant[70] describes in detail the same process of disembedding which took place in England with the enclosure of 'farm, fen and forest', and the effects on the environment of the transition 'from peasant control for the purpose of subsistence to capitalism control for the purpose of profit'. A transition instrumental to the emergence of a mechanistic view of nature,[71] and mapping onto a similar transition from local justice(s) to universal/izing Justice.

Paraphrasing Ivan Illich's[72] distinction between universal peace and people's peace, we can say that a universal 'market' justice tends to make cultures alike whereas [justice] is that condition under which each culture flowers in its own incomparable way. Justice cannot be exported: its attempted export means war and poverty. It follows that any articulation of justice which is abstracted from its local, embodied context has a disruptive, 'belligerent' potential. Critical resistance groups – such as La Via Campesina or the Climate Justice Now! Coalition – counter this abstraction by opposing 'food sovereignty'[73] to 'food security', energy sovereignty to industrial energy production and distribution. This fundamentally implies people's control over their means of subsistence, and rejects the professional satisfaction of needs through global markets, which 'naturalizes' the global capitalist industry, and its production, distribution and exchange methods, processes and social power relations: indeed its very vision of man, unencumbered, self-interested, alienated. Recalling Waltzer, 'every substantive account of...justice is a local account... Justice is rooted in the distinct understandings of places, honors...things of all sorts, that constitute a shared way of life. To override those understandings is (always) to act unjustly'.[74]

In the end, climate justice(s) – besides and beyond identifying historical responsibility and redressing, financially, historical and present wrongs[75] – must open towards people's sovereignty, autonomy, self-coping, in a wider perspective where people's equity fosters sovereign ecological cultures.

Notes

* Acknowledgment: I am greatly thankful to Ruth Thomas-Pellicer, for her support, for patiently reading through previous drafts and for offering tremendous help for improvements.

1 And to an extent to the 'moral' responsibility of rich countries to act urgently to prevent adverse effects of climatic changes. For the economy of this article, I will emphasize on the distributional focus.

2 The Ad Hoc Working Group on Long-term Cooperative Action Under the Convention.

3 'What's missing from the climate talks? Justice!' Joint Press Release, 14 December 2007.

4 Focusing sharply, for example, on critiques of market instruments such as carbon trading.

5 Which is comprised primarily of the climate regime proper (UNFCCC, the Kyoto Protocol and related organs and bodies), but also of other UN agencies and institutions such as UNEP, UNDP, other international organizations such as the World Bank etc.

6 The four questions are: 1) What is a fair allocation of the costs of preventing the global warming that is still avoidable? 2) What is a fair allocation of the costs of coping with the social consequences of the global warming that will not in fact be avoided?3) What background allocation of wealth would allow international bargaining (about issues 1 and 2) to be a fair process? 4) What is a fair allocation of emissions of greenhouse gases (over the long-term and during the transition to the long-term allocation)? See Shue, H. (1993) 'Subsistence Emissions and Luxury Emissions', *Law & Policy*.

7 In particular through the principle of common but differentiated responsibility, which has been operationalized in terms of differential commitments, differentiated contributions to various climate mitigation and adaptation funds etc. See, inter alia, De Lucia, V., (2009) 'Common but Differentiated Responsibility for the Global Environment', in B. Mukherjee, R. Ray and S. K. Basu (eds) (2009) *Environment - Gathering Crises*. Sikha Books, Kolkata, India.

8 See Global Commons Institute (1996) 'Draft Proposals for a Climate Change Protocol based on Contraction and Convergence: A Contribution to Framework Convention on Climate Change', *Ad Hoc Group on the Berlin Mandate*, 6th September 1996 AGBM/1.9.96/14, Global Commons Institute (2001) References for Contraction and Convergence, 11 August, http://www.gci.org.uk/refs/C&CUNEPIIIg.pdf and Meyer, A. (2004) 'Briefing: Contraction and Convergence, Engineering Sustainability', 157(4): 189-192.

9 Baer, P., T. Athanasiou and S. Kartha, (2007) 'The Right to Development in a Climate Constrained World The Greenhouse Development Rights Framework', *Heinrich Böll Stiftung*, publication series on ecology, Volume 1.

10 To which extent they can be utilized to avoid climate action.

11 Mapping which in neoclassical environmental economics goes under various names/practices: contingent valuation, willingness to pay, hedonic pricing, value of a statistical life, cost-benefit analysis etc.

12 It is often the case that environmental and development activists and NGOs, when protesting against corporate and/or State 'attacks' on the environment or on local populations, condemn lack of compensation. But as the People's Coalition on Food Sovereignty (http://www.foodsov.org/html/takeaction06.htm) reported about evictions occurring in the State of West Bengal, India in 2006, 'The communities do not want compensation. They want to retain their land, because this is their life'.

13 See for example World Rainforest Movement (2001) *The Bitter Fruit of Oil Palm: Dispossession and Deforestation*, http://www.wrm.org.uy/plantations/material/OilPalm.pdf and

CDMWatch/ SinksWatch (2004) *How Plantar sinks the World Bank's rhetoric: Tree Plantations and the World Bank's sinks agenda,* June 2004.

14 See Illich, I. (1990) 'Needs', unpublished manuscript available at http://www.davidtinapple.com/illich/1990_needs.PDF accessed on October, 10 2008.

15 Development was already framed in these terms in the Inaugural Address of Harry Truman in 1949.

16 Extending and expanding the commodification of labor (man) and nature highlighted by Polanyi as the final step in the process leading to the great transformation. See Polanyi, K. (2001) *The Great Transformation: the Political and Economic Origins of our Time.* Boston: Beacon Press.

17 On this see Hornborg, A. (2001) *The Power of the Machines: Global Inequalities of Economy, Technology and Environment.* Altamira Press.

18 Lohmann, L. (2008) 'Carbon Trading, Climate Justice and the production of Ignorance', *Development,* 51: 359–365.

19 In general see Gramsci, A. (2007) *Selections from the Prison Notebooks.* London: Lawrence and Wishart; and Gramsci, A. (2007) *Quaderni del Carcere, Edizione Critica dell'Istituto Gramsci,* a cura di Valentino Gerratana, Einaudi.

20 Litowitz, D. (2000) 'Gramsci, hegemony, and the law', *Brigham Young University Law Review,* 2: 515–551.

21 Cox, R. W. (1981) 'Social Forces, States and World Orders', Millennium: *Journal of International Studies,* 10(2): 126–55.

22 In this section I am drawing particularly on Havelock, J. (1978) *The Greek Concept of Justice: from its Shadow in Homer to its Substance in Plato.* Harvard University Press; Burke, K. (1969) *A Grammar of Motives.* University of California Press; and Ong, W. J. (2002) *Orality and Literacy.* Routledge.

23 Functioning, for example, so as to facilitate activities such as seeding after a river flooding the fields, harvesting before night frost would set in etc.

24 ethos. Dictionary.com. *The American Heritage® Dictionary of the English Language, Fourth Edition.* Houghton Mifflin Company, 2004, http://dictionary.reference.com/browse/ethos (accessed: September 15, 2008).

25 *Ethe* is the plural forms of *ethos.*

26 See Toledo,V. M. (1999) 'Indigenous Peoples and Biodiversity', paper presented at the *Congrés de Biodiversitat,* Institut d'Estudis Andorrans.

27 The whole paragraph draws on Brown, D. (2008) 'Procedural Justice and the Work of the IPCC', *Climate Ethics,* http://climateethics.org/?p=32 , posted Tuesday, March 4th, 2008 at 12:36 am.

28 Lohmann, L. (2008) 'Carbon Trading, Climate Justice and the production of Ignorance'.

29 Though the IPCC reports 'only' mirror the predominant scientific production. In particular I refer to the policy dimension, hence the reports of IPCC's Working Group III, on mitigation.

30 Cox, R. W. (1981) 'Social Forces, States and World Orders'.

31 For example: is the best policy option carbon trading or is it carbon taxes? Should improvements in energy efficiency be mandated or incrementally achieved through price incentives? Should technology be funded by the State or through private investments?

32 The full list of members of the Gold Standard includes, to date, 60 NGOs.

33 This certification (as many others that exist) was a response to the mounting evidence of frauds, ecological damages and social misery consequent a large number of carbon offset projects. To reassure carbon consumers the Gold Standard was developed. See http://www.cdmgoldstandard.org/.

34 Either in its Waxman-Markey form or in its Senate version, the Boxer-Kerry bill.

35 In the words of Steve Cochran, director of our national climate campaign. See http://www.edf.org/page.cfm?tagID=337.

36 See NRDC's CAP 2.0 Program at http://www.nrdc.org/globalWarming/cap2.0/files/uncapped.pdf.

37 See http://www.climatenetwork.org/about-can.

38 See http://www.climatenetwork.org/about-can/three-track-approach.

39 The Future in our Hands.

40 Originally a Swiss organization. See http://www.myclimate.org._Private consumers are encouraged and enabled to calculate their carbon footprint, and then purchase emissions offsets through a series of projects in developing countries, which are supposed to have a 'double dividend': climate mitigation and sustainable development. For cogent critiques of the private offset markets (as well as the CDM), see, inter alia, Bachram , H. (2004) 'Climate Fraud and Carbon Colonialism: The New Trade in Greenhouse Gases', *Capitalism Nature Socialism*, 15(4); Lohmann, L. (2006) *Carbon Trading: a critical conversation on climate change, privatization and power*, and Haya, B. (2007) 'Failed Mechanism: How the CDM is subsidizing hydro developers and harming the Kyoto Protocol', report for *International Rivers*, October 2007.

41 See http://unfccc.int/resource/docs/2008/smsn/ngo/043.pdf and http://unfccc.int/resource/docs/2009/smsn/ngo/135.pdf.

42 Fueling the ongoing processes of commodification/marketization: of the air, through pollution trading, of the land and forests through REDD schemes, CDM, and possibly biochar policies and the so called 'Payment for Environmental Services' in the near future.

43 Gill, S. (1993) 'Epistemology, ontology and the 'Italian school'', in Stephen Gill (ed.), *Gramsci. Historical Materialism ami International Relations*. Cambridge.

44 See Gill, S. (1993) 'Epistemology, ontology and the 'Italian school''; and Gramsci, A. (1972).

45 Paterson, M. (2007) 'Climate governance and the legitimation of a finance-led regime of accumulation', *Paper for conference on Pathways to Legitimacy? The Future of Global and Regional Governance*, Centre for the Study of Globalisation and Regionalisation, University of Warwick, September 2007.

46 More climate justice in this context.

47 As well as through organic intellectuals.

48 Vidal, J. (2008) 'Clash over plan to save tropical forests', 21 August, http://www.guardian.co.uk/environment/2008/aug/21/forests.carbonoffsetprojects/print.

49 See http://unfccc.int/resource/docs/2009/smsn/ngo/135.pdf.

50 See http://www.climatenetwork.org/climate-change-basics/mechanisms-for-achieving-kyoto-targets .

51 Haug (1985) as quoted in Brand, U. (2007) 'The Internationationalization of the State as the Reconstitution of Hegemony', *IPW Working Papers* No. 1/2007.

52 I.e. UNCR's hegemonic proposition.

53 Robinson, W. I. and J. Harris (2000) 'Towards A Global Ruling Class? Globalization and the Transnational Capitalist Class', *Science & Society*, 64(1): 11–5411.

54 In this section I am rely heavily on Havelock (1978); and Ong (2002).

55 The Greek word for justice, *dike*, has in fact originally the meaning of 'custom, usage', pinning justice to a place, a community, into which it is embedded both in space and time. As a contingent embodiment of people's habits, it translates into a conception of 'what is done', albeit in the normative sense of 'what is right'.

56 Underlying universal/izing justice.

57 Laws, Book VI.

58 See Constantini, C. (2008) 'Equity's Different Talks', *Cardozo Electronic Law Bullettin*, 14.

59 'Man is the measure of all things, of the things that are as to 'how' they are, and of things that are not, as to how they are not', Kerferd, G. B. (1981) *The Sophistic Movement*, Cambridge: Cambridge University Press.

60 Sandel, M. J. (1998) *Liberalism and the limits of justice*, 2nd Edition. Cambridge: Cambridge University Press.

61 Which is quite different from relativism, as it does not depend on the individual whim, but rather is an evolving socially shared, 'intersubjective' convention. Its contingency resides in its fluidity across time and space, and in its not acquiring an immutable, ontological status.

62 Latour, B. (1993) *We Have Never Been Modern*, trans. C. Porter. Cambridge, Mass: Harvard University Press.

63 And, conversely, '[f]or each Nature there exists a corresponding state of Society'.

64 The emphasis falls on a clear radicalization of peoples with places, and the appreciation of reality is 'totalizing', rather than artificially broken into the nature versus culture dualism. See Latour, B. (1993) *We Have Never Been Modern*.

65 Latour, B. (1993) *We Have Never Been Modern*. In line with Toledo'a analysis of the mutually sustaining relation between cultural and biological diversity.

66 Respectively of GHGs in the atmosphere and of capitalist profits.

67 Such as the CDM.

68 Such as Oil palm, Eucalyptus, Pine.

69 With the benefits of 'development' accruing in the form of industrial agriculture, jobs creation, earning of foreign currency through commodity export or the carbon markets (CDM or private offset), See, among others, World Rainforest Movement, CDM Watch.

70 Merchant, C. (1989) *The Death of Nature: Women, Ecology and the Scientific Revolution*. HarperOne.

71 And which ultimately justified the pillaging of time/space for the purpose of capitalist accumulation.

72 Illich, I. (1980b) 'The De-Linking of Peace and Development', Opening address on the occasion of the first meeting of the Asian Peace Research Association, Yokohama.

73 For example, La Via Campesina: 'Food sovereignty is the right of peoples to healthy and culturally appropriate food produced through ecologically sound and sustainable methods, and their right to define their own food and agriculture systems', see the *Declaration of the Forum for Food Sovereignty, Nyéléni 2007*.

74 Walzer, J. P. (1983) *Spheres of Justice. A defense of Pluralism and Equality*. Basic Books.

75 This is the concept of climate debt, which, however, should and can only be, a 'stepping stone', and should include, besides financial reparations also *restitutio ad integrum* where possible.

22

Resistance Makes Carbon Markets

Matthew Paterson

Introduction

While critiques of carbon markets have developed powerful exposés of scams in the CDM and voluntary offset markets, less well recognized is that their opposition and critique, and the occasional associated political resistance, has in fact been important in shaping the character of those markets in the first place. Proponents of carbon markets would have us believe there is no real distinction between the abstract ideas of the free market economists who, since Ronald Coase, have dreamed up schemes for pollution markets, and actually existing carbon markets, as if the abstract ideas have been smoothly rolled out. Occasionally critics fall into a similar trap. But the fact that many have criticized such markets from the outset, and a much broader range of people are distinctly uneasy with the idea of commodifying the atmosphere, or paying others to offset their emissions, has in fact shaped what those markets look like in practice.[1]

As we worry and criticize carbon markets as they develop, then we should understand that critique in two contexts at least. One is that previous critiques and opposition to carbon market development have affected the specific character of carbon markets, with the effect of making them potentially more effective environmentally and to avoid the worst excesses of their socially unjust dynamics. The other context is the political dilemma this poses – that even while resistance may indeed be fertile, it is unlikely to achieve the intended goals of many critics themselves, that of overturning the basic tendency to market mechanisms as means to respond to climate change.

I illustrate these points using two examples. The first is the debate about forestry in the CDM as it was being negotiated between 1997 and 2001, which resulted in the virtual exclusion of forestry from the CDM. The second is the development of certification systems in the offset markets, mostly by environmental and development NGOs, and their emerging effects on that market.

Kyoto, the CDM, and Forestry

Throughout the period before 1997, when the Kyoto Protocol was agreed, questions of how and whether industrialized countries would be able to have 'flexibility' in meeting their commitments were central to negotiations. Already in the UNFCCC negotiations, a proposal originally by Norway in 1991, and leapt upon by the United States, that countries should be able to meet their commitments 'individually or jointly', had been accepted. In the period after the FCCC came into force, in late 1994, what this term might mean was routinely discussed. A central question was whether such 'joint implementation' could involve developing countries. The latter, led by India and China in particular, stated consistently that such an approach was impossible because 'joint implementation' implied that both parties involved had emissions limitations commitments to implement, and the FCCC is clear that developing countries have no such commitments. They opened the door however, in accepting a pilot phase for 'activities implemented jointly' (the wording not implying jointly held commitments, but the substance was the same), which started in 1995. In this, companies and governments in the Annex I countries would be able to develop mitigation projects in the South. There were no credits for these activities; it was intended as a 'learning by doing' period.[2]

The 'learning' occurred rather quickly however – within a year US negotiators in particular were already insisting on some such mechanism in the agreement to be signed in Kyoto. Investing in East European countries as they underwent 'transition' to capitalism was all very well, but projects in the South would be so much cheaper in terms of the emissions reduced, that it would be economically irrational not to include them. During 1996 and into early 1997, US negotiators piled on the pressure for 'flexibility', proposing and insisting on emissions trading in Kyoto (resisted for a long time by the EU and Southern countries), turning the AIJ pilot phase into Joint Implementation (albeit just among Annex I countries – so mostly involving projects in the former Soviet bloc).

It looked like developing countries had successfully resisted the idea that they might be included in such projects. But at the last minute, Brazil, which has pursued an independent line from other developing countries on many issues in the climate negotiations, proposed a 'clean development mechanism'. This was intended as a compensation mechanism – that Annex I countries who failed to meet their targets to reduce their emissions would have to pay into a fund which would invest in emissions reductions in developing countries. The US negotiators jumped on the proposal and said in effect 'why not have this as a carrot not a stick?' A quick re-write ensued and the CDM – 'Kyoto's surprise' – was created.[3]

But what in practice the CDM would become was negotiated over the next 4 years, in the run-up to what would be called the 'Marrakech Accords', which are the operational rules for the Kyoto Protocol. In fact, the forestry negotiations were so complex that the rules were only finalized a couple of years after Marrakech, in December 2003. In the Protocol itself, the CDM is only described in very general institutional terms – that it is subject to the authority of the FCCC parties, that projects need to deliver 'real, measurable, and long-term benefits related to the mitigation of climate change', and the like.[4] Nothing is said about what types of projects can or cannot be included, how emissions foregone are supposed to be calculated, who gets to decide on such calculation techniques, whether there should be limits to rich countries' recourse to the mechanism to meet their commitments, or similar details. These were all negotiated over the following years.

As the conversations about putting the mechanism into practice developed, there was much conflict precisely about these aspects of the mechanism. Central was (and is) the question of methodology, of how to make a claim that any given project reduces emissions compared to what they would have been absent the project. And in trying to think through this question, forestry became a key debate. This was in part because of the technicalities of measuring the absorption of carbon by sinks such as forests, which are completely different to those associated with measuring emissions. Not only are you measuring counterfactuals – what emissions would be without the project compared to what they are with it – which is the basic problem with all offset systems like the CDM, but you are doing so for the uptake of CO_2 from the air by tree species which vary widely in their growth rates, and in ecological conditions which vary enormously, so the uncertainty range increases drastically. Also, the permanence of the projects is very much more in doubt, since guaranteeing the continued growth of forests, and thus the uptake of carbon, over a long period of time, is considerably more problematic than that for an energy efficiency project or a wind farm.[5] This sort of debate pervaded the Marrakech negotiations in general, not just to do with the CDM – as some countries wanted to increase their ability to use sinks in meeting their commitments (and strategically, to try to get the US to be able politically to sign), thus weakening commitments generally. But it had a particular resonance in relation to the CDM.

Mainstream environmental NGOs such as Friends of the Earth, as well as more radical organizations like the World Rainforest Movement, and the informal protest networks arising out of anti-globalization movements which mobilized in particular at the Hague COP in 2000, all raised a range of concerns about including sinks in the CDM. The principal concerns of these movements was that forestry projects in the CDM would act as a series of 'low-hanging fruit', enabling cheap offsets which would undermine the effectiveness of the Kyoto targets being negotiated. This would then create longer-term problems for developing countries by using up all the easy, cheap, abatement options, and thus operate as a sort of 'carbon colonialism'. The projects would also be ecological disasters by creating incentives to develop large eucalyptus monoculture plantations, and would be social disasters in displacing indigenous peoples, peasant farmers, and others, from their lands. The Plantar Project in south-eastern Brazil, a US-funded project under the AIJ pilot phase (see above), was widely publicized as a textbook case of this sort of problem. Finally, they were subject to huge uncertainties in measuring the uptake of carbon by forests in different ecological contexts, and thus particularly prone to the problems of 'climate fraud'.[6]

NGOs built alliances with governments in the South, who worried that the CDM in general, but CDM sink (i.e. forestry) projects in particular, would in practice enable Northern countries to get out of their emissions reductions obligations. They also had significant support, at least until the US pulled out of Kyoto in March 2001, from most European countries.

As a consequence of the sustained pressure, and political alliance building, such concerns, along with the technical problems in measuring sinks compared to measuring emissions, resulted in rules which have made such projects extremely difficult to get approved in the CDM processes, and only a small handful of CDM projects are in forestry as a consequence. 'Avoided deforestation' projects were excluded entirely, and the rules for other types of forestry projects (deforestation, reforestation, afforestation) were written to be very tight. Countries could have a maximum of 1% of their 1990 emissions mitigated by such projects.[7] In fact, the numbers have come in well below this amount: in 2007 under 1% of the CERs issued were in forestry projects (as compared to around 30% of projects in the voluntary carbon markets).[8]

Of course, carbon traders and project developers don't like the virtual exclusion of forests from the CDM. They include this question in their recurrent attacks on the CDM institutions for their (alleged) rigidity. And forestry is back on the CDM agenda for the post-Kyoto period, in the way that the debates about Reduced Emissions from Deforestation and Forest Degradation (REDD) are being articulated. It is not yet clear what way these negotiations will go, but there is great pressure from business actors to link REDD to the CDM given the market opportunities this creates for the carbon market industry. It is also difficult in political terms to see how industrialized countries will provide significant financial support for REDD activities in developing countries without some way of getting credit for it in relation to their emissions reductions obligations, and the CDM is (for them) the obvious means of doing this. This re-opens a Pandora's box whose lid had largely been kept shut in the negotiations for Marrakech.

The Dynamics of Certification Schemes

The second example of where resistance makes carbon markets is the development and dynamic of certification systems in the offset markets, both the CDM and the voluntary markets. There are now around 15 such certification systems, mostly developed by mainstream environmental NGOs like WWF, which is behind the Gold Standard, one of the best-known schemes. These systems have arisen precisely because of the widespread worry (fed by the exposés by critics of carbon markets) that offsets (including the CDM) may have little benefit in terms of carbon abatement and serve merely to assuage the consciences of Western consumers or the pursuit of green PR by corporations. The certification systems aim in many cases to add extra hurdles to those in the CDM markets, or to create possibilities for different types of projects in the voluntary markets (community-based agro-forestry, for example, in the Social Carbon standard). While their boosters clearly exaggerate the effects, there is nevertheless some evidence that these standards are having an impact on practices in the offset markets, and that there may be a 'race to the top' as the more stringent standards gain in market share and the prices projects certified by them can garner.

Boosters of voluntary carbon markets are keen to let us know that these markets precede the regulatory ones; that the first voluntary carbon market transaction was in 1989 when AES, a US electricity company, invested in a forestry plantation (of pine and eucalyptus, precisely the sort critics of carbon markets worry about) in Guatemala to offset the emissions from its new coal-fired power plant in Connecticut. But these markets started to get going in the late 1990s and only really became sizeable in the last few years, growing from US$22m in 2003 to US$331 in 2007.[9]

Early on, there was no way of checking that the investment did what the company making the investment claimed it did. Companies like the CarbonNeutral Company, originally founded as Future Forests, could invest in projects proclaiming their virtue and there was no way an outside observer could judge the reasonableness of their claims. But as these markets developed, two processes have intertwined to change this picture.

First, the principal demand for voluntary market offsets has become corporate. In the public eye, carbon offsets are about enabling individuals feeling guilty about their consumption levels to purchase offsets for their flights (a particular focus in the marketing of offsets) or more generally, hence the criticism of offset markets as 'the new indulgences'.[10] The whole paraphernalia of carbon footprint counters, strategic links between airlines and offset providers, and so on, is where these markets are visible on an everyday level. But these sources provide only a tiny fraction of the demand for offset market. Individuals purchased only 5% of the offsets sold in 2007, while NGOs bought 13%, governments 0.4% and businesses the rest.[11] So the driving force behind demand is not individual guilt tripping, but the desire of corporations for good PR. Amongst various reasons corporations give for buying offsets, PR and Corporate Social Responsibility (which often amounts to the same thing) predominate.[12] So the desire of a bank to be able to claim it is Carbon Neutral, or is working towards carbon neutrality, is the central driving force behind this market.

Given the scandals of the last decade, from Enron, WorldCom, and the collapse of LTCM onwards, and of course particularly acutely in the current crisis which has been triggered by financial dodgy-dealings, but also the increasing ability of social movements to expose corporate abuses from human rights to environmental destruction, the visibility of large firms like HSBC is very high. Their desire for PR is a double-edged sword – if you make bold claims, it is difficult to evade the glare of those checking up on you. Indeed many companies prefer simply to keep their head below the parapet – if you put it above, then you get shot at. The experience of BP's re-branding as 'beyond petroleum' is an object lesson in the risks of making such bold claims. So when buying offsets, many large corporations are averse to taking risks with their image.

The second process has been the development of third-party certification schemes for the offset markets. One of the first was the Gold Standard, established by WWF in 2003. As we saw above, one of the key worries in offset markets is about forestry. The Gold Standard sought to overcome this problem by excluding forestry by fiat. For a project to be certified under the scheme, it can only be in renewable energy or energy efficiency. This certification scheme arose precisely out of the sorts of worries about the potential in offset markets for 'climate fraud and carbon colonialism' identified by critics of carbon markets.[13] Others have likewise been developed by NGOs seeking to respond. But the response can take different forms, as in the Social Carbon standards attempt to design a forestry standard which escapes the problems of large-scale plantations, displacement of communities, and lack of social benefits of such projects.

Most of the standards have been organized by groups of corporations – in particular the banks involved in carbon trading who are wholesaling credits (as in the Voluntary Carbon Offset Standard) or by groups which represent the businesses involved in carbon offsetting, such as the Voluntary Carbon Standard (VCS) developed by the Climate Group and the International Emissions Trading Association (IETA), or the (even more arms-length) ISO14064 standard. But while this clearly raises the question of corporate capture of this verification process, those organizations are in effect responding to the same set of concerns – that offset projects may be scams in terms of their emissions reductions or cause other PR problems to do with human rights or other abuses.

It is notable that the main flurry of these standards coming on-stream was in 2007 – so relatively late in the emergence of the voluntary market, and after the steady stream of stories about scams in the offset markets appeared in the public eye, highlighted by critics of carbon markets and campaigning (or sometimes just opportunistic) journalists. Ecosystem Marketplace, an organization boosting carbon markets, called 2007 'the year of the standard'.[14] By 2007, 87% of projects were being verified by a third-party verifier according to some sort of standard.[15] Between 2006 and 2007, what was also noticeable was a dramatic drop (from 23% to 2% of all projects which used a standard) in the number of projects being verified according to the retailer's own standard, and a noticeable decline in the proportion going through the Chicago Climate Exchange, who operate their offset projects according to principles not accepted by any other voluntary market operators.[16] In contrast, the proportion going to projects often regarded as providing *relatively* plausible claims about additionality (such as the Gold Standard, VCS or VER+) went up. Projects using those standards also appear to gain a price premium – a project using the Gold Standard in particular sells for about twice the average price in the voluntary market.[17]

Now none of this should be taken as a simple example of successful industry self-regulation, and that carbon offsets are not a problem at all. The scams identified by critics of the offset markets continue, and exposés of problems such as climate fraud or carbon colonialism are likely to be necessary for years to come. But it does point to two things. One is that the carbon market in 2005 is not necessarily the same thing as the carbon market in 2008; the sorts of problems that will arise will evolve as the market does itself. But more importantly, the evolution of this market, and the way that it is regulated or regulates itself, is driven precisely by the exposés of the problems of offsetting *per se*. Whether driven by NGOs attempting to improve the sorts of projects investors engage in, or avoid the worst sorts of projects, or companies seeking to avoid exposés and seek good PR, it is the opposition to carbon offsetting which is playing an important role in shaping what sorts of offsets can be pursued, how they need to be justified, and so on.

Conclusions

So, resistance makes markets. The character that carbon markets have – the operational rules, informal norms, changing dynamics – is produced at least in part by the reactions to opposition to carbon markets as a means of responding to climate change. Resistance is indeed, thus fertile. The process here is very much a microcosm of a more general phenomenon identified by writers such as Hardt and Negri, or in a rather different way, Naomi Klein.[18] That is, the overarching projects of 'global neoliberal governance' (or Empire, if you prefer) have been generated as strategies by powerful global forces to overcome, co-opt, or destroy opposition to its power. At times (as in the many examples given by Klein in *The Shock Doctrine*) the strategy is to eliminate opposition, often physically through torture and murder. But at other times the approach is more subtle, involving an appropriation of the themes and ideas of critics and a twisting to enable further accumulation of capital and extension of global power. The appropriation of many ideas in the 1970s 'counterculture' for capitalist purposes, from Ben and Jerry's to 'flat hierarchies', to the 'new economy', is a classic example.[19]

Carbon markets could precisely be interpreted as a sort of 'disaster capitalism', in Klein's terms, where climate change was articulated as an 'urgent disaster' in order to legitimize the opening up of new frontiers for investment and financial trading, all the while deploying climate as the legitimizing signifier. But while this is clearly a strategy by big, globally organized business to serve its interests, it is nevertheless the case that the impetus to do so was to appropriate the energy and desire to act on climate change, and forestall and respond to criticism of corporate capitalism for failing to act.[20]

But the effects here are of course contradictory. For many involved in opposing carbon markets, the point is to get rid of them, not make them work better. We are thus in the awkward position of shaping the form of something we may prefer didn't exist.

And of course we should not over-emphasize the power of resistance. Carbon markets in general are still expanding rapidly and show little sign of slowing down overall, despite the financial crisis and generalized calls for re-regulation of finance and a 'green new deal'. New schemes are likely to come on stream in New Zealand, Australia, and perhaps Canada in the next couple of years, and cap and trade still seems to be the centrepiece of any legislation likely to come out of Washington. In the 'post-Kyoto' negotiations, the CDM still has the aura of an unassailable institution about it – many of the developing countries that in the run-up to Kyoto and then the negotiations to Marrakech were highly hostile to the CDM, now see it rather as an opportunity to attract investment and technology. So resistance does clearly not result in the aim of many critics – to cause the abandoning of carbon markets as a response to climate change.

But this is not new in the history of capitalism. Oppositional movements from at least the late 19th century have seen their protests and resistance turned into new sorts of legitimacy for capitalist and state elites, and often new sorts of means for corporations to make money. From Bismarck's welfare state, to the New Deal and the Fordist-Keynesian management system from the 1930s onwards, reform has been driven by responses to protest and movement activism. Often this simultaneously shores up the legitimacy of the system while benefiting corporations directly – the key winners in the Fordist-Keynesian system were arguably the car manufacturers, oil companies, road builders, and construction companies who made money from the organization of economic development by the state. Carbon markets are not much different to that – the winners are different (the city firms like Barclays Capital dominating the trading, the auditors like SGS dominating the consulting and verification systems), but the alliance between legitimacy and profitability are similar.[21]

So this puts critics in an awkward position. There is no easy way out of this dilemma. Short of an eco-socialist transformation, any project to reduce emissions dramatically will entail some of response from which some bits of big business benefit – no response to climate change can oppose all business and expect to win. This is the Gramscian logic of hegemony at work – within capitalist society no political project can become hegemonic without involving some powerful sections of capital. And renewable interests on their own are not powerful enough to overcome the opposition of coal and oil interests. From this sort of political point of view, the genius of carbon markets is that such a set of capitalist interests have been engineered – substantial parts of the City of London, albeit rhetorically and we should be sceptical of their rhetoric – that have interests in reducing emissions. More stringent targets = higher carbon price = more hedging by regulated firms = more arbitrage and demand for derivative products, and thus profits for city firms. The Carbon Market and Investors Association, a group of city firms who split from the International Emissions Trading Association because the latter caved in too often to the interests of regulated firms, now argues openly for stringent targets to reduce emissions.[22]

So it is unlikely, in my view, that resistance will overcome the tendency to carbon markets as a key element in the response to climate change. But it will shape the character of those markets, potentially reduce some of their worst effects, and generate support for the policies and social change which are essential and for which carbon markets may well be simply a distraction. In the context of the financial crisis in particular, it is clear that this opens up certain opportunities for critics and for resistance. Already, the reframing of carbon markets through the notion of 'subprime carbon' has emerged.[23] And critics have not been slow to point out that many of the firms closely associated with the financial collapse – Merrill Lynch, most prominently – have also been closely involved in carbon trading and other private climate governance schemes developed by the financial industry, like the Carbon Disclosure Project. Highlighting the possibilities for arbitrage, speculation, development of derivatives instruments, and so on, within carbon markets, and thus the vulnerability to now-discredited financial strategies, has become a new weapon in the armoury of critics. There is little evidence that such strategies however are causing a 'roll-back' in the impetus to develop carbon markets – indeed, if financiers are restricted in other areas as they are re-regulated, then there is every likelihood that they will push hard to keep new markets expanding, and that mainstream politicians will want to accommodate them. But there is good reason to believe that it will affect how new carbon markets are designed, just as previous critiques and opposition have affected carbon markets as they have emerged so far. One specific direction is that it opens up space to make claims about dealing with climate change in a way which goes beyond the fetishization of carbon markets. The notion of the 'green new deal' is one such approach; deliberately alluding to previous periods of crisis and reform, both to connect climate change to the other crises produced by contemporary capitalism, while opening up space to show how markets alone cannot deliver the decarbonization that the world economy needs to engage in.[24]

Notes

1   This argument draws variously on different sorts of criticisms (Marxist, ecological, feminist, institutionalist, poststructuralist) of (neo)classical economics which insist that actually existing markets cannot be understood as abstract ideal types but as fundamentally social and political institutions, assembled through various sorts of constructions and impositions of power. The ideas presented here draw on two particular approaches to this question – those emphasising questions of class and financial power in the construction of markets to benefit powerful groups, and those focusing more on the particular ways markets are 'assembled', in particular in how economics as a profession constructs markets. For good examples of the former perspective, see Mansfield, B. (ed.) (2008) *Privatization: Property and the Remaking of Nature-Society Relations*. Oxford: Blackwell, or Heynen, N., J. McCarthy, S. Prudham, and P. Robbins (eds) (2007) *Neoliberal Environments: False promises and unnatural consequences*. London: Routledge. For an example of this argument regarding climate change, see Bumpus, A. and D. Liverman, (2008) 'Accumulation by Decarbonization and the Governance of Carbon Offsets', *Economic Geography*, 84(2): 127-155. On the latter, see in particular Callon, M. (1998) *The Laws of the Markets*. Oxford: Blackwell, or MacKenzie, D. (2006) *An Engine, Not a Camera: How financial models shape markets*. Cambridge MA: MIT Press. MacKenzie has turned his lens to carbon markets; see MacKenzie, D. (2009) 'Making things the same: Gases, emission rights and the politics of carbon markets', *Accounting, Organizations and Society*, 34(3-4): 440-455. The specific contribution here is to emphasise the role played by specific political strategies of resistance in constructing markets, a point which tends to be neglected by authors like MacKenzie or Callon, or relegated to an after-the-fact protest by eco-Marxist writers like those in the Mansfield or Heynen et al books.

2   This history draws in particular on Paterson, M. (1996) *Global Warming and Global Politics*. London: Routledge, ch. 3.

3   Werksman, J. (1998) 'The clean development mechanism: Unwrapping the Kyoto surprise', *Review of European Community and International Environmental Law*, 7: 147-58.

4   United Nations, Kyoto Protocol to the United Nations Framework Convention on Climate Change, New York: United Nations, 1998, Article 12.5 (a).

5   For a discussion of some of these technical aspects, see Yamin F. and J. Depledge (2004) *The International Climate Change Regime: A Guide to Rules, Institutions and Procedures*. Cambridge: Cambridge University Press, pp. 180-182.

6   For a detailed account of these NGO positions in the period running up to Marrakech, see Bäckstrand, K., and E. Lövbrand (2006) 'Planting Trees to Mitigate Climate Change: Contested Discourses of Ecological Modernization, Green Governmentality and Civic Environmentalism', *Global Environmental Politics*, 6(1): 50-75.

7   Bäckstrand, K. and E. Lövbrand. (2006) 'Planting Trees to Mitigate Climate Change: Contested Discourses of Ecological Modernization, Green Governmentality and Civic Environmentalism', p. 59

8   Capoor, K. and P. Ambrosi (2008) *State and Trends of the Carbon Market 2008*, Washington DC: World Bank, p. 29. Their pie chart has agro-forestry at 0.1% of the CERs. It also has biomass at 5%, some of which may involve forestry, and 'other' at 1%. It is unclear if this figure is just for 2007 projects, or is a cumulative figure. Nevertheless, the proportion of projects in forestry is very small.

9   Hamilton, K., M. Sjardin, T. Marcello, and G. Xu (2008) *Forging a Frontier: State of the Voluntary Carbon Markets 2008*. Washington DC: Ecosystem Marketplace, p. 6.

10 Smith, K. (2007) *The Carbon Neutral Myth: Offset Indulgences for your Climate Sins*. Amsterdam: Carbon Trade Watch. For a fuller analysis of the dynamics of legitimisation and delegitimisation of carbon markets, see Paterson, M. (2010 forthcoming) 'Legitimacy and accumulation in climate change governance', *New Political Economy*.

11 Hamilton, K., M. Sjardin, T. Marcello, and G. Xu (2008) *Forging a Frontier: State of the Voluntary Carbon Markets 2008*, p. 66.

12 Hamilton, K., M. Sjardin, T. Marcello, and G. Xu (2008) *Forging a Frontier: State of the Voluntary Carbon Markets 2008*, p. 67.

13 The title of Heidi Bachram's now classic article, see Bachram, H. (2004) 'Climate Fraud and Carbon Colonialism', *Capitalism, Nature, Socialism*, 15(4): 5-20.

14 Hamilton, K., M. Sjardin, T. Marcello, and G. Xu (2008) *Forging a Frontier: State of the Voluntary Carbon Markets 2008*, pp. 47-58.

15 Hamilton, K., M. Sjardin, T. Marcello, and G. Xu (2008) *Forging a Frontier: State of the Voluntary Carbon Markets 2008*, p. 55.

16 In particular, the CCX is explicitly not interested in the question of 'additionality'. That is, it is happy to pay people for carbon credits from practices already being undertaken. The majority of CCX credits are being generated by paying farmers in North America who *are already* practicing no-till agriculture for their credits.

17 New Carbon Finance, *Voluntary Carbon Index, First Edition*, 15 September 2008, p. 2, http://www.newcarbonfinance.com/?p=about&i=freereports, accessed 13 November 2008.

18 Hardt M. and A. Negri (2000) *Empire*. Cambridge MA, Harvard University Press; Hardt M., and A. Negri (2004) *Multitude: guerre et démocratie à l'âge de l'Empire*. Montréal: Boréal; Klein, N. (2007) *The Shock Doctrine: the rise of disaster capitalism*. Toronto: Knopf.

19 On the appropriation of the counterculture, see in particular Frank, T. (2001) *One market under God: extreme capitalism, market populism and the end of economic democracy*. London: Secker & Warburg.

20 This appropriation is often bodily in character – many of the people in carbon trading firms started out in NGOs or thinktanks working on climate change. They take the energy and passion into the marketplace, helping both legitimise the market and shape its character.

21 This of course is Gramsci's process of 'passive revolution', whereby ruling elites adapt the form that their hegemony takes by taking on parts of claims made by subordinate groups in order to neutralise their broader revolutionary goals. See Showstack Sassoon, A. (1982)

'Passive Revolution and the Politics of Reform', in *Approaches to Gramsci*. London: Writers and Readers, pp. 127-148.

22 See the association's statements of aims at http://www.cmia.net/membership.php, viewed 22 June 2009.

23 See for example Friends of the Earth US, *Subprime Carbon? Re-thinking the world's largest new derivatives market*, Washington DC: Friends of the Earth, March 2009.

24 See Green New Deal Group, A Green New Deal: Joined-up policies to solve the triple crunch of the credit crisis, climate change and high oil prices, London; New Economics Foundation, 2008.

23

Green Capitalism, and the Cultural Poverty of Constructing Nature as Service Provider*

Sian Sullivan

People differ not only in their culture but also in their nature, or rather, in the way they construct relations between humans and non-humans.[1]

Loss

We hear a lot these days about loss. In April 2009, the International Monetary Fund (IMF) estimated that banks, insurance instruments and pension funds have 'lost' some US $4.1 trillion from the global economy.[2] The amounts lost to taxpayers via government removal of the toxic assets littering the financial sector are so huge as to be almost meaningless. According to the IMF, UK taxpayers have already lost over £1.2 trillion to Britain's financial sector,[3] while in North America the Inspector General of the Troubled Asset Relief Program (TARP) stated recently that potential government/taxpayer assistance could total $23.7 trillion.[4] Meanwhile, the International Union for the Conservation of Nature (IUCN) asserts that the wildlife crisis actually is worse than the economic crisis, with almost 900 species lost already in an analysis of some 45,000, and no fewer than 16,928 of these currently threatened with extinction.[5] Habitat loss to 'development' is a major cause of these extinctions. Greenpeace reports of the Brazilian Amazon that 'one acre [is] lost every 8 seconds', the hamburger-cattle sector identified here as the major driver of clear-felling in this landscape.[6]

Crisis Capitalism and the Creation of 'Value'

Notwithstanding the complexities beneath these alarming figures, they do seem to signal some sort of crisis, both of capitalism, and of 'the environment'. Intuitively it makes sense to think that these crises might be connected in two key ways. First, that economic exploitation and the profit motive, in driving production and transformed consumption of 'natural resources', is causing and contributing to ecological crisis. And second, that the ecological crisis arising from these pressures is itself generating crisis in the global economy, through making manifest the material limits to economic production and consumption. This is the so-called Limits to Growth argument of the 1970s,[7] which posited resource limits to economic growth, and the need to sensibly distribute resources as well as reducing production and consumption to avert both economic and ecological crises.

But this intuitive view – that ecological loss is entwined with and also signals economic crisis – seems to be somewhat naïve. To look at these connections another way is to see that capitalism thrives on crisis. This is its engine of innovation and creativity. As with the Kafkaesque derivatives markets that in part have pushed the international finance market into such recent toxicity,[8] capitalism makes a virtue of crisis. If the risk of loss or hazard can be priced, and this financial value captured via trade and speculation, then economic growth – the unassailable good of capitalist 'culture' – will be maintained, to the presumed benefit of everyone.

It also is in times of crisis that new forms of capitalist value, new frontiers of accumulation, and new enclosures and dispossessions, are created. In *The Shock Doctrine*, Naomi Klein forcefully argues that various crisis events, from natural disasters to terrorist attacks, in fact are central to the creation of the openings required for incursions of corporate capital investment, thinly masked by the seemingly liberating guise of instituting free markets and democracy.[9]

In this zeitgeist of crisis capitalism, the environmental crisis itself has become a major new frontier of value creation and capitalist accumulation. Referred to by terms such as 'market environmentalism',[10] 'green neoliberalism'[11] and 'green capitalism',[12] the understanding is that if we just price the environment correctly – creating new markets for new 'environmental products' based on monetized measures of environmental health and degradation – then everyone *and* the environment will win. If nature can be rationally abstracted and priced into assets, goods and services, then environmental risk and degradation can be measured, exchanged, offset and generally minimized. At the same time, the new financial values accruing to the declining 'stock' of nature's assets, goods and services might in and of themselves attract more financial value via speculative trade on stock exchanges. Indeed, stock exchanges focusing only on new environmental products now are arising, the Climate Exchanges in London and Chicago being key examples. These have been established for the sole purpose of brokering and trading the new commodity/currency of tradeable carbon – itself created as the vehicle via which climate-change-causing carbon emissions can be measured and ostensibly reduced.

'An Ecosystem at your Service?'[13]

Behind this monetization of environmental crisis is a logic and language that transforms the global environment – Nature – into a provider of services for humans. This conceptual capture, and the economic rationalization of nature's value that it permits, is facilitating the creation of markets for the exchange of 'ecosystem services' in the form of Payments for Ecosystem Services (PES).

Arguably this construction and discourse is justifying *right now* what in time might be considered a critical, cultural transformation in how relationships between humans and the non-human world are conceived, valued, managed and governed globally.

Conservation biologists have been labelling nature as service provider by using the language of ecosystem services since the 1970s.[14] As noted above, this is a decade which also saw the first globalizing statements of concern regarding the ecological limits to economic growth and the emergence of environmentalist discourses requiring development to be ecologically, as well as economically, 'sustainable'.[15] Some years later, Robert Costanza and colleagues brought the concept of ecosystem services firmly into economics by estimating their annual value globally to be \$16-54 trillion.[16] The ensuing alliance between environmental economists and environmental campaigners has emphasized 'convergence between commercial interest and environmental imperative' in demonstrating 'the business case for sustainable development'.[17] At the same time, assertions of the monetized values for defined ecosystem services has led to the corresponding conclusion that currently they are not being valued for what they are worth, and that somehow they should be paid for. As Jean-Christophe Vié, Deputy Head of IUCN's Species Programme, stated recently:

> It's time to recognize that nature is the largest company on Earth working for the benefit of 100 percent of humankind – and it's doing it for free.[18]

In recent years, two phenomena have conspired to push these concerns and concepts together to generate a utopian win-win scenario of both mitigating environmental degradation *and* facilitating economic growth through pricing the ecological services provided by nature. The first is the 2005 publication of the influential United Nations Millennium Ecosystem Assessment (MEA), which highlights human-generated change of the biosphere and overwhelmingly uses the language of ecosystem services in speaking of the non-human world. These are further categorized into provisioning services (food, water, timber, fibre, etc.), regulating services (floods, droughts, land degradation and disease), supporting services (such as soil formation and nutrient cycling), and non-material cultural services (recreational, spiritual, religious, etc.).[19] Through combining the quantification skills of ecological science and economics, the MEA proposes that breaking nature down into these increasingly scarce services,[20] quantifying their functionality, and assigning a price to them, will assist conservation by asserting their financial value; at the same time as fostering economic growth by creating new tradeable assets.[21]

The second is the creation of a multi-billion dollar market in a new commodity – carbon – intended to mitigate (i.e. minimize) climate change by providing the possibility of profitably exchanging one of the gases contributing to anthropogenic global warming. As noted above, this is generating a market-based context for approaching the broader environmental concerns of the MEA. Like Adam Smith's putative economic 'invisible hand',[22] the assumption is that both good environmental governance and the equitable distribution of environmental services will derive from the correct pricing of quantified environmental goods and services, combined with the self-regulating market behaviour that will emerge from their market exchange.

In this case, the financial price attributed to carbon is allocated to, and therefore captured by, heavy industry emitters. It is they who gain tradeable carbon credits (i.e. the currency representing carbon), for example, under the European Union's Emissions Trading Scheme.[23] Some (currently minimal) scarcity is built into the market by allocating credits at a level below what major installations require to cover their emitting levels, so as to meet the emissions reducing targets set by the Kyoto Protocol of the UN Framework Convention on Climate Change (UNFCC). Once these credits enter the international financial system their future value can be speculated on (as with any other currency or commodity, including derivatives) and significant profits can ensue. In the wake of this, a veritable ecosystem of economists, stockbrokers and financial advisers has emerged to service trade in this new commodity, as epitomized by the Europe Climate Exchange in the City of London. This is 'the leading marketplace for trading carbon dioxide (CO_2) emissions in Europe and internationally',[24] and basically a stock exchange for the currency of tradeable carbon credits. Interestingly, the website of the Europe Climate Exchange provides very little information connecting this exchange with environmental impacts through the reduction of atmospheric CO_2. Such presentation seems to emphasize that this is a product with a great deal to do with trade, finance and profit, that operates at a rather large remove from the materiality of global climate and eco-systems.

The Ecosystem Marketplace

Of course, payments for the environmental services produced by nature's labour do not go to the environment itself, but to whoever is able to capture this newly priced value. A key logic is that such payments will act as compensation for economic opportunity costs in contexts where environmental-use practices are altered so as to conserve ecosystem services. As stated by Conservation International, '[t]he payment for ecosystem services concept helps address the destruction of Earth's habitats, landscapes and ecosystems by assigning a value to these services, and compensating the people, communities and countries whose actions enhance or protect ecosystem services and the costs that work incurs'.[25]

This might take the form of relatively simple direct payments for transformed behaviour to maintain a particular and clearly defined environmental good. In water management, for example, the water available to those living downstream can be directly negatively affected by water-users upstream, and PES schemes may be established to alter upstream behaviour so as to maintain downstream water quality and access. Paradigmatic here is the case of Vittel (Nestlé Water) in north-east France, who came to a financial agreement to compensate farmers for altering their nitrate-based fertilizing practices upstream which were contaminating the aquifer producing the bottled mineral water sold by the company.[26] In this case the key parameters were relatively clear to define. They included the environmental good (uncontaminated water), the potential 'servicers' of that good (nitrate-using farmers), the environmental problem (contamination by nitrate-based fertilizers), and the purchaser of the environmental good (Vittel). Further critical factors are embodied here with implications for the applicability of such initiatives elsewhere and over broader geographical scales, such as between contexts in the urban industrialized North and the rural 'underdeveloped' South. The wealth of the purchasing company and the continued market value of their product, provided economic sustenance for their interest in pursuing the ecosystem services exchange. The land constituting the source area for the water is enclosed as private property under clear tenure arrangements, permitting the establishment of relatively direct contracts between service purchasers and providers. And Vittel was able to collaborate with a professional and well-funded prolonged (four-year) period of research on the connections between farming practices, water quality and potential collaborative alternatives, prior to the long-term establishment of a PES scheme. Even with these factors, the initiative cost Vittel some 24.25 million Euros to develop in its first seven years (an estimated 980 Euros per hectare per year),[27] and it took some ten years following the initial four-year period of research for the scheme to become operational.

Increasingly, PES involves the creation of derived environmental 'products' that are agreed by sellers and buyers to represent some sort of measure of environmental health or degradation. An example might be the creation of schemes financed as commercial deals by private investors whereby new products representing a defined environmental good are sold both to fund conservation practice and to generate a return to investors. The Malua Wildlife Habitat Conservation Bank (MWHCB), also referred to as the Malua BioBank, in Sabah, Malaysia,[28] might be considered a paradigmatic example here. In this scheme a collaboration between private investors and the Sabah government has created saleable 'Biodiversity Conservation Certificates', each representing 100m^2 of rainforest restoration and protection. Over a 50-year license of conservation rights to the BioBank from the Sabah government,[29] the sale of certificates is intended to 'make rainforest rehabilitation and conservation a commercially competitive land use'.[30] It is projected that the initial US$10 million of private investment committed for the rehabilitation of the Malua Forest Reserve over an initial six years will be recovered from the sale of these certificates and also will endow a trust fund (the Malua Trust) to fund the long-term conservation management of the BioBank over the remaining 44-year period of the license. In this case, investment is via the Eco Products Fund, LP, a private equity investment vehicle managed by the international asset brokers New Forests Inc.[31] and Equator Environmental, LLC (whose self-defining phrase is 'creating value by investing in ecosystems'[32]). As a member of the collaborative Clinton Global Initiative[33] between governments, the private sector, NGOs and 'other global leaders', the Eco Products Fund commits US$1 million over 6-10 years towards finding ways, globally, '[t]o realize value from illiquid environmental assets such as carbon, water, and biodiversity, and to use innovative financial structures to represent the value of these critical services in the marketplace'.[34] In the case of the Malua BioBank, any profits from the sale of biodiversity certificates are to be shared between the forest management license holder and the investor. The purchase of certificates does *not* constitute an offset against rainforest impacts elsewhere, and as such is designed to constitute a simple purchase of conservation. It is projected that by the end of the licensing period the initial endowment 'will be fully capitalized and this funding can be used either to renew the conservation rights to the Malua Forest Reserve or to establish a conservation bank on another property with high biodiversity value'.[35] Within-country 'conservation banks' and 'species banks', involving the creation and trading of 'credits' representing biodiversity values on private land, also are proliferating, particularly in the US.[36]

While purchase of the Malua BioBank's biodiversity certificates is not designed to offset environmentally damaging activities due to the transformation of landscapes through economic development elsewhere, much of the anticipation regarding the new pricing of ecosystem services revolves around exactly this. Thus the attribution of new prices to conserved land already owned by commercial companies might be mobilized so as to offset environmental degradation caused through resource extraction elsewhere. Even more attractively, companies might be able to trade newly priced marketable ecosystem services on appropriated land that they now own, thereby capturing new financial value from the new construction of nature as service provider. Mining conglomerate Rio Tinto, for example, are exploring with the IUCN 'opportunities to generate marketable ecosystem services on land owned or managed by the company'.[37] These might include 'potential biodiversity banks in Africa, as well as the opportunity to generate marketable carbon credits by restoring soils and natural vegetation or by preventing emissions from deforestation and degradation'.[38] Environmental credits rewarded to businesses for ecosystem improvement activities also might be 'banked' against future environmental liabilities' or sold to other land developers 'to compensate for the adverse environmental impacts of their projects',[39] with a new generation of 'commercial conservation asset managers' required to broker these exchanges and revenues.

These new forms of ecosystem value thus become conventional business opportunities for investment: the ensuing transformation of ecosystem services into marketable assets provides 'new trading opportunities' such that buyers and sellers of these services can generate profit that 'does not imply the loss of natural assets'.[40] Large corporations, investors and investment brokers now are moving to claim slices of emerging ecosystem markets, and the potential finance flows accruing from newly priced species, ecosystems, services and environmental products.

The new global multi-billion dollar trade in carbon, in particular, is providing a market-based model, embraced by both business and major environmental organizations, for *pricing* and exchanging environmental products across the environmental spectrum under the rapidly proliferating arenas of PES and the proposed programme administered by the United Nations Environment Programme (UNEP) for Reducing Emissions from Deforestation and Degradation (REDD). A critical component of the logic underlying these approaches is an assumption that environments, emissions and effects in very different locations somehow are equivalent and therefore substitutable, such that they allow negative impacts in one location to be *offset* against environmental investments in another. So the REDD programme proposes equivalence between carbon emitted in the fossil-fuel fumes of cars and industry etc., with that stored in living and decomposing biomass in the myriad configurations of long-evolved and diverse assemblages of species. Emissions therefore can be offset against newly priced carbon stored in standing forests, principally in 'developing countries'. An accompanying logic is that the new financial value accruing to standing forests will act to reduce the carbon emissions produced by their potential transformation into different landscapes which currently might be more economically profitable (to some people at least); examples might include the clear-felling of the Amazon for hamburger-cattle, soya or oil production.

But significant questions remain. Are the molecules of CO_2 emitted through fossil-fuel burning really equivalent to the carbon stored in complex terrestrial ecosystems whose assemblages have evolved over many millennia? Do such offsetting schemes actually reduce environmental impacts (e.g. levels of CO_2 emissions), or do they instead provide incentives to continue to profit from these emissions and their trade? And as discussed below, how does trade in derived environmental products relate to and affect the peoples, livelihoods and lifeworlds located in the landscapes from which these products are derived?

Nevertheless, new markets for ecosystem services and other ecological products now are proliferating, with an accompanying array of brokers advertising ecological wares online. Websites and companies abound with names such as 'Ecosystem Marketplace',[41] 'Species Banking'[42] and 'Climate Change Capital'.[43] At the same time, the major global conservation NGOs such as Conservation International (CI), The Nature Conservancy, and the World Wide Fund for Nature (WWF) are embracing PES as a critical tool for generating and distributing the finance needed for conservation activities. A CI glossy brochure called *Nature Provides*, published in August 2009, thus announces the forthcoming launch of ARIES – Artificial Intelligence for Ecosystem Services – described as a 'web-based technology...offered to users worldwide to assist rapid ecosystem service assessment and valuation at multiple scales, from regional to global'.[44] This alliance between investment capital, business and environmental organizations is being fostered by the world's oldest and largest global environmental organization – the International Union for the Conservation of Nature (IUCN) – a network of governments, donor agencies, foundations, member organizations and corporations (www.iucn.org). An onlooker at the four-yearly IUCN World Conservation Congress in Barcelona in October 2008, for example, would be forgiven for thinking that multinational corporations now are the planet's conservationists. At this event, the World Business Council for Sustainable Development (WBCSD) was particularly visible. This is a network of the Chief Executive Officers of some 200 corporations, whose mission statement is 'to provide business leadership as a catalyst for change toward sustainable development, and to support the business license to operate, innovate and grow in a world increasingly shaped by sustainable development issues'.[45] The image in Figure 1, taken at the prominent

WBCSD stand at the 2008 World Conservation Congress, is suggestive of its planetary reach and ambition. It depicts the brand logos of many of the world's largest multinationals, stretching across an abstract earth, smoothed of difference, diversity and inequality. This is a world good for capital. But is it also good for cultural and ecological diversity?

Figure 1: The world according to the World Business Council for Sustainable Development: a smooth earth populated by corporate logos. From the WBCSD display at the 2008 World Conservation Congress of the International Union for the Conservation of Nature. Photo: Sian Sullivan.

A Unifying Language?

Recently, the UNEP and the IUCN described ecosystem services as a 'unifying language' in global environmental policy.[46] This indeed may be the desire. Significant questions remain, however, with serious relevance for the distribution of power and voice in global decision-making. Who is creating and writing this language and for whom? What are the ontological and epistemological assumptions built into the construction of nature as service provider – i.e. what is understood to be the nature of nature? And what are thereby legitimated as appropriate methods for claiming 'nature knowledge'? How are human/non-human relationships being structured, both materially and conceptually, in the process of creating and instituting this 'unifying language'? And what knowledges and experiences are being othered and displaced through the parlance and practice of ecosystem services markets?

Some of these questions can be approached through the brief descriptions of PES concepts and schemes outlined above. The construction and monetization of nature as service provider clearly produces a range of significant transformations. Through PES the non-human world in all its diversity and mystery becomes the provider of services for humans. People dwelling in areas now valued for the ecosystem services they provide to people in other locations become the necessary custodians and providers of these services, with recompense from service-users being dependent on services received. This may be a double-edged sword for people living in newly priced service-providing landscapes, especially in the Global South. Continuing a long history of displacement for environmental conservation,[47] food-producing practices and cultures may be restructured and constrained in the process of shifting from direct production for subsistence and livelihoods to producing environmental service-oriented landscapes. And finally, those numerate in the labyrinthine abstractions accompanying the creation of new ecological commodities and markets – accountants, brokers, bankers and assisting ecological scientists – become the expert mediators and managers of monetary value for both.

All these transformations emphasize conceptual difference rather than continuity between human and non-human worlds. Nature somehow is backdrop to, rather than co-creator of, human activity. At the same time they reinforce somewhat Hegelian master-servant relationships between human and non-human realms, extended further to those between 'experts' on, and inhabitants of, newly priced service-providing landscapes.[48] Nature serves culture; and those dwelling in landscapes newly monetized for their provision of ecosystem services are themselves constructed as servers for visions of the appropriate nature of these landscapes, as perceived by policy and technical experts who, while globally mobile, frequently are based in distant urban locations.

These transformations are critical for cultures as well as for environments worldwide. I opened this chapter by noting the ways in which economic and ecological crisis narratives revolve around assertions of loss. To complete the picture, the 2009 United Nations Educational, Scientific and Cultural Organization (UNESCO) Atlas of the World's Languages in Danger announces the loss of 233 known languages, with a further 574 classified as 'critically endangered'.[49] If language is a key lexicon through which culture is expressed, shared and made meaningful, then the loss of languages equates with the demise of cultures. The causes are complex interactions of marginalization, 'acculturation' to modern monetary and capitalist culture, and direct displacement. The outcome is a subtle 'culturecide': the death of collective identities through displacement by a dominant and globalizing culture that has amongst its norms and values certain disciplining assumptions about the nature of reality. These include rather strict conceptual separations between culture and nature (echoed by that between mind and body, male and female, civilized and wild and so on) – separations which tend to privilege the first part of each of these binaries; together with the elevation of monetized exchange as the key measure and mediator of value. As indicated by the global loss of languages, the peoples, cultures and epistemologies that are othered in this capitalist structuring of values can become rather 'disposable',[50] in part through constructing them as poor, marginal, and often as environmentally problematic.

As an extension of a globalizing capitalist culture which has these assumptions at its heart, it is difficult not to see the unifying language of ecosystem services as part and parcel of these processes of cultural displacement in the realm of human/non-human relationships, understandings and values. In part this is because the proliferating freedoms and futures espoused by free-market environmentalism simultaneously close off possibilities for other freedoms and futures in how relationships between human and non-human worlds are practiced and expressed. Many forms of value, appreciation, understanding and experience of non-human worlds simply are incommensurable with economic pricing mechanisms, and are displaced or closed off completely in the process of pricing for monetized exchange.[51] Where money and capital are the measures of wealth, economically marginalized indigenous cultures, as well as those who choose to live by different values within highly industrialized nations,[52] are seen only as materially poor and thus requiring intervention to foster economic development. A recent UN Food and Agriculture Organization report thus focuses on the desire to better capture the ecosystem services provided by dryland ecosystems globally, in part through shifting the livestock-based livelihoods of 'the poor' who dwell in such lands.[53] As I have noted elsewhere,[54] the 'poor' in these contexts include peoples as diverse as Maasai of East Africa, Raika pastoralists of India's Rajasthan, and Quechua-speaking highland herders in Peru: a global fabric of rich and different cultures sustained through mixed farming practices of which livestock constitute a major part. Importantly, such peoples may not define themselves and their land-entwined lifeworlds as 'poor', as indicated by Maasai in the strong statement that 'the poor are not us'.[55]

A particular irony here is that many of the endangered languages noted above are those of so-called indigenous cultures; of people who retain and can trace some form of coherent connection with the landscapes with which their lineages are entwined. Often these connections seem to be in landscapes that currently are highly valued for their biodiversity and other environmental riches. At risk of essentializing or romanticizing, perhaps it might be that the complexities of indigenous cultural engagement with these landscapes have something to do with their current conservation value. It might also signal that disappearing languages and their associated cultures have something relevant to say and teach about other possibilities for what it means to be and become human today, in dynamic relationship with non-human worlds.

Cultured Lands

Despite a problematic past in service to colonial endeavours, anthropology has relevance here as an academic discipline that at least makes some effort to understand and enter into culturally unfamiliar experiences and conceptions of being human. With Damara or ≠Nū Khoen people living in the dry, open lands of north-west Namibia, I have been privileged to witness, experience and learn some very different ways of relating with the non-human world. Here, for example, the process of acquiring food and other substances, while a pragmatic effort to procure resources, at the same time also required constant conversation and exchange with the ancestors and other non-human presences populating the land. Non-human worlds were alive to be spoken to, and variously remonstrated with and celebrated through words, song, dance and gift-giving. People were not separate and alienated from the non-human world; they were co-creators with it.

To illustrate this, let me relate one story here.[56] Figure 2 is an image taken in 1995 at a place called | Giribes, which are large open grassy plains to the north-west of a larger settlement called Sesfontein or! Nani | aus. We had driven there early in the morning, and the sun was starting to burn. I had my notebook and plant press at the ready, and was keen to get going with the resource-use documentation – the knowledge collection, if you like – that I hoped to do that day. But the first thing that these three people did – they are Nathan ≠Ûina Taurob on the right, his daughter and her partner – was to move some way away from the car, sit down and start talking out at the land. I remember feeling slightly bemused and impatient at the time, anxious to get on with the 'real work' of resource collection and documentation. But I was curious enough to ask what they were doing.

The answer I received was that this was *aoxu* – the practice of connecting with and giving something away to their ancestors remaining in this landscape and to the spirits of the land, to ask for safe passage and for success in finding the foods they wished to gather. They were giving away tobacco – ≠Nū Khoen, particularly of Sesfontein/!Nani | aus, have long been known regionally for the pungent tobacco they grow in small gardens – and also the leaves of *tsaurahais* or *Colophospermum mopane*, valued locally for their healing properties. The direction they are facing is to the north – towards the settlement of Purros. This is the land where Nathan ≠Ûina grew-up; it is the land (*!hūs*) that he knew and loved, and with which his heart as a healer (*!gaiaob*) was connected. Nathan and his family were no longer able to live there, but in the 1990s they continued to return to these areas, sometimes for several weeks at a time. Most of this movement was completely invisible to the various formal administrations of the region. And some of it meant moving into tourism concessions, run by commercial enterprises, to which they officially no longer had access.

Figure 2: Nathan ≠Úina Taurob and family greet and gift the spirits of the land in |Giribes plains, north-west Namibia. Photo: Sian Sullivan.

It took a fairly prolonged period of unlearning of my own encultured assumptions regarding the nature of reality to reach some understanding of what might be going on here. From this and other experiences, I know now that it is possible for human beings to embody an implicit ethos of reciprocity in relationship with the other sentient beings making up what we now call biodiversity. In this way of doing things, all 'resource-use practice' simultaneously is a conversation, a negotiation and an exchange that binds people into multilayered and multifaceted direct reciprocal arrangements with ancestors, spirit and with other species. It is not just about something that is taken to be consumed; it also is about something that is returned, through direct material and energetic exchanges with the non-human world. Human beings can thereby communicate with and serve the known and unpredictable manifestations of the non-human world, and in doing so affirm reciprocal moral obligations as well as make moral sense of phenomena that cannot be completely knowable or ultimately controlled. Infusing this is an epistemic and ontological orientation to non-human worlds that embraces continuity with, rather than separateness between, these realms, and that encourages movements with, rather than ownership and management over, dynamic ecosystem processes. I perceive also that this practice and logic is encountered in remaining shamanic cultures worldwide – cultures that interestingly also seem to be those who have lived in maintaining relationships with currently much sought after biodiversity. There is depth, diversity and coherence in the understandings of, and communications with, an animated non-human world embodied by many of the world's now disappearing cultures.[57] But these are ways that seem opaque to a modern world whose cosmovision rests instead on fetishized commodities, financial transactions, private property, competition and hunger for growth.

International PES policy developments such as REDD assert the need for 'ensuring effective participation' of indigenous peoples and local communities,[58] and many such communities may see participation in these schemes as a means of generating income and gaining footholds in global economic structures. Others, however, express resistance to 'being participated' on the programmatic terms laid out by these schemes. A recent declaration of Confederation of Indigenous Nationalities of the Ecuadorian Amazon (CONFENIAE) thus states that:

> We reject the negotiations on our forests, such as REDD projects, because they try to take away our freedom to manage our resources and also because they are not a real solution to the climate change problem, on the contrary, they only make it worse.[59]

Such resistance denotes a missed opportunity. This is not in terms of local peoples coming on board in these narrowing trajectories for determining value for the global environment. It is in terms of missed opportunities for listening to and learning from different ways of conceptualizing and enacting relationships with the non-human world.

Serving Nature?

Green capitalism and market environmentalism are rapidly becoming the dominant policy and political choices linking environmental health with economic development. In this paradigm the creation and capture of market value for the services provided for humans by the non-human world is considered the most efficient and sustainable means of mitigating global environmental problems while maintaining and even enhancing economic growth. In this article I ask some questions of this significant conceptual reframing of nature as service provider. What might this discourse say of the ways in which our collective relationship with the non-human world is construed and constructed? What is othered and excluded in the process, and what significance does this have for understanding both the phenomenon of nature and for the cultural and epistemological inclusiveness of contemporary environmental agendas? And finally, what potential does the understanding of nature as service provider really have for kindling health in the earth's psychosocial- and eco-systems?

Gretchen Daily and colleagues represent a common optimism in claiming that '[t]he main aim in understanding and valuing natural capital and ecosystem services is to make better decisions, resulting in better actions relating to the use of land, water, and other elements of natural capital'.[60] Such a statement, however, is devoid of political and epistemological context. It affects an illusion of solution through ecological modernization[61] and linear progress.[62] At the same time, and in common with most international environment and development initiatives, it uses a depoliticized language that excises the significance of 'for who' and 'by whom' questions in this new governance arena.[63]

The core idea underlying these initiatives is that so-called environmental services have not been correctly valued to date. Of course I would agree that capitalist culture has tended to ride roughshod over both biological and cultural diversity. But it seems to me that *pricing* something financially is not the same thing as *valuing* it.

We are critically impoverished as human beings if the best we can come up with is money as the mediator of our relationships with the non-human world. Allocating financial value to the environment does not mean that we will embody practices of appreciation, attention, or even of love in our interrelationships with a sentient, moral and agential[64] non-human world. Instead, it lowers 'the moral tone of social life' and, through doing so, it furthers damage to both humans and ecosphere because 'the pricing of everything works powerfully as a device for making morality and love... seem irrelevant'.[65]

We are bearing witness to another significant and accelerating wave of enclosure and primitive accumulation to liberate natural capital for the global market. Commodification now extends from genes to species and to ecosystems, i.e. to all the domains of diversity that are delineated by the Convention on Biodiversity (www.cbd.int). The continued capture and monetized exchange of the non-human world in the form of Payments for Ecosystem Services (PES) seems set to have an impact on global human/non-human relationships as significant as that which began with the transformation of land into individualized property in England from the Tudors onwards: formalized throughout Europe through escalating Enclosure Acts and accompanying property law, and exported globally via European colonial adventure.[66] We know from history that this past revolution in capital creation, accumulation and investment had major social and environmental implications, reducing diverse cultures to labour in the service of capital, and disembedding peoples' relationships with landscapes in the process.[67]

It seems clear that collectively we are in need of some radically different ways of valuing the global environment. But is it possible to turn instead for training and inspiration to those who, in many different contexts, and often against the odds, seem to have both valued and *served* nature's 'services'? And through doing so is it possible to (re)claim and (re)learn communicative relationships with non-human worlds: worlds which express the same moral, creative, mysterious and playful agencies that humans also embody? Perhaps it might be that ways of relating with and valuing non-human worlds that are othered by modernity and capitalist culture, in fact are those offering openings into possibilities for dwelling that are less hungry and more sustainable, at the same time as perhaps being more meaningful, poetic and *enjoyable*. But to hear and learn this requires an act that capitalist and developmentalist culture almost by definition cannot countenance: which is stopping to listen to, and perhaps even embodying, such alternatives.

Notes

* This chapter was first published as Sullivan, S. (2009) 'Green capitalism, and the cultural poverty of constructing nature as service provider', *Radical Anthropology* 3: 18-27, http://www.radicalanthropologygroup.org/journal_03.pdf.

1 Latour, B. (2009) 'A disputatio: nature vs culture, *Anthropology Today*, 25(1-2): 2.

2 Conway, E. (2009a) 'IMF puts losses from financial crisis at $4.1 trillion', *The Telegraph*, 21 April, http://www.telegraph.co.uk/finance/financetopics/recession/5194711/IMF-puts-losses-from-financial-crisis-at-4.1-trillion.html, accessed 10 August 2009.

3 Conway, E. (2009b) 'IMF puts UK banking bail-outs at £1,227bn', *The Telegraph*, 31 July, http://www.telegraph.co.uk/finance/newsbysector/banksandfinance/5949751/IMF-puts-UK-banking-bail-outs-at-1227bn.html, accessed 10 August 2009.

4 Kuhnhenn, J. (2009) 'Banks report using govt. assistance for loans', *Associated Press*, 19 July, http://news.yahoo.com/s/ap/20090719/ap_on_bi_ge/us_banks_bailout, accessed 10 August 2009.

5 IUCN (2009) 'Wildlife crisis worse than economic crisis', http://www.iucn.org/about/work/programmes/species/?3460/Wildlife-crisis-worse-than-economic-crisis-IUCN, accessed 10 August 2009.

6 Greenpeace (2009) *Slaughtering the Amazon*, http://www.greenpeace.org/usa/press-center/reports4/slaughtering-the-amazon, accessed 12 August 2009.

7 Meadows, D.H., D.L. Meadows, and J. Randers (1972) *The Limits to Growth: A Report for the Club of Rome's Project for the Predicament of Mankind*. London: Pan.

8 Taibbi, M. (2009) 'The big takover', *The Rolling Stone*, 19 March 2009, Online. http://www.rollingstone.com/politics/ story/26793903/the_big_takeover, accessed 23 April 2009.

9 Klein, N. (2008) *The Shock Doctrine: The Rise of Disaster Capitalism*. London: Penguin, but see critique, e.g. Norberg, J. (2008) 'Defaming Milton Friedman', *Reason*, October 2008, Online. http://www.reason.com/news/show/128903.html, accessed 10 August 2009.

10 Anderson, T.L. and D.R. Leal (1991) *Free-Market Environmentalism*. London: Palgrave Macmillan.

11 Goldman, M. (2005) *Imperial Nature: The World Bank and Struggles for Social Justice in the Age of Globalization*. London: Yale University Press.

12 Heartfield, J. (2008) *Green Capitalism: Manufacturing Scarcity in an Age of Abundance*. London: Mute Publishing Ltd.

13 Thank you to Simon Fairlie of *The land is ours* campaign (www.tlio.org), who suggested this as the title for an article I wrote recently for *The Land*; Sullivan, S. (2009) 'An ecosystem at your service?', *The Land*, Winter: 21-23. Thank you also to Mike Hannis for bringing to my attention various sources used here.

14 Bormann, F.H. (1976) 'An inseparable linkage: conservation of natural ecosystems and the conservation of fossil energy', *BioScience*, 26: 754-760; also see Ehrlich, P.R. (1982) 'Human carrying capacity, extinctions and nature reserves', *BioScience*, 32: 331-333.

15 cf. IUCN/UNEP/WWF (1980) *World Conservation Strategy: Living Resource Conservation for Sustainable Development*, http://www.iucn.org/dbtw-wpd/edocs/ WCS-004.pdf.

16 Costanza, R., R. d'Arge, S. de Groot, M. Farber, B. Grasso, K. Hannon, S. Limburg, R. Naeem, J. O'Neill, R. Paruelo, R. Raskin, P. Sutton, and M. van den Belt, (1997) 'The value of the world's ecosystem services and natural capital', *Nature*, 387: 253-260.

17 As critiqued in Crompton, T. and T. Kasser (2009). *Meeting Environmental Challenges: The Role of Human Identity*, http://assets.wwf.org.uk/downloads/meeting_environmental_challenges_the_role_of_human_identity.pdf, p.2, accessed 24 June 2009.

18 IUCN/UNEP/WWF (1980) *World Conservation Strategy: Living Resource Conservation for Sustainable Development.*

19 MEA (2005) *Millennium Ecosystem Assessment: Ecosystems and Human Well-being,* Washington D.C.: Island Press, p.3.

20 Sagoff, M. (2008) 'On the economic value of ecosystem services', *Environmental Values,* 17: 239-257.

21 Ruffo, S. and P.M. Kareiva (2009) 'Using science to assign value to nature', Guest Editorial, *Frontiers in Ecology and the Environment,* 7(3).

22 Smith, A. (1977) (1776) *An Inquiry into the Nature and Causes of the Wealth of Nations.* Chicago: University of Chicago Press.

23 EU ETS (2009) Emisision Trading System (EU ETS), http://ec.europa.eu/environment/climat/emission/ index_en.htm, accessed 10 August 2009.

24 ECX. (2009) Europe Climate Exchange: About ECX, http://www.ecx.eu/About-ECX, accessed 10 August 2009.

25 Conservation International. (2009) *Nature Provides: Ecosystem Services and Their Benefits to Humankind,* Arlington: CI, p. 4, http://www.conservation.org/Documents/ CI_Ecosystemservices_Brochure.pdf, accessed 10 August 2009.

26 See analysis in Perrot-Maître, D. (2006) 'The Vittel payments for ecosystem services: a "perfect" PES case?', London: International Institute for Environment and Development (IIED), http://www.katoombagroup.org/documents/tools/TheVittelpaymentsforecosystem services2.pdf, accessed 18 August 2009.

27 Perrot-Maître, D. (2006) 'The Vittel payments for ecosystem services: a "perfect" PES case?', p.18.

28 www.maluabank.com.

29 Via the regional state organisation Yayasun Sabah, www.ysnet.org.my.

30 MWHCB Inc. 2009 Frequently asked questions, http://www.maluabank.com/faq.html, accessed 18 August 2009.

31 www.newforests-us.com.

32 www.equatorllc.com.

33 www.clintonglobalinitiative.org.

34 Clinton Global Initiative 2009 Eco Products Fund, L.P. (2007) http://www.clintonglobalinitiative.org//Page.aspx?pid=2646&q=271968&n=x, accessed 27 April 2009.

35 Clinton Global Initiative 2009 Eco Products Fund, L.P., MWHCB Inc.

36 Fox, J. and A. Nino-Murcia (2005) 'Status of species conservation banking in the United States', *Conservation Biology,* 19: 996-1007.

37 Bishop, J. (2008) 'Building biodiversity business: notes from the cutting edge', *Sustain,* 30: 10-11, p.10.

38 Bishop, J. (2008) 'Building biodiversity business: notes from the cutting edge'.

39 Bishop, J. (2008) 'Building biodiversity business: notes from the cutting edge'.

40 Bishop, J. (2008) 'Building biodiversity business: notes from the cutting edge'.

41 www.ecosystemmarketplace.com.

42 www.speciesbanking.com.

43 www.climatechangecapital.com.

44 Bishop, J. (2008) 'Building biodiversity business: notes from the cutting edge'; Conservation International. (2009) *Nature Provides: Ecosystem Services and Their Benefits to Humankind,* p.6.

45 WBCSD (2009) Mission statement, http://www.wbcsd.org/templates/Template WBCSD5/layout.asp?type=p&MenuId=NjA&doOpen=1&ClickMenu=LeftMenu, accessed 18 August 2009.

46 UNEP/IUCN (2007). 'Developing international payments for ecosystem services: towards a greener world economy', http://www.unep.ch/etb/areas/pdf/IPES_ IUCNbrochure.pdf, accessed 23 September 2008, p. 2.

47 Igoe, J., Brockington, D. and R. Duffy (2008) *Nature Unbound: Conservation, Capitalism and the Future of Protected Areas*, London: Earthscan; also see Dowie, M. (2009) *Conservation Refugees: The Hundred-Year Conflict between Global Conservation and Native Peoples*, Cambridge Massachusetts: MIT Press.

48 Hegel, G.W.F. (1977/1807) *Phenomenology of Spirit*, trans. Miller, A.V., Oxford: Clarendon Press.

49 UNESCO 2009 UNESCO Interactive Atlas of the World's Languages in Danger. http://www.unesco.org/culture/ich/index.php?pg=00206, accessed 10 August 2009.

50 cf. Giroux, H. (2006) *Stormy Weather: Katrina and the Politics of Disposability*, Boulder Colorado, Paradigm Publishers.

51 Spash, C. (2008) 'Ecosystems services valuation', *Socio-economics and the Environment in Discussion, CSIRO WorkingPaper Series* 2008-03, http://www.csiro.au/files/files/pjpj.pdf, accessed 21 February 2009.

52 e.g. see Chapter 'Planning for permaculture' by M. Hannis, this volume.

53 FAO (2006) *Livestock's Long Shadow*, Rome: FAO, http://www.fao.org/docrep/010/a0701e/a0701e00.htm, accessed 26 September 2008.

54 Sullivan, S. (2009) 'An ecosystem at your service?', *The Land*, Winter: 21-23.

55 Anderson, D.M and Broch-Due, V. (1999) *The Poor Are Not Us*. Oxford: James Currey.

56 Sullivan, S. (2008) 'Bioculturalism, shamanism and economics', *Resurgence*, 250, Online. http://www.resurgence.org/magazine/article2631-Bioculturalism-Shamanism-Economics.html.

57 Some of my favourite texts that extend this point are Knight, C. (1991) *Blood Relations: Menstruation and the Origins of Culture*. London: Yale University Press; Narby, J. (1998) (1995). *The Cosmic Serpent: DNA and the Origins of Knowledge*. London: Victor Gollancz; Ingold, T. (2000) *The Perception of the Environment: Essays in Livelihood, Dwelling and Skill*. London: Routledge; Brody, H. (2001) *The Other Side of Eden: Hunter-gatherers, Farmers and the Shaping of the World*. London: Faber and Faber; Lewis-Williams, J.D. and Pearce, D.G. (2004) *San Spirituality: Roots, Expressions, and Social Consequences*. London: Altamira Press; and Griffiths, J. (2006). *Wild: An Elemental Journey*. London: Penguin Books.

58 See for example Angelsen, A., S. Brown, C. Loisel, L. Peskett, C. Streck, and D. Zarin (2009) *Reducing Emissions From Deforestation and Degradation (REDD): An Options Assessment Report*, Meridian Institute, for the Government of Norway, http://www.redd-oar.org/links/REDD-OAR_en.pdf, accessed 18 August 2009.

59 CONFENIAE 2009 *CONFENIAE on REDD: Ecuadorian Indigenous Peoples' Statement*, http://colonos.wordpress.com/2009/08/05/confeniae-on-redd-ecuadorian-indigenous-peoples-statement/, accessed 18 August 2009.

60 Daily, G.C., S. Polasky, J. Goldstein, P. Kareiva, H.A. Mooney, L. Pejchar, T.H. Ricketts, J. Salzman, and R. Shallenberger, (2009) 'Ecosystem services in decision making: time to deliver', *Frontiers in Ecology and the Environment*, 7, 21-28: 23.

61 Hajer, M.A. (1995) *The Politics of Environmental Discourse*. Oxford: Clarendon Press.

62 Gray, J. (2002) *Straw Dogs: Thoughts on Humans and Other Animals*. London: Granta Books.

63 cf. Ferguson, J. (1994) *The Anti-Politics Machine: 'Development', Depoliticization, and Bureaucratic Power in Lesotho, Minnesota*. University of Minnesota Press.

64 Plumwood, V. (2006) 'The concept of a cultural landscape: nature, culture and agency in the land', *Ethics and the Environment*, 11: 115-150.

65 Read, R. (2007) 'Economics is philosophy, economics is not science', *International Journal of Green Economics* 1(3/4): 307-325, p. 315.

66 Federici, S. (2004) *Caliban and the Witch: Women, the Body and Primitive Accumulation*. Brooklyn: Autonomedia.

67 Polanyi, K. (2001) (1944) *The Great Transformation: The Political and Economic Origins of Our Time*. Uckfield: Beacon Press.

IV

ALTERNATIVES

Part IV of the book hopefully gives us hope. The point that each paper in this part makes is that there are real alternatives to carbon markets, many of which already exist in local communities around the world. If indeed carbon markets often deliver quite perverse outcomes – as many contributions to this book have shown – then what are the alternatives? What can we practically do to mitigate climate change and create a real sustainable, low carbon future? The point that this section makes is that a sustainable future is in our hands. The first two papers offer a more 'political' answer to carbon markets, suggesting that, rather than managing complex carbon markets, we need to think about the 'ecological debt' we have created and consider climate change from a point of view of justice. The remaining papers collected in this part of the book offer a range of very practical insights into how communities already live in sustainable ways. What these contributions seem to be saying is that we cannot depend on governments or markets to save us from climate change. If something has to be done, then it has to be done by us – each and one of us.

24

Repaying Africa for Climate Crisis: 'Ecological Debt' as a Development Finance Alternative to Emissions Trading*

Patrick Bond

Introduction

Carbon trading is under attack, but is there an alternative strategy to transfer resources to the Global South to support a different model of development? Is it reasonable to make calls for 'ecological debt' or 'climate compensation' in the form of a fund that would justifiably exceed $100 billion/year within a decade, without tendentious reliance upon emissions trading brokers, offset salesmen, futures and options, 'additionality' requirements, corruption, and the 'commodification of the air' associated with the Kyoto Protocol and its likely successor regime?

There is a fairly simple financial choice facing those advocating global climate governance: the North would pay hard-hit South sites to deal with climate crisis either through 'Clean Development Mechanism' (CDM) projects and declining overseas development aid – both entailing plenty of damaging side effects – or instead, pay through other mechanisms that must provide financing quickly, transparently and decisively, to achieve genuine income compensation plus renewable energy to the masses. The Kyoto Protocol – and its potential Copenhagen COP successor – is all about the former choice, because the power bloc in Europe and the US put carbon trading at the core of their emissions reduction strategy, while the two largest emitters of carbon in the Third World, China and India, are the main beneficiaries of CDM financing.

What that means is that problems caused when Al Gore's US delegation brought pro-corporate compromises to Kyoto in 1997 – deceitfully promising US sign-on to Kyoto in exchange for carbon trading – will now amplify and haunt this debate for a long time to come. For what we have witnessed since Kyoto came into effect in February 2005 is a climate-reduction stalemate by a coalition of selfish, fossil-fuel addicted powers. Terribly weak targets may get a mention (or even no mention, as at the Bali 2007 Conference of Parties), but market mechanisms will be invoked as the 'solution' so as to appease polluting capital and associated governments, especially Washington. Some of the less principled environmental NGOs and opportunistic Third World elites (and even campaigners like Wangari Maathei) will sign up, as has become a habit in such global governance gambits.

Market mechanisms – especially carbon trading and offsets – allow corporations and governments generating greenhouse gases to seemingly reduce their net emissions. They can do this, thanks to the Kyoto Protocol, by trading for others' 'certified emissions reductions' (e.g. CDM projects in the Third World) or emissions rights (e.g. Eastern Europe's 'hot air' that followed the 1990s economic collapse). The pro-trading rationale is that once property rights are granted to polluters for these emissions, even if given not auctioned (hence granting a generous giveaway), a 'cap' can be put on a country's or the world's total emissions. It will then be progressively lowered, if there is political will. So as to minimize adverse economic impact, corporations can stay within the cap even by emitting way above it, by buying others' rights to pollute. But critics[1] argue that the carbon market isn't working, for several reasons:

- the idea of inventing a property right to pollute is effectively the 'privatization of the air';

- greenhouse gases are complex and their rising production creates a non-linear impact which cannot be reduced to a commodity exchange relationship (a ton of CO_2 produced in one place accommodated by reducing a ton in another, as is the premise of the emissions trade);

- the corporations most guilty of pollution and the World Bank – which is most responsible for fossil fuel financing – are the driving forces behind the market, and can be expected to engage in systemic corruption to attract money into the market even if this prevents genuine emissions reductions;

- many of the offsetting projects – such as monocultural timber plantations, forest 'protection' and landfill methane-electricity projects – have devastating impacts on local communities and ecologies, and have been hotly contested;

- the price of carbon determined in these markets is haywire, having crashed by half in a short period in April 2006 and by two-thirds in 2008, thus making mockery of the idea that there will be a sufficient market mechanism to turn the society towards renewable energy;

- there is a serious potential for carbon markets to become an out-of-control, multi-trillion dollar speculative bubble, similar to exotic financial instruments associated with Enron's 2002 collapse (indeed, many Enron employees populate the carbon markets);

- as a 'false solution' to climate change, carbon trading encourages merely small, incremental shifts, and thus distracts us from a wide range of radical changes we need to make in materials extraction, production, distribution, consumption and disposal; and

- the idea of market solutions to market failure ('externalities') is an ideology that rarely makes sense, and especially not following the world's worst-ever financial market failure, and especially not when the very idea of derivatives

– a financial asset whose underlying value is several degrees removed and also subject to extreme variability – was thrown into question.

Most scientists insist that at least an 80% drop in emissions will be necessary within four decades, with the major cuts (of at least 45%) before 2020. To achieve this, carbon markets won't work, as the leading US climate scientist, James Hansen, concluded in leading the intellectual opposition to Barack Obama's cap and trade scheme. Obama's legislation – the Waxman-Markey bill that passed the US House of Representatives in June 2009 (with similar Senate legislation bogged down at the time of writing in September 2009) – was so profoundly flawed that the more ambitious wing of environmental civil society argued it should be scrapped, especially because of the legislation's destruction of Environmental Protection Agency powers to regulate carbon pollution.[2] Even the financial speculator George Soros criticizes cap and trade:

> The cap and trade system of emissions trading is very difficult to control and its effects are diluted. It is pretty much breaking down because there is no penalty for developing countries not to add to their pollution. You count the saving but you don't count the added pollution going on. As a world, I don't think we are getting our act together on climate change at the moment... [CDMs] are not effective: you buy credits in third world countries that don't have a cap on emissions and you can get carbon credits whether you can sell them or not... It is precisely because I am a market practitioner that I know the flaws in the system.[3]

To be sure, one wing of civil society – e.g., campaigners Avaaz, the World Wildlife Federation and the Climate Action Network – are apparently asking that Copenhagen 'seal the deal' no matter such flaws, which can be partially explained by the fact that some in the latter group have substantial conflicts of interest as carbon-traders themselves. According to Michael Dorsey, professor of political ecology at the US's Dartmouth College, these include CAN board member Jennifer Morgan of the Worldwide Fund for Nature, who took leave for two years to direct work on Climate and Energy Security at carbon trading firm E3G, Kate Hampton, formerly of Friends of the Earth, who joined Climate Change Capital as head of policy while simultaneously advising the EU on energy and the environment, working for the California Environmental Protection Agency, and acting as president of International Carbon Investors and Services, and several others.

Dorsey concludes: 'After more than a decade of failed politicking, many NGO types...are only partially jumping off the sinking ship – so as to work for industries driving the problem. Unfortunately, many continue to influence NGO policy from their current positions, while failing to admit to or even understand obvious conflicts of interest'.[4]

Critics condemn carbon trading for these and many other reasons, and term the emissions trade a 'false solution'. In contrast, central to a genuine solution to climate crisis is the task of raising the world's standards of living in a manner not characterized by the fossil fuel addiction of industrial society. Climate-related finance will be required, and this might logically begin with the North's payment of ecological debt to the South for excess use of environmental space and for the damage done to many ecosystems already, and in future when vast damages are anticipated especially in Africa.

Ecological Debt Defined

According to the Quito group Accion Ecologica: 'ecological debt is the debt accumulated by Northern, industrial countries toward Third World countries on account of resource plundering, environmental damages, and the free occupation of environmental space to deposit wastes, such as greenhouse gases, from the industrial countries.'[5] The term came into professional use in 1992 at the Earth Summit of the United Nations in Rio de Janeiro of 1992, with NGOs promoting the concept through an 'Alternative Treaty'. An initial voice was the Institute of Political Ecology in Santiago, Chile, and contributed to world consciousness about CFC damage to the ozone layer. A Colombian lawyer, José María Borrero, wrote a 1994 book on the topic, and further research and advocacy was provided by the Foundation for Research on the Protection of the Environment. By 1999 Friends of the Earth International and Christian Aid agreed to campaign against Ecological Debt, especially in relation to climate damage.

The leading scientist in the field, Autonomous University of Barcelona's Joan Martinez-Alier, calculates ecological debt in many forms: 'nutrients in exports including virtual water, the oil and minerals no longer available, the biodiversity destroyed, sulphur dioxide emitted by copper smelters, the mine tailings, the harms to health from flower exports, the pollution of water by mining, the commercial use of information and knowledge on genetic resources, when they have been appropriated gratis ('biopiracy'), and agricultural genetic resources.' As for the North's 'lack of payment for environmental services or for the disproportionate use of Environmental Space,' Martinez-Alier criticizes 'imports of solid or liquid toxic waste, and free disposal of gas residues (carbon dioxide, CFCs, etc).' According to Martinez-Alier:

> The notion of an Ecological Debt is not particularly radical. Think of the environmental liabilities incurred by firms (under the United States Superfund legislation), or of the engineering field called 'restoration ecology', or the proposals by the Swedish government in the early 1990s to calculate the country's environmental debt.[6]

The sums involved are potentially vast. As Martinez-Alier puts it, 'tropical rainforests used for wood exports have an extraordinary past we will never know and ongoing biodiversity whose destruction we cannot begin to value.' However, 'although it is not possible to make an exact accounting, it is necessary to establish the principal categories and certain orders of magnitude in order to stimulate discussion... If we take the present human-made emissions of carbon, [this represents] a total annual subsidy of $75 billion is forthcoming from South to North.'[7] Leading ecofeminist Vandana Shiva[8] and former South Centre director Yash Tandon[9] estimate that wild seed varieties alone account for $66 billion in annual biopiracy benefits to the US. Examples of biopiracy in

Africa, according to a 2005 study commissioned by Edmonds Institute, African Centre for Biosafety, include:

- three dozen cases of African resources – worth $billions – captured by firms for resale without adequate 'Access and Benefit Sharing' agreements between producers and the people who first used the natural products;
- diabetes drug produced by a Kenyan microbe and Libyan/Ethiopian treatment;
- antibiotics from Gambian termite hill and giant West African land snails;
- antifungal from Namibian giraffe and nematocidal fungi from Burkina Faso;
- infection-fighting amoeba from Mauritius;
- Congo (Brazzaville) treatment for impotence;
- vaccines from Egyptian microbes;
- South African and Namibian indigenous appetite suppressant Hoodia;
- drug addiction treatments, multipurpose kombo butter from Central, W.Africa;
- beauty, healing treatment from Okoumé resin in Central Africa;
- skin and hair care from the argan tree in Morocco, Egyptian 'Pharaoh's Wheat', bambara groundnut and 'resurrection plant';
- endophytes and improved fescues from Algeria and Morocco;
- groundnuts from Malawi, Senegal, Mozambique, Sudan and Nigeria;
- Tanzanian impatiens; and
- molluscicides from the Horn of Africa.[10]

A partial ecological debt accounting was published by environmental scientists in early 2008, and counted $1.8 trillion in concrete damages over several decades.[11] According to co-author Richard Norgaard, ecological economist at the University of California, Berkeley, generated a crucial finding: 'At least to some extent, the rich nations have developed at the expense of the poor, and, in effect, there is a debt to the poor. That, perhaps, is one reason that they are poor. You don't see it until you do the kind of accounting that we do here'.[12] The study included factors such as greenhouse gas emissions, ozone layer depletion, agriculture, deforestation, overfishing, and the conversion of mangrove swamps into shrimp farms, but the researchers did not (so far) succeed in calculating other damages, e.g. excessive freshwater withdrawals, destruction of coral reefs, bio-diversity loss, invasive species, and war.

Another route into the intellectual challenge of calculating ecological debt was taken by the World Bank (2006) in its estimates of tangible wealth (in the book *Where is the Wealth of Nations?*). In addition to resource depletion and rent outflows, there are also other subsoil assets, timber resources, non-timber forest resources, protected areas, cropland and pastureland to account for. The 'produced capital' normally captured in GDP accounting is added to the tangible wealth. In the case of Ghana, to consider one example, that amounted to $2,022 per person in 2000. The same year, the Gross National Saving of

Ghana was $40 and education spending was $7. These figures require downward adjustment to account for the consumption of fixed capital ($19), as well as the depletion of wealth in the form of stored energy ($0), minerals ($4) and net forest assets ($8). In Ghana, the adjusted net saving was $16 per person in 2000. But given population growth of 1.7%, the country's wealth actually shrunk by $18 per person in 2000. Notwithstanding the World Bank's conservative counting bias,[13] Africa shows evidence of net per capita wealth reduction, largely traceable to the extraction of nonrenewable resources that is not counterbalanced by capital investment from those firms doing the extraction (Table 1).

	Income per capita before adjustment ($)	Change in wealth per capita after adjustment ($)
Benin	360	-42
Botswana	2925	814
Burkina Faso	230	-36
Burundi	97	-37
Cameroon	548	-152
CapeVerde	1195	-81
Chad	174	-74
Comoros	367	-73
Rep of Congo	660	-727
Côte d'Ivoire	625	-100
Ethiopia	101	-27
Gabon	3370	-2241
The Gambia	305	-45
Ghana	255	-18
Kenya	343	-11
Madagascar	245	-56
Malawi	162	-29
Mali	221	-47
Mauritania	382	-147
Mauritius	3697	514
Mozambique	195	-20
Namibia	1820	140
Niger	166	-83
Nigeria	297	-210
Rwanda	233	-60
Senegal	449	-27
Seychelles	7089	904
South Africa	2837	-2
Swaziland	1375	8
Togo	285	-88
Zambia	312	-63
Zimbabwe	550	-4

Table 1: African countries' adjusted national wealth, 2000[14]

African Leaders United?

How is Africa reacting? Generally the leadership of African countries has cooperated with those doing the resource extraction and utilizing Africa's ecological space, with only complaints by exploited communities, by workers subject to safety/health violations and exploitation, and by environmentalists. However, finally in mid-2009, the African Union's leadership on climate issues became a force to be reckoned with, thanks to Ethiopian prime-minister Meles Zenawi, who also chaired the New Partnership for Africa's Development and thus was invited to G20 gatherings along with the South African government. Sometimes considered a US proxy power in the Horn of Africa – thanks to the disastrous, Washington-sponsored 2007 invasion of neighboring Somalia – Zenawi is rather more complex. He was once a self-described Marxist but became a tyrant whose troops killed scores of students and other democrats. It is ironic, thus, for Zenawi to lead the ecological debt charge, reportedly demanding a minimum of $67 billion – and up to $200 billion – annually from the North by 2020.[15]

Ironic or tragic, nevertheless this voice must be heard, considering how much Africa will be devastated by the climate crisis. The most shocking probable outcome of climate change is, according to the UN Intergovernmental Panel on Climate Change director R.K. Pachauri, 'that there could be a possible reduction in yields in agriculture of: 50% by 2020 in some African countries... In Africa, crop net revenues could fall by as much as 90% by 2100, with small-scale farmers being the most affected.'[16] The Climate Change Vulnerability Index, calculated in 2009 'from dozens of variables measuring the capacity of a country to cope with the consequences of global warming', listed 22 African countries out of 28 across the world at 'extreme risk', whereas the United States is near the bottom of the world rankings of countries at risk even though it is the leading per capita contributor to climate change.[17]

There is no question that those most responsible should pay reparations, now that there is near-universal awareness of the damage being done by rising greenhouse gas emissions, and by the ongoing stubborn refusal by the rich to cut back. The amounts can be debated, for of course $67 billion/year for Africa is way too low, given how many incalculably valuable species will be lost, how much devastation to individuals and communities is already underway, how many economies will falter, how much ecology is threatened.

The question is not mainly a technical one, but related to power. Behind African elites' considerations are the threat to repeat their performance in Seattle in 1999 and Cancun in 2003, when denial of consent in World Trade Organization negotiations were the proximate cause of the summits' collapse on both occasions. On September 3, 2009, Zenawi issued a strong threat from Addis Ababa about the upcoming Copenhagen conference: 'If need be we are prepared to walk out of any negotiations that threatens to be another rape of our continent.'[18]

To gather that power, Zenawi established the Conference of African Heads of State and Government on Climate Change: chairpersons of the AU and the

AU Commission, representatives of Ethiopia, Algeria, the Democratic Republic of Congo, Kenya, Mauritius, Mozambique, Nigeria, Uganda, Chairpersons of the African Ministerial Conference on Environment and Technical Negotiators on climate change from all member states. They met at the AU Summit in Sirte, Libya in July 2009, agreeing that Africa would have a sole delegation to Copenhagen with a united front and demands for compensation. According to AU official Abebe Hailegabriel, 'Trillions of dollars might not be enough in compensation. Thus there must be an assessment of the impact before the figure.' Added AU head Jean Ping, 'Africa's development aspirations are at stake unless urgent steps are taken to address the problems of climate change. Climate change will fundamentally affect productivity, increase the prevalence of disease and poverty...and trigger conflict and war'.[19]

The most important African negotiator – and largest CO_2 emitter (responsible for more than 40% of the continent's CO_2) – is South Africa.[20] Long seen as a vehicle for Western interests in Africa, Pretoria's negotiators have two conflicting agendas: increasing Northern payments to Africa (a longstanding objective of the New Partnership for Africa's Development, which requested $64 billion per annum in aid and investment concessions during the early 2000s, much of which would be channeled through Pretoria as an aspiring good-governance gate-keeper); and increasing its own CO_2 outputs through around 2050, when the Long-Term Mitigation Scenario – South Africa's official climate cap – would come into effect (only at that point are significant emissions declines offered as a scenario). In the meantime, Pretoria has earmarked more than $100 billion for emissions-intensive coal and nuclear fired electricity generation plants due to be constructed during 2010-15, which would amplify Africa's climate crisis, requiring more resources from the North for adaptation. But the South African ruling class does not, officially, see itself as an ecological creditor. As the environment minister, Buyelwa Sonjica put it in September 2009: 'We expect money. We need money to be made available... we need money as of yesterday for adaptation and mitigation.'[21]

South African negotiators are also amongst leaders of the G77 on this issue, and are on record from August 2009 demanding that 'at least 1% of global GDP should be set aside by rich nations', according to one report, so as to help developing countries conduct research, improve flood control, protect their coastlines, create seed banks and take other steps to cope with the severe storms and droughts linked to climate change. The money also could help poor countries obtain technology to reduce their carbon emissions. Alf Wills, a top South African environmental official, summed up the position going into Copenhagen: 'No money, no deal.'[22]

There are other allies, especially Bolivia, whose submission to the UNFCCC in 2009 made the ecological debt demand explicitly:

> The climate debt of developed countries must be repaid, and this payment must begin with the outcomes to be agreed in Copenhagen. Developing countries are not seeking economic handouts to solve a problem we did not cause. What we call for is full payment of the debt owed to us by developed countries for

threatening the integrity of the Earth's climate system, for over-consuming a shared resource that belongs fairly and equally to all people, and for maintaining lifestyles that continue to threaten the lives and livelihoods of the poor majority of the planet's population. This debt must be repaid by freeing up environmental space for developing countries and particular the poorest communities. There is no viable solution to climate change that is effective without being equitable. Deep emission reductions by developed countries are a necessary condition for stabilising the Earth's climate. So too are profoundly larger transfers of technologies and financial resources than so far considered, if emissions are to be curbed in developing countries and they are also to realize their right to development and achieve their overriding priorities of poverty eradication and economic and social development. Any solution that does not ensure an equitable distribution of the Earth's limited capacity to absorb greenhouse gases, as well as the costs of mitigating and adapting to climate change, is destined to fail.[23]

Bolivia's government is generally driven by left-leaning popular forces in the rural and urban social movements. Other countries that have expressed similar sentiments include Venezuela, Paraguay, Malaysia and Sri Lanka. In Africa, where most countries do not have such strong movements, what is the state of play around civil society's ecological debt demands?

Civil Society Reactions

The threat of a walkout at Copenhagen was contemplated with interest by civil society groups, both in Africa and across the world. The former became increasingly active in August 2008 when Africa chapters of Jubilee South converged in Nairobi to debunk limited 'debt relief' by Northern powers and to plan the next stage of financial campaigning. Nairobi-based Africa Jubilee South co-coordinator Njoki Njehu concluded, 'Africa and the rest of the Global South are owed a huge historical and ecological debt for slavery, colonialism, and centuries of exploitation.' Njehu says Jubilee's challenge as it rebuilds is to link issues as diverse as food sovereignty, climate change, trade and EU Economic Partnership Agreements and continuing debt bondage. 'From the initial 13 countries that participated in the Jubilee South founding conference in Johannesburg in 1999, the Africa Jubilee South network has grown to organizations and movements from 29 countries.'[24]

A year later in Nairobi, the Africa Peoples Movement on Climate Change (a Nairobi-based initiative hosted by Ibon and the Kenyan Debt Relief Network) pronounced:

- We reject the principle and application of Carbon Trading, which is a false solution based on inventing a perverse property right to pollute, a property right to air;
- We demand that human rights and values be placed at the centre of all global, national and regional solutions to the problem of climate change;
- We call on colleagues in the social and economic justice movement globally to rigorously campaign against the undemocratic corporate led agendas which will dominate the deliberations and processes at COP 15;

– We emphasize that ecological, small holder, agro-biodiversity based food production can ensure food and seed sovereignty and address climate change in Africa;

– We support the call by African leaders for reparations on Climate Change and support the initiative of the upcoming AU ministers of environment meeting and call for African governments to embrace more people centred alternatives for the African peoples;

– We urge African governments to engage civil society groups positively and collaborate with them to build common national and international responses on the problems of climate change.[25]

Another node of ecological debt organizing is the World Council of Churches (WCC), whose Central Committee adopted a formal statement in September 2009 on the North's 'deep moral obligation to promote ecological justice by addressing our debts to peoples most affected by ecological destruction and to the earth itself.' It is useful to consider the WCC's analysis because it does not stop at the debt, but attacks the mode of production itself:

> We call for the recognition, repayment and restitution of ecological debt in various ways, including non-market ways of compensation and reparation, that go beyond the market's limited ability to measure and distribute... This warrants a re-ordering of economic paradigms from consumerist, exploitive models to models that are respectful of localized economies, indigenous cultures and spiritualities, the earth's reproductive limits, as well as the right of other life forms to blossom. And this begins with the recognition of ecological debt.[26]

The WCC Central Committee made several requests, including that the environmental justice and faith community:

– Urges Northern governments, institutions and corporations to take initiatives to drastically reduce their greenhouse gas (GHG) emissions within and beyond the United Nations Framework Convention on Climate Change (UNFCCC), which stipulates the principles of historical responsibility and 'common, but differentiated responsibilities' (CDR), according to the fixed timelines set out by the UNFCCC report of 2007.

– Urges WCC member churches to call their governments to adopt a fair and binding deal, in order to bring the CO2 levels down to less than 350 parts per million (ppm), at the Conference of Parties (COP 15) of the UNFCCC in Copenhagen in December 2009, based on climate justice principles, which include effective support to vulnerable communities to adapt to the consequences of climate change through adaptation funds and technology transfer.

– Calls upon the international community to ensure the transfer of financial resources to countries of the South to keep petroleum in the ground in fragile environments and preserve other natural resources as well as to pay for the costs of climate change mitigation and adaptation based on tools such as the Greenhouse Development Rights (GDR) Framework.

– Demands the cancellation of the illegitimate financial debts of Southern countries, most urgently for the poorest nations, as part of social and ecological compensations, not as official development assistance.[27]

It is evident at this writing (October 2009) that the COP15 – or its immediate successors – will not make the urgent progress required in three areas: first, cutting emissions to the levels at which climate disaster can be averted (at least 45% by 2020); or second, replacing the 'false solution' of carbon trading with genuine emissions cuts; or third, on providing restitution and reparations to Third World peoples, or even canceling their illegitimate debts. To be sure, in September 2009, the European Union acknowledged its responsibility to begin paying ecological debt, but only up to $22 billion annually to fund adaptation, roughly 1/7th of what EU environment commissioner Stavros Dimas observed would be required by 2020 ($145 bn). Some of that would be subtracted from existing aid. The EU damage estimates were considered far too conservative, as China's mitigation and adaptation costs alone would be $438 bn annually by 2030, according to Beijing. According to one report, the EU view is that emissions trading should be the basis of 'much of the shortfall', because 'The international carbon market, if designed properly, will create an increasing financial flow to developing countries and could potentially deliver as much as €38bn per year in 2020.'[28] As noted above, however, this strategy is replete with fatal flaws.

Because of the influence of big capital and pro-market ideology on Northern governments in the Kyoto process to date, not only will emissions continue rising and the ecological debt *not* be properly paid, carbon trading will *not* be dropped as a central EU and US strategy. As a result, critical narratives will become more common, and in turn will force serious advocates of environmental justice to raise very important strategic issues about how to get the North to repay the ecological debt.

Conclusion: Repaying the Debt?

Existing North-South redistributive processes are not effective. Northern foreign aid to the South goes only a small way towards ecological debt repayment. It is a far lower sum (and falling) than military spending (which is rising), and in any case 60% is 'phantom aid', according to the Johannesburg-based agency Action Aid.[29] Aid is also a tool of imperialism. Other North-South payments to Africa are yet more dubious, including the debt relief promised in 2005. In spite of enormous hype at the Gleneagles G8 meeting, the International Monetary Fund calculates that notwithstanding a lower debt stock, the actual debt repayments of the lowest-income African countries stayed stable from 2006-08 and then increased 50% in 2009 as a percentage of export earnings.[30] So although there was debt cancellation, it was on unrepayable debt, with debilitating debt servicing charges for low-income African countries still preventing local accumulation and provision of social services, much less financing preparations for climate change adaptation.

There are important debates about who should pay what share. But in general, it is important to note that ecological debt results from the unsustainable production and consumption systems adopted by elites in the Northern countries, which are to some extent generalized across the Northern populations. Hence even poor and working-class people in the North, often through no fault of their own, are tied into systems of auto-centric transport or conspicuous consumption, which mean that they consume far more of the Earth's resources than do working-class people of the South.

Hence, recalling the WCC position in favour of a 'Greenhouse Development Rights' framework, it is worth considering that a per capita 'right to pollute' – and to trade pollution rights – will have some of the same dubious outcomes. The bigger questions which GDRs pose are whether environmental justice can be measured merely in terms of formal 'equality'; whether environmental justice is instead historical, political-economic, and grounded in social struggles of those adversely affected; and in turn, whether environmental justice should not aim higher, for a broader, deeper eco-social transformation? The WCC hints at such a perspective, but the GDR approach may foreclose it by reducing the challenge to incremental reformism. When it comes specifically to GDRs as a methodology for calculating debt liabilities and beneficiaries, Larry Lohmann of The Cornerhouse (a British development institute) critiques the model's tacit endorsement of a long-discredited concept of 'development' that condescendingly sees 'resilience' as 'far beyond the grasp of the billions of people that are still mired in poverty', and that singles out for special climate blame 'subsistence farming, fuel wood harvesting, grazing, and timber extraction' by 'poor communities' awaiting Northern tutelage in capital flows, social networking, carbon trading and methods for holding policymakers accountable.[31]

Is a rights-based approach to environmental services preferable, as a strategy for demanding and properly redistributing ecological debt payments from North to South? South Africa's 'Free Basic Services' provide insights into the possibilities and limitations of rights discourses for redistributing wealth from North to South. In 2000 (just after Nelson Mandela left the presidency), the ruling party's municipal campaign platform highlighted this promise: 'African National Congress-led local government will provide all residents with a free basic amount of water, electricity and other municipal services, so as to help the poor. Those who use more than the basic amounts will pay for the extra they use.' But as can be shown in excruciating detail, it was the failure to move beyond individualized nuclear-family household units and tokenistic amounts of free basic water (6 kl/household/month) and electricity (50kWh/household/month) that led to many 'service delivery protests' during subsequent years, contributing to South Africa's standing as the country with the most per capita social unrest. Attempts to gain justice in these cases through the court system – even the Constitutional Court in September 2009 – proved extremely frustrating.[32]

Juridical approaches to ecological debt may not be optimal, although interesting precedents have emerged. In November 2008 a San Francisco court began considering an ecological debt and reparations lawsuit – under the Alien Tort Claims Act – filed by Larry Bowoto and the Ilaje people of the Niger Delta against Chevron for involvement in 1998 murders reminiscent of those that took the life of Ken Saro-Wiwa and eight other Ogoni leaders on November 10, 1995. The first judgments went against Bowoto *et al* but appeals are in process. In June 2009, Shell Oil agreed in a similar lawsuit to an out-of-court settlement with reparations payments of $15.5 million. Although representing just four hours' worth of Shell profits, it was considered a crucial step in establishing liability and disincentivizing corporate exploitation of people and nature. In late 2009, further reparations lawsuits were expected in the New York Second District Court by victims of apartheid who initially requested $400 billion in damages from US corporations which profited from South African operations during the same period. Supreme Court justices had so many investments in these companies that in 2008 they bounced the case back to a lower New York court to decide, effectively throwing out an earlier judgment against the plaintiffs: the Jubilee anti-debt movement, the Khulumani Support Group for apartheid victims, and 17,000 other black South Africans. When Judge Clara Scheindlin replaced the late John Sprizzo, the case suddenly was taken seriously and in March 2009 moved a step closer to trial when she rejected the corporations' attempt to have it dismissed.[33]

Beyond these kinds of tort actions, will courts start declaring climate-related ecological debt a valid concept? Environmental rights to protection from climate change were explored in a court case filed by Friends of the Earth, Greenpeace and the cities of Boulder, Colorado, Arcata, Santa Monica and Oakland in California, against the US Export-Import Bank and Overseas Private Investment Corporation, which had invested, loaned or insured $32 billion in fossil fuel projects from 1990–2003 without regard to the US National Environmental Policy Act (NEPA). At present, only US cities have formal standing to sue for damages from climate change under NEPA, in the wake of a 2005 federal ruling. The out-of-court-settlement in February 2009 meant that the defendants will in future incorporate CO_2 emissions into planning, but there are prospects for further suits that extend into identification and payment of damages.[34]

There are quite obvious limits to prospects for court relief under the Alien Tort Claims Act or NEPA, the two most advanced areas. Hence it would be consistent to also proceed with more immediate strategies, as well as direct action tactics. As Al Gore expressed it in 2007, 'I can't understand why there aren't rings of young people blocking bulldozers and preventing them from constructing coal-fired power plants'.[35] Arguing that 'Protest and direct action could be the only way to tackle soaring carbon emissions,' the US National Aeronautic and Space Administration's leading climate scientist, James Hansen, 'The democratic process is supposed to be one person one vote, but it turns out that money is talking louder than the votes. So, I'm not surprised that people are getting frustrated. I think that peaceful demonstration is not out of order, because we're running out of time.'[36] Hansen himself moved to direct action in 2009, demonstrating at coal-fired power plants in Coventry, England and West Virginia (where at the latter site he was arrested).

But the most effective examples of direct action come from the Global South, especially the Niger Delta. In January 2007, at the World Social Forum in Nairobi, many other groups became aware of this movement thanks to eloquent activists from the Delta, including the Port Harcourt NGO Environmental Rights Action (ERA). In separate disruptions of production (including armed interventions), the Movement for the Emancipation of the Niger Delta prevented roughly 80% of the country's oil from being extracted, although a cease-fire was called in mid-2009.[37] The strategy is consistent with the grassroots, coalface and fenceline demands of civil society activists in the Oilwatch network (headquartered at ERA) to *leave the oil in the soil, the coal in the hole, the tar sand in the land.* Activists from Accion Ecologica popularized this approach in their struggle to halt exploitation of the oil beneath the Yasuni park in the Ecuadoran Amazon. The German state development agency GTZ conceded to a $50 mn/year grant, although Yasuni may become a pilot carbon trading case unless Ecuadoran environmental and indigenous rights activists can resist.[38] By October 2009, the hope lay in Ecuadorean Foreign Minister Fander Falconi's public announcement that he would review the Yasuni to reconsider any relationship with the carbon market.

The legacy of keeping oil in the soil includes Alaskan and Californian environmentalists who halted drilling and even exploration. In Norway, the global justice group, ATTAC, took up the same concerns in an October 2007 conference and began the hard work of persuading wealthy Norwegian Oil Fund managers that they should use the vast proceeds of their North Sea inheritance to repay Ecuadorans some of the ecological debt owed via a Yasuni grant. In Australia, regular blockades of Newcastle coal transport (by rail and sea) by the activist group, Rising Tide, correspond to Gore's injunction.

Canada is another Northern site where activists are hard at work to leave the oil in the soil. In a November 2007 conference in Edmonton, the University of Alberta Parkland Institute addressed the need to halt development of tar sand deposits (which require a liter of oil to be burned for every three extracted and devastate local water, fisheries, and air quality). Institute director Gordon Laxer laid out careful arguments for strict limits on the use of water and greenhouse gas emissions in tar sand extraction; realistic land reclamation plans (including a financial deposit large enough to cover full-cost reclamation up-front); no further subsidies for the production of dirty energy; provisions for energy security for Canadians (since so much of the tar sand extract is exported to the U.S.); and much higher economic rents on dirty energy to fund a clean energy industry (currently Alberta has a very low royalty rate). These kinds of provisions would strictly limit the extraction of fossil fuels and permit oil to leave the soil only under conditions in which much greater socio-ecological and economic benefit is retained by the broader society.[39]

There are many other examples where courageous communities and environmentalists have lobbied successfully to keep nonrenewable resources (not just fossil fuels) in the ground for the sake of the environment, community stability, disincentivizing political corruption, and workforce health and safety. For many victims, the extraction of these resources is incredibly costly in terms of local land use, water extraction, energy consumption, and political corruption, and requires constant surveillance and community solidarity. The adverse balance of forces noted at the outset should be restated: the climate negotiators and corporations of the Global North will consistently fail to make sustained emissions cuts; to depart from the ineffectual, dangerous carbon trading modus operandi; and to offer adequate reparations for the ecological debt. This will, in turn, require national states to take stronger actions, such as Zenawi has threatened, or as Bolivia's Evo Morales has done in his ecological debt statement, or as Ecuador's Rafael Correa did in defaulting on odious foreign debt in early 2009. But most of all, it will require people of conscience across the world to become involved, and to offer solidarity and activism aimed at leaving fossil fuels in the ground.

Notes

*   Patrick Bond thanks colleagues – especially Professor Dennis Brutus – at the University of KwaZulu-Natal Centre for Civil Society in Durban: http://www.ukzn.ac.za/ccs. This paper was first presented to the Gyeongsang University Institute for Social Studies (supported by the Korea Research Foundation's grant KRF 2007-411-J04602).

1   The analysis generated by Larry Lohmann is probably the most sophisticated, e.g., Lohmann, L. (2006) *Carbon Trading: A critical conversation on climate change, privatization and power.* Uppsala: Dag Hammarskjold Foundation, http://www.dhf.uu.se/pdffiler/DD2006_48_carbon _trading/carbon_trading_web_HQ.pdf and more recently, Lohmann, L. (2010) 'Uncertainty markets and carbon markets: Variations on Polanyian themes', *New Political Economy;* Lohmann, L. (2009a) 'Climate as investment', *Development and Change;* see also Lohmann's Chapter 2 in this volume; Lohmann, L. (2009) 'Regulatory challenges for financial and carbon markets,' *Carbon & Climate Law Review,* 3(2); and Lohmann, L. (2009) 'Toward a different debate in environmental accounting: The cases of carbon and cost-benefit', *Accounting, Organisations and Society,* 34(3-4).

2   See, e.g, the groups Biological Diversity, Climate SOS and the Sustainable Energy and Economy Network, http://www.biologicaldiversity.org/action/toolbox/ ACESA/sign-on_letter.html; http://www.climatesos.org and http://www.seen.org.

3   Wheelan, H. (2007) 'Soros slams emissions trading systems: Market solution is "ineffective" in fighting climate change', Responsible Investor, 18 October, http://www.responsible-investor.com/home/article/soros_slams_emissions_cap_ and_trading_systems/.

4   Bond, P. (2009) 'A timely death?', *New Internationalist,* January, http://www.newint.org/ features/2009/01/01/climate-justice-false-solutions/.

5   Ecologica, A. (2000) 'Trade, climate change and the ecological debt,' Unpublished paper, Quito.

6   Martinez-Alier, J. (2003) 'Marxism, Social Metabolism and Ecologically Unequal Exchange', Paper presented at Lund University Conference on World Systems Theory and the Environment, 19-22 September.

7   Martinez-Alier, J. (2003) 'Marxism, Social Metabolism and Ecologically Unequal Exchange'.

8   Shiva, V. (2005) 'Beyond the WTO Ministerial in Hong Kong', ZNet Commentary, 26 December.

9   http://www.globalpolicy.org/socecon/develop/devthry/well-being/2000/tandon.htm.

10   McGown, J. (2006) 'Out of Africa: Mysteries of Access and Benefit Sharing', Edmonds Washington, the Edmonds Institute and Johannesburg, the African Centre for Biosafety.

11   Srinivasan, U., S. Carey, E. Hallstein, P. Higgins, A. Ker, L. Koteen, A. Smith, R. Watson, J. Harte, and R. Norgaard (2008) 'The debt of nations and the distribution of ecological impacts from human activities', *Proceedings of the National Academy of Sciences of the United States of America,* 105(5), http://www.pnas.org/ content/105/5/1768.

12   The Guardian (2008) 'Rich countries owe poor a huge environmental debt', 21 January, http://www.guardian.co.uk/ science/2008/jan/21/environmental.debt1.

13   The Bank's estimates are conservative for at least three reasons: a minimalist definition based upon international pricing in 2000 (not potential future values when scarcity becomes a more crucial factor, especially in the oil industry); the partial calculation of damages to the local environment, to workers' health/safety, and especially to women in communities around mines; and the Bank's use of average – not marginal – cost resource rents also probably leads to underestimations of the depletion costs.

14   World Bank (2006) *Where is the Wealth of Nations?,* Washington, DC, p.66.

15   McLure, J. (2009) 'Ethiopian leader chosen to represent Africa at climate summit,' Addis Ababa, 1 September.

16   Pachauri, R.K. (2008) 'Summary of testimony provided to the House Select Committee on Energy Independence and Global Warming,' US Congress, Washington DC, globalwarming.house.gov/tools/assets/files/0342.pdf.

17   Agence France Press (2009) 'Albania to Zimbabwe: the climate change risk list', 2 September.

18   Ashine, A. (2009) 'Africa threatens withdrawal from climate talks', *The Nation,* 3 September.

19   Bond, P. (2009) 'Don't play games with humanity's future', *The Mercury,* 2 September.

20   Bond, P., R. Dada and G. Erion (2009) *Climate Change, Carbon Trading and Civil Society,* Pietermaritzburg: University of KwaZulu-Natal Press.

21   Sapa (2009) 'SA not "compromising anything" at climate change negotiations – Sonjica,' 15 September.

22   Sapa (2009) 'SA on climate change: "No money, no deal"', 5 August.

23   Republic of Bolivia (2009) 'Submission to the Ad Hoc Working Group on Long-term Cooperative Action under the UN Framework Convention on Climate Change,' La Paz, April.

24   Bond, P. and D. Brutus (2008) 'Ecological debt and our centre's survival', ZCommentaries, 21 August, http://www.zcommunications.org/zspace/commentaries /3594.

25   Africa Peoples Movement on Climate Change (2009) 'Confronting the Climate Crisis: Preparing for Copenhagen and Beyond', Nairobi, 30 August.

26   World Council of Churches Central Committee (2009) 'Statement on eco-justice and ecological debt,' Geneva, 2 September.

27   World Council of Churches Central Committee (2009) 'Statement on eco-justice and ecological debt,'.

28   Chaffin, J. and E. Crooks (2009) 'EU sets out €15bn climate aid plan', *Financial Times,* 8 September.

29   Action Aid (2005) 'Real Aid: An Agenda for Making Aid Work', Johannesburg.

30   International Monetary Fund (2009) 'The implications of global financial crisis for low-income countries', Washington, DC.

31   Lohmann, L. (2009) personal correspondence.

32   Bond, P. (2010) 'South Africa's "rights culture" of water consumption: Breaking out of the liberal box and into the commons?', in B. Johnston, L. Hiwasaki, I. Klaver, and V. Strang

(eds), *Water, Cultural Diversity & Global Environmental Change: Emerging Trends, Sustainable Futures?*. Paris: UNESCO.

33   Bond, P. and K.Sharife (2009) 'Apartheid reparations and the contestation of corporate power in Africa', *Review of African Political Economy*, 119.

34   Friends of the Earth (2009) 'Landmark global warming lawsuit settled', Washington, DC, 6 February.

35   Kristoff, N. (2007) 'The big melt', *New York Times*, 16 August.

36   Adam, D. (2009) 'Leading Climate Scientist: "Democratic Process Isn't Working"', *The Guardian*, 18 March.

37   Mistilis, K. (2009) 'Niger Delta standoff', *Pambazuka News*, 442, 16 July, http://pambazuka.org/en/category/comment/57769.

38   Gallagher, K. (2009) 'Paying to keep oil in the ground', *The Guardian,* 7 August.

39   Laxer, G. (2007) 'Freezing in the dark?', Presentation, Parkland Institute, University of Alberta, Edmonton, 7 November.

25

Rethinking the Legal Regime for Climate Change: The Human Rights and Equity Imperative

Philippe Cullet

Introduction

The existing climate change regime has a strong equity basis. This includes, in particular the differential obligations that developed and developing countries have taken with regard to emissions reduction. Equity will in all likelihood remain a key component of any future climate change deal because differential treatment remains a condition for developing country participation. While equity has been a key principle of climate change instruments for the past two decades, the same cannot be said with regard to human rights aspects of climate change. The increasing certainty about the causes and the consequences of climate change as well as the increasingly visible links between climate change and most other environmental issues has made the link between ongoing climate change and human rights much more evident. Yet, human rights play at best an extremely marginal role in the existing climate change regime.

The existing climate change regime needs to be revisited for several reasons. Firstly, the commitments taken by developed countries until now are insufficient to effectively mitigate climate change. Secondly, the existing conceptual framework for differential treatment based largely on a division between developed and developing countries is increasingly incapable of providing results that are just. This is due to the fact that there is little that can justify putting together countries about to be submerged by rising sea levels and countries whose economies are completely dependent on oil extraction. Similarly there is little in common between a small land-locked least developed country like Malawi and giants like China and India. Thirdly, the state-centric framework to achieve differentiation has never been the best possible policy instrument. It has, however, become increasingly less appropriate in the past 15 years that have witnessed very fast growth in some countries but of a kind that has seen inequalities between the poor and the rich increasing markedly. A new framework that recognizes not only the distinction between rich and poor between countries but also within countries is thus necessary. This is where the inclusion of a human rights perspective can make a significant difference.

Limitations of the Existing Regime and Evolving Situation

The principle of common but differentiated responsibilities underlies the whole climate change regime.[1] It provides a strong basis for climate change mitigation commitments that are not simply based on a mechanical application of the principle of sovereign legal equality. The latter would not yield results that are substantively equal. In this sense, the climate change regime is among the most progressive instruments dealing with a global issue.

Yet, this is insufficient because the present regime is not conceived broadly enough. A number of additional issues need to considered in the future to ensure that the overall problem of climate change is effectively addressed, to ensure that each country's contribution to any mitigation effort is proportional to its responsibility and capacity, to ensure that countries of the South suffering disproportionately more from ongoing climate change are given the priority they deserve and to ensure that climate change is not addressed as an issue that affects countries as if everyone in a given country was equal in front of climate change.

One of the ways in which a broader framework can be given to the equity dimension of climate change is by giving a more prominent role to the notion of vulnerability. Vulnerability is an apt starting point because it has influenced the climate change regime from the outset.[2] It emphasizes the fact that countries and people are not similarly placed when it comes to making choices that influence their contribution to climate change or when it comes to the impacts of climate change. It also constitutes an acknowledgment of the fact that while climate change is a global issue, different parts of the globe are and will be affected in different ways by its negative impacts. This is true of countries since countries like the Maldives and Nepal do not and will not suffer the same kind of problems because of climate change. This is also true of people since different individuals and communities do not and will not be affected in the same way. A number of reasons explain these differences, ranging from geographical factors that sees people living in low-lying areas more likely to be affected by rising sea levels than those living higher up to socio-economic factors that see the rich being able to adapt much more easily, for instance by relocating to safer locations.

In legal terms, the existing regime is limited in several ways. Firstly, its equity framework is built in large part around mitigation issues which indirectly gives more importance to long-term concerns over ongoing adaptation threats. Secondly, the climate change regime is built around a convenient but limited framework that largely divides the world between developed and developing countries. This does not provide a framework to recognize the vast gaps that exist between different countries such as Tuvalu and Saudi Arabia and the lack of common interests these countries have in the face of climate change. Thirdly, the framework adopted in 1992 may not be anymore adapted to changed circumstances two decades later. Whereas issues of historical responsibility have changed little over the past two decades, the same is not true of overall greenhouse gas emissions. Since climate change is global in nature a simple

North-South division may not be sufficient to address the problem at stake. Fourthly, the whole climate change regime is not adapted to the challenges that need to be addressed. The traditional state-centric international law framework is reductionist because it does not give effective space to people's concerns. This requires a different paradigm for the future. The inclusion of a human rights dimension to the climate change regime constitutes a first way to reorient the regime.

Another issue that needs to be addressed in future is the premise on which emissions reduction commitments are allocated. At present the regime assumes that countries have a certain entitlement to pollute and the main question is to allocate these entitlements among countries. This is convenient in terms of negotiations but does not include any acknowledgment that climate change mitigation must start not from the comfort zone of existing pollution levels but from the needs of the global environment together with the needs of the regions and people likely to suffer most. This includes rethinking the ways in which we conceive air, the atmosphere and the right to pollute in legal terms.

These different issues and the need for new ways to think about climate change policy are illustrated by the case of India. On the one hand, India remains without any possible doubt a developing country. India's position in the ranking of the Human Development Index at number 128 just ahead of several least developed countries like Laos and Cambodia reflects the reality that the majority of Indians experience. On the other hand, India has experienced fast economic growth in recent years. Additionally it has increasingly sought to flex its political muscle on the world stage by seeking recognition as a major power. In terms of climate change, as in many other dimensions, India is today two countries. The India that shines has standards of living that often match those of developed countries with a concomitant negative environmental impact in terms of climate change. In India the majority of the population has made little progress since 1990. Thus, 77 per cent of the population has an income of less than \$2 a day.[3] In fact, while there has been some reduction in the percentage of people in 'extreme poverty', the overall number of poor and vulnerable people has increased from 733 to 836 millions between 1993-94 and 2004-05.[4]

From an equity perspective, India must be analyzed from these two different perspectives. On the one hand, from the perspective of climate change, an international problem requiring the collaboration of all states to address it, India has a duty to contribute to efforts to mitigate climate change. In fact, India is already contributing to climate change mitigation through its involvement in the Clean Development Mechanism like other developing countries. Yet, progressively, more needs to be done. Additionally, from the perspective of a big country that shows no signs of overall vulnerability, it is increasingly difficult to justify that India should hide behind the veil of its 'developing country' status since it has little in common with countries like Malawi or the Maldives in terms of vulnerability.

On the other hand, the overwhelming majority of India's population is as vulnerable as the average inhabitants of other developing countries, including in many cases people in least developed countries. India's rank of 94 on the Global Hunger Index (out of 118 countries listed) reflects this other reality.[5] Equity, as realized through differential treatment in international law cannot justify the imposition of emissions reduction or stabilization commitments in a way that would increase the vulnerability of the already vulnerable majority of the population. This would go against the idea of progressive of realization of fundamental rights.

New Bases for the Climate Change Regime

This section focuses on some of the many issues that need rethinking in the continuous search for an effective climate change regime. It highlights the need for a new understanding of differentiation. It also emphasizes the primacy of human rights and vulnerability as a necessary foundation of further measures on climate change. Further, it argues that air and the atmosphere should be recognized as a common heritage to ensure that the benefits of climate mitigation are not appropriated by private actors but rather ploughed back into renewable energy or other measures that are sustainable and primarily benefit the most vulnerable. Finally, it argues that a new basis for allocating entitlements must be found to ensure that the poor and vulnerable are not indirectly dispossessed of something that is in essence humankind's primary survival resource.

Rethinking Differential Treatment for Future Emissions Reduction Commitments

The basis for differentiation remains as strong as it was at the time of the negotiation of the Climate Change Convention. Indeed, on the whole it is the same small number of countries that contribute most to climate change in per capita terms. At the same time, there is still a majority of countries whose contribution to climate change is negligible, starting with all least developed countries. These countries are also the most vulnerable to the impacts of climate change.

Yet, rapid economic development in some part of the world over the past decade has altered the balance of overall contributions that countries make. In particular, the share of big developing countries like India and China in global GHG emissions has increased since 1990. This is due to the fact that their emissions have been growing at least 4 per cent per year, faster than any other region of the world.[6] Since the climate change legal regime is primarily about achieving a global environmental benefit, any substantial increase in emissions is to be taken into account, wherever the additional emissions are generated.

The need to rethink differential treatment for the future is due to the fact that the legal regime must reflect the changes that have taken place since the early 1990s in the position of some developing countries, must reflect the increasingly central role that climate change plays among environmental issues and must reflect the fact that climate change is much more than an environmental and economic issue but as well a core human right issue.

295

First, it is increasingly difficult to attribute emissions on the basis of the fiction of legal equality of states alone. On the one hand, the direct or indirect contribution of each individual country varies, according to wealth and other factors. On the other hand, questions arise concerning the responsibility of a country for all emissions arising from its territory. The case of special economic zones (SEZ) is a telling example. Where companies invest under particularly beneficial conditions and where they export all the products they manufacture, equity requires that emissions be at least partly allocated to the actors that take advantage of the lax legal regimes that increase profits on products that are marketed in wealthier parts of the world. Beyond SEZs, a number of other situations may call for similar treatment, for instance, where deforestation is undertaken to use the cleared land to produce cash crops that are mostly exported. In this case, it is necessary to find new ways to allocate responsibility for climate change. These should take into account not only countries' contributions but also that of actors that directly benefit in economic terms from greenhouse gas emitting activities. This would constitute a useful application of the polluter pays principle. The issue can therefore not be reduced to a simple dichotomy between taking and not taking commitments. It is also not a simple case of whether developing countries as a block (the G77 group) should or not take on commitments under the Kyoto Protocol.

Second, differential treatment is not in itself an instrument that seeks to favour developing countries. It so happens that under most existing treaties, differentiation has been approved based on countries' classification as developed or developing. Yet, since there is no generally agreed definition of which country is a developing country and since the decision is often left to self-identification, this is in itself no effective guide. Further, the simple division in two groups is only for convenience's sake but is increasingly itself inequitable since it does not take into account the complete lack of congruence between the respective situations of Malawi and South Korea or Vanuatu and India. The real purpose of differential treatment, which is to foster substantive equality and a partnership among all countries in solving problems of a global nature, cannot be equated with the division of the world between developed and developing countries. There are thus a number of situations where developing countries should either be individually targeted for preferences or at least clubbed in smaller groups so that, for example, small island states that are going to disappear as a side-effect of climate change do not have to be put in the same category as OPEC countries that have become much wealthier because of the growth of the global carbon economy.[7]

Third, differential treatment goes beyond the granting of preferences based on differences in levels of economic development. In fact, differential treatment in environmental treaties primarily seeks to foster the overall environmental goals of the agreement by fostering the participation of countries that may have little incentive to participate. Thus, in the case of climate change, developing countries as a whole would have had little incentive in 1992 to join a global legal regime to address a problem they had hardly contributed to cause.[8]

The implication is that differential treatment in the context of any subsequent commitment period under the Kyoto Protocol needs to be much more closely tailored to the overall environmental goals of the regime while providing a much-needed equity angle. This means that differentiation must be an instrument that takes into account both the contribution of each country to the problem, its capacity to mitigate and adapt and the vulnerability of its population. In the case of a country like India, this also requires going beyond a simplistic decision on commitments versus no commitments. What differential treatment calls for is that big countries like India and China whose emissions grow faster than any other regions of the world take up their responsibilities as member of the international community and more specifically as aspiring military and political global powers. At the same time, the focus of differential treatment on equity clearly bars the imposition of any commitment that would harm the majority of the vulnerable population of these countries. Mechanisms thus need to be devised at the international and national levels to ensure that the burden of any commitments falls exclusively on polluting industries, on the people whose lifestyle makes a significant contribution to climate change and on the government to ensure that climate change friendly policies are implemented. In other words, commitments should go alongside with new forms of international technology transfers and new forms of resource redistribution at the national level.

It is clear that countries like India cannot simply curb their economic growth in a bid to satisfy the North. These countries must, nevertheless urgently reorient their growth and find alternative economic development paths. One of the possible solutions is to rely on technology transfers where the North provides the more environmentally friendly technologies it has already developed to ensure that economic growth in developing countries is not hampered by taking climate change friendly measures. This could include, for instance, wind and solar energy technologies. Another solution lies in focusing on renewable energy, something that can easily be fostered by reallocating resources away from carbon intensive energy sources. In other words, addressing climate change does not have to be a costly proposition in terms of economic growth. It may in fact provide an excellent opportunity to rethink failed economic development strategies. Thus, climate change does not provide a basis for promoting just any energy source that is not harmful from a climate change point of view. Current efforts to suggest that nuclear energy is an apt alternative to carbon-based energy do not take into account the fact that nuclear energy has no justification from an environmental point of view. Indeed, while the actual production of energy may be harmless in terms of greenhouse gas emissions, nuclear energy is unacceptable from the point of view of its other impacts, particular because there is no environmentally acceptable solution to nuclear waste at present and because of a number of side-effects of nuclear power generation on human health are either unknown or not in the public domain.[9]

With regard to resource redistribution, two main points can be made. Firstly, one option may be for some developing countries like India and China to take on commitments with a view to ensure that climate change is effectively averted. This would give a strong signal that the world cannot tolerate more emissions and that further economic development strategies need to be rethought throughout the world. The commitments taken by such countries in the name of the global environment benefit that is climate change mitigation and reduced costs of climate change adaptation should be borne in part or entirely by developed countries under the CBDR principle. Secondly, any form of compensation that is provided by developed to developing countries with commitments should be carefully targeted. It must benefit the poor and in priority the poorest and the most vulnerable. This is a matter of equity and human rights since both focus on the situation of the most disadvantaged. Resources made available should be invested primarily in mitigation and adaptation measures for the poor since they are the most vulnerable and least able to adapt, as well as in measures that put the poor at the centre of any new economic development strategies. Together, this will ensure that differentiation contributes to global and local environmental benefits as well as to poverty alleviation and the realization of human rights. This new framework is imperative to redirect climate change law towards being more environmentally friendly and more equitable.

The Human Rights Imperative

Links between climate change and human rights can be identified at different levels. Yet, human rights have not been a significant dimension of climate change policy debates. This can be partly ascribed to the fact that while climate change is in essence an environmental problem, it requires much more significant changes in strategies of economic development than other environmental problems. Additionally the link between GHG emissions and economic growth has ensured that debates have given significant attention to economic, trade and financial aspects of climate change. Another less obvious reason is that the addition of a human right dimension to climate change has the potential to completely change the way in which law and policy is conceived in this area. Indeed, the human rights consequences of climate change are potentially so severe that they will overwhelmingly prevail over economic and related considerations if human rights are effectively taken into consideration in climate change law and policy. Nevertheless, human rights must be placed at the centre of law and policy on climate change. This is a precondition for ensuring the legitimacy of climate change law and ensuring that measures taken on environmental grounds do not have negative human rights consequences.

Human rights concerns arise both in the context of mitigation and adaptation. With regard to climate change mitigation issues arise for developing countries with regard to taking on emission stabilization or reduction commitments. Indeed, commitments are only justifiable if their consequences are completely offset for the majority of the poor. This is a direct consequence of the principle that countries can take progressive measures to realize socio-economic rights but they cannot backtrack.[10] It goes further than this since climate change commitments should also not lead to any reduction in the measures currently taken to progressively realize human rights. Thus, it would not be enough to take measures to reduce GHG emissions in the generation of electricity. At the same time, measures must be taken to increase access to electricity for the majority of villagers who do not have access at present. This may require a reduction in consumption from the wealthier individuals and economic actors or the installation of alternative, CO2 free sources of electricity in villages.

Conversely, the realization of the human rights to life, health, food, water and environment for the majority of the poor should be put at the centre of climate change policies. In other words, any shift away from a carbon-based economy must be conceived in priority with the realization of human rights in mind.

In the context of adaptation, human rights consequences are easier to identify since there is an immediate connection between ongoing climate change-related damages and the realization of human rights. Again, since the poor are the most vulnerable to climate change, they are also the most affected by ongoing damages. Thus, food shortages and floods induced by climate change invariable affect the poor first and need to be given priority.

While the environmental law regime does not directly address the human rights dimension of climate change yet, UN human rights organs have started giving the issue consideration.[11] This has, for instance, resulted in resolutions of the Human Rights Council focusing on climate change.[12] These make the link between ongoing impacts of climate change and the realization of human rights, recognizing the specific problems that the poor face in this context as well as low-lying countries. They also link climate change concerns and the right to development, thus squarely emphasizing the development dimension of climate change.

Much more needs to be done to ensure that human rights are effectively put at the centre of climate change law and policy. Currently, the emphasis is largely on adaptation to ongoing climate change and the human rights impacts of events that are most likely directly linked to global warming. This is important but needs to be supplemented by a much broader agenda. Indeed, the realization of human rights also has a big contribution to make to climate change mitigation and adaptation. Thus, the realization of the human right to a clean environment – a right not yet formally recognized in international law but existing in nearly 120 countries – could be directly linked to greenhouse gas reduction measures. This is in fact what all countries will eventually have to do. While reducing greenhouse gas emissions is fundamental to addressing climate change, these gases are the same that cause local air pollution, one of the most severe environmental problems in many parts of the world. In this sense, a human rights perspective that starts from the realization of human rights has the potential to be much more powerful than a simple look at the ways in which climate change damages affect the realization of human rights. A general though unspecific recognition of this need is visible in the 2009 resolution which recognizes that human rights have 'potential to inform and strengthen international and national policy-making in the area of climate change'.[13]

New Legal Constructs for a More Equitable and Environmentally Sustainable Regime: Recognizing Air as a Common Heritage

Air was for the longest time the object of little interest by lawyers, economists or policy makers. Indeed, while air is the first basic element that allows us to survive, it was for all practical purposes beyond appropriation. This situation changed relatively quickly over the course of the twentieth century with the introduction of aviation that led states to assert control over their airspace.[14] At the same time, the question of air pollution led to the realization that while air may be beyond legal control, humankind was able to impact on air in various negative ways. Yet, a treaty like the Convention on Long-range Transboundary Air Pollution does not address the question of air pollution from the point of view of states' right to pollute.[15] As a result, it proposes a series of measures to reduce air pollution without trying to ascribe entitlements or addressing the status of air or the atmosphere. Beyond airspace, which cannot be directly compared with air or the atmosphere, the only other dimension that states have addressed is that of outer-space where the consensus is that it is a common heritage of humankind.[16]

In the context of the climate change regime, the international community has agreed that the climate and its adverse effects are a common concern of humankind.[17] This implies an acknowledgment that the climate can only be addressed through common action of all states but it does not indicate whether states or individuals are in a position to lay specific claims on air or on air pollution. The Kyoto Protocol does not address this issue directly either. However, the Protocol indirectly provides the most polluting nations on Earth specific polluting entitlements. In other words, while no legal claims to air or the atmosphere are staked by any state, an indirect appropriation takes place. This is problematic because science has clearly showed that the global sink that is the atmosphere can only absorb a limited amount of carbon. Above a certain limit, consequences which are extremely harmful will most likely take place. In other words, the polluting rights indirectly given to developed countries under the Kyoto Protocol constitute entitlements that affect all nations on Earth.[18]

The approach taken in the Kyoto Protocol is problematic. The starting point for regulating emissions is 'grandfathering' (see also Chapter 19 in this volume), which indirectly rewards industries that have done least to cut back pollution before the adoption of the new regime. Grandfathering also rewards countries that industrialized early because their high level of pollution becomes the baseline against which reductions are debated. Countries that have lagged in industrial development suffer the double disadvantage under a grandfathering scheme of having lower levels of economic development and lower pollution levels that entitle them to similarly lower polluting entitlements. Both equity and environmental concerns call for a different type of response to climate change. In terms of equity or environmental conservation, the shortcomings of grandfathering call for giving the climate change regime new bases. One of the starting points for a differently conceived regime is to rethink the legal status of air and the atmosphere.

The Kyoto Protocol is in principle a treaty focusing on an environmental problem. Yet, in reality because of the nature of the problem being addressed, the real focus has been on economic development and the impacts that addressing climate change will have on economic growth. The debate has thus been framed mostly as an economic development issue within the broader context of environmental quality. This is unfortunate because it sidelines increasingly important impacts of air pollution on human health and thus the realization of the human right to health. More generally, the current regime fails to take into account the human impacts of air pollution and thereby fails to directly acknowledge that vulnerability is not just an issue in terms of the impacts of climate change but also in terms of the causes of climate change. For instance, the urban poor in developing countries are much more likely to be affected by air-related health issues than the middle classes.

Since air pollution cannot be regarded as being limited to a dichotomy between environmental quality and economic growth, the legal status of air must be conceived in a broader perspective. Given that there is only one atmosphere, it follows that it needs to be managed as such. Individual control over air is physically impossible and would go against the need for a global solution. Air, the atmosphere and the global climate should thus be seen as a common heritage of humankind that needs to be commonly conserved and managed. The most obvious starting point for developing this concept is the notion of common heritage developed in the context of the law of the sea.[19] Common heritage status implies first of all that no sovereign claims can be made on the area or resource covered.[20] It also prohibits unilateral appropriation and requires international cooperation in the exploitation of resources, for instance, by giving an international body the necessary authority.[21]

The introduction of common heritage status for air and the atmosphere would make a significant contribution to policy debates on the future climate change regime. Indeed, it would provide a new solid basis for rethinking the allocation of emissions reduction commitments and for regulating the use of flexibility mechanisms according to priorities focused on differential treatment and vulnerability rather than in terms of economic efficiency and the indirect allocation of individual property rights over a global heritage.

Common heritage status would, for instance, lead to setting a new framework for the CDM. At present the CDM provides essentially economic benefits to project partners. The CDM policy framework itself does not indicate how these benefits should be used. As a result, they can simply be used to foster the partner's business. Since benefits accruing through CDM projects are linked to climate change mitigation, under a scheme where air is a common heritage, there is no reason for project partners to receive unconditional benefits. Indeed, there are a number of social and environmental priorities that must be addressed in the context of climate change. The resources raised in the name of climate change mitigation should thus be used for activities that specifically contribute to addressing the global heritage since no one should be able to acquire direct or indirect rights to pollute something which is vital for survival for all living beings. The use of CDM proceeds to address issues related to the global good is even more important in a context where governments often claim that they have insufficient resources to implement effective environmental and social policies.

Turning the air, the atmosphere and the global climate into a common heritage will no doubt be fiercely resisted by a number of actors who have and still benefit immensely from the absence of clear concepts determining who is entitled to 'use' air and 'pollute' the atmosphere. Yet, this is in fact but a small extension of the notion of public trust, a concept widely used.[22] Interestingly, the Indian Supreme Court already declared more than a decade ago that air is a public trust in India.[23] The notion of public trust implies that the state has to act as a trustee on behalf of all individuals, must take a long-term view of its protection and must ensure socially equitable and environmentally sustainable access to and use of the resource.[24] It also implies that the state is not in a position to trade away or sell pollution rights or carbon credits in its role of trustee.[25] These safeguards include fostering the realization of human rights and ensuring that no violations of existing protection level takes place as well as the respect for environmental law in general and not just of climate change law.

Towards New Forms of Entitlements on Air
The basis for today's climate change law is, on the whole, the grandfathering of existing emission patterns. In political terms, this can be easily explained since any other formula would affect existing polluters more than the economic actors or the countries that contribute less to climate change. Yet, this is an ineffective way to address climate change. Indeed, while a baseline based on existing energy use puts the burden on developed countries and on polluting industries, it does not provide any compensation mechanism to non-industrialized countries and to people who have not benefited from the standards of living achieved while causing climate change.

As long as existing levels of economic development and existing pollution patterns constitute the basis for regulation, climate change law will largely reflect the priorities of the economically and politically more powerful states. An equitable and effective climate change regime needs to be based on a different paradigm that takes into account a broader variety of factors. The starting point is the common benefit that a healthy global environment represents for the whole of humankind and for life on Earth in general. Basic principles of environmental law, such as precaution and equity are thus at the centre of efforts to define entitlements. Today, environmental protection is conceived by all states as encompassing human rights, social and economic aspects. This is implies that it is not only the realization of the right to a clean environment recognized in nearly than 120 countries that is at stake but also the realization of all human rights.

This broad framework leads to the development of a regime, which does not give economic growth and economic development the kind of importance they have under the Climate Change Convention and Kyoto Protocol. It is human development and not economic development, which should be the starting point for a climate change regime. Human development gives primacy to human rights and environmental considerations but does not per se deny the necessity of economic development. In fact, the link between economic development and the realization of human rights, in particular socio-economic rights is well established. This is important because it recasts economic development as a tool for the realization of the human rights of the poor and marginalized. In this context, the success or failure of policies and laws is rated according to their impact on the poor.

In terms of climate change the first step would be to move away from a system that allocates polluting rights based on past or present emissions. Indeed, any such scheme rewards long-term polluters – developed countries – and provides incentives to the few countries among developing countries such as some East Asian countries, India and China to increase their pollution levels as fast as they can so that their own emissions levels will be grandfathered the day they take on commitments under the Kyoto Protocol. This is unjustifiable in environmental terms and inequitable for the majority of developing countries and all least developed countries that will be made to suffer the consequences of their lower levels of economic development twice over.

The most widely proposed alternative to grandfathering allocations is one based on per capita entitlements.[26] The basis for an equitable climate change policy should indeed take into account that every single human being has a right to a certain quantity of emissions. These include subsistence emissions including emissions related to the growing of food or the use of firewood to cook meals or purify water.[27] This also includes livelihood emissions, which relate to everyone's right to benefit from the fruits of economic and technological development, for instance, by having access to electricity. Thus, there should be a basic human entitlement to a certain level of emissions. This level needs to take into account the needs of the global environment and may thus imply reduced emissions by the minority of the world's population that directly or indirectly emits much more than what the global atmosphere can support.

This entitlement is to be conceived from two related but distinct perspectives. At the international level, it provides a new way to allocate emission rights, which is fairer to countries that have not benefited from the fruits of economic growth. At the national level, it provides a similar mechanism whereby the poor and marginalized that do not have access to the amenities that their wealthier counterparts benefit from, obtain a right to benefit from existing resources. In other words, the developed world and the minority of wealthy citizens within each country each have a debt to the poorer segments of the community.

While the basis for entitlements should be on a per capita basis, this cannot be the only criterion. Two reasons, at least, call for a more selective approach. Firstly, a per capita entitlement may have the negative impact of fostering population policies, which may not otherwise be in the interest of the concerned countries. Secondly, an equitable legal framework should also take into account that some countries have low population density because their environment is already degraded to such an extent that population has failed to grow over time. Since these countries usually happen to be among the poorest as well, recognition of their situation must also be taken into account.

The entitlement proposed here must differ from a Kyoto Protocol entitlement in an additional respect. The debt that rich countries and rich people within each country have accumulated towards the poor cannot be redeemed by simply stabilizing emissions or reducing them.[28] The entitlement scheme must be based on the premise that the only way in which emissions can be accessed from the poor that do not use their quota is by accepting a duty to invest an equivalent amount of money towards developing non-carbon development paths. If that is not undertaken, the entitlement system will simply end up being another market mechanism through which the poor will sell their entitlements but without any policy framework imposing the necessary changes for effectively mitigating climate change in the long term. Thus, any future CDM should only fund projects that provide zero-carbon emissions so that the CDM itself becomes a vehicle for technology transformation and not just a cheap compliance mechanism that, at best, does nothing for the poor and at worst contributes to harming them further where already discredited development options are reintroduced in the guise of climate change friendly policies.

The new entitlement framework is thus conceived as a mechanism through which the poor and vulnerable can demand new technologies or emissions convergence. In other words, this entitlement framework imposes on the rich parts of the world (rich countries and rich segments of the population) to either reduce their own emissions or invest in ways and means so that the poor do not follow the rest of the world in increasing their own emissions as economic development eventually reaches them. In India, where the richest classes produce 4.5 times more CO_2 than the poorest class and almost 3 times more than the all-India average, this convergence is also required.[29] A number of different initiatives could be taken. For instance, in a situation where, in India, only 31 per cent of rural households use electricity, there is untold potential for emissions increase if the poor are provided with the same kind of amenities that the rich benefit from.[30] The entitlement framework based on human rights indicates that the poor also have in principle a right to the lifestyle that the rich enjoy. As a result, the only way to ensure that poverty eradication does not harm the global environment more, while at the same time providing alternative economic development paths for the rich and poor alike is for the rich to invest in new ways to deliver development benefits. For instance, electricity generation in India could easily be focused on local solutions, in particular solar energy. Similarly, technological research should focus on new forms of public transport rather than on private vehicles with a lower negative climate change impact. Simply improving or changing the fuel on which private vehicles run may have a positive contribution on the global environment. However, as witnessed in the case of Delhi and its shift to CNG on a large-scale, this neither solves the environmental pollution caused by vehicles per se nor addresses the huge social and other problems caused by increasing reliance on private modes of transportation.[31]

Conclusion

Climate change must be addressed in earnest urgently. While the existing climate change regime made a noteworthy attempt at starting the process of effectively addressing the problem, much more needs to be done in the future. The task of addressing the environmental problem of climate change is complicated by the fact that it affects all aspects of development.

The international community understood that the special nature of climate change required special measures. This explains why the climate change regime provides one of the strongest recognition of the need for differential measures among environmental treaties with the adoption of different commitments for developed and developing countries under the Kyoto Protocol.

This now needs to be supplemented by much stronger measures because the scientific understanding of climate change has evolved dramatically since the early 1990s. While differential treatment will remain the basis of any broadly acceptable regime it needs to be strengthened and to be adapted to evolving realities. The limitations of a regime that focuses mostly on states also needs to be addressed by incorporating a human rights dimension. Finally, it is time for the international community to consider the broader implications of a climate change legal regime and adopt new principles clarifying the legal status of air and the atmosphere. Indeed, air and the atmosphere need to be clearly separated from any sovereign claims that states may have over natural resources.

Notes

1   See generally Cullet, P. (2003) *Differential Treatment in International Environmental Law.* Aldershot: Ashgate.

2   Article 3(2), Framework Convention on Climate Change, New York, 9 May 1992 [hereafter Climate Change Convention].

3   National Commission for Enterprises in the Unorganised Sector, Report on Conditions of Work and Promotion of Livelihoods in the Unorganised Sector 6 (2007).

4   Id. 7.

5   International Food Policy Research Institute, The Challenge of Hunger 2007 - Global Hunger Index (2007).

6   Central Pollution Control Board, Newsletter (October 2002) available at http://www.cpcb.nic.in/News%20Letters/Archives/Climate%20Change/ch9-CC.html.

7   Least developed countries are, for instance, frequently targeted for preferential measures. See e.g., Article 4(9) and 12(5), Climate Change Convention (n 2 above) and Article 66, Agreement on Trade-Related Aspects of Intellectual Property Rights, Marrakech, 15 April 1994, reprinted in 33 International Legal Materials 1197 (1994).

8   See e.g., Bodansky, D. (1993) 'The United Nations Framework Convention on Climate Change: A Commentary', *Yale Journal of International Law*, 18: 451, 463.

9   See, e.g., Katz, A. (2008) 'Chernobyl: The Great Cover-up', *Le Monde diplomatique*, http://mondediplo.com/2008/04/14who.

10 Article 2(1), International Covenant on Economic, Social and Cultural Rights, New York, 16 December 1966.

11 See, e.g., Report of the Office of the United Nations High Commissioner for Human Rights on the Relationship Between Climate Change and Human Rights, UN Doc. A/HRC/10/61 (2009).

12 UN Human Rights Council Resolution 7/23, 'Human Rights and Climate Change', 28 March 2008, UN Doc. A/HRC/7/78 and UN Human Rights Council Resolution 10/4, 'Human Rights and Climate Change', 25 March 2009, UN Doc. A/HRC/10/29.

13 UN Human Rights Council Resolution 10/4, 'Human Rights and Climate Change', 25 March 2009, UN Doc. A/HRC/10/29, preamble, last paragraph.

14 Article 1, Convention on International Civil Aviation, Chicago, 7 December 1944, 15 *UNTS* 295.

15 Convention on Long-range Transboundary Air Pollution, Geneva, 13 November 1979, 1302 *UNTS* 217.

16 Article 11, Agreement Governing the Activities of States on the Moon and Other Celestial Bodies, New York, 18 December 1979, reprinted in 18 *International Legal Materials* 1434 (1979).

17 Preamble, Climate Change Convention (see above).

18 See e.g., Lohmann, L. (2006) *Carbon Trading: A Critical Conversation on Climate Change, Privatisation and Power*. Uppsala: Dag Hammarskjold Foundation.

19 See United Nations General Assembly, Declaration of Principle Governing the Sea-Bed and the Ocean Floor, and the Subsoil Thereof, Beyond the Limits of National Jurisdiction, GA Res. 2749 (XXV), 17 December 1970, Resolutions Adopted by the General Assembly During its 25th Session, 15 Sept. - 17 Dec. 1970, GAOR 25th Sess., Supp.28 (A/8028).

20 See e.g., Holmila, E. (2005) 'Common Heritage of Mankind in the Law of the Sea', 1 *Acta Societatis Martensis* 187, 195.

21 See Part XI, United Nations Convention on the Law of the Sea, Montego Bay, 10 December 1982 and Agreement Relating to the Implementation of Part XI of the United Nations Convention on the Law of the Sea of 10 December 1982, New York, 28 July 1994, 1833 *UNTS* 3.

22 In the case of water, see e.g., for California, *National Audubon Society v. Department of Water and Power of the City of Los Angeles*, Supreme Court of California, 17 February 1983, 658 P.2d 709 and for South Africa, Section 3, National Water Act (1998).

23 *M.C. Mehta* v. *Kamal Nath*, 1997 1 SCC 388. For a similar example in the United States, *see e.g.*, Article 1(27), Pennsylvania Constitution.

24 Id.

25 *Cf.* Takacs, D. (2008) 'The Public Trust Doctrine, Environmental Human Rights, and the Future of Private Property', *New York University Environmental Law Journal*, 16: 711, 733.

26 See, e.g., International Institute for Sustainable Development, Per Capita Emission Rights (1998), http://www.iisd.org/didigest/sep98/sep98.2.htm.

27 *Cf.* Shue, H. (1993) 'Subsistence Emissions and Luxury Emissions', *Law & Policy*, 15: 39.

28 *Cf.* Halme, P. (2007) 'Carbon Debt and the (In)Significance of History', *Trames*, 11(4): 346.

29 Greenpeace India (2007) Hiding Behind the Poor: A Report by Greenpeace on Climate Justice.

30 Anonymous (2007) 'What Equals Effective', *Down to Earth*, 16(14): 62.

31 See, e.g., Kumar, N. and A. Foster (2007) 'Have CNG Regulations in Delhi Done Their Job?', *Economic & Political Weekly*, 42(51): 48.

26

Low Impact Development*

Larch Maxey and Simon Dale

What is Low Impact Development?

Low Impact Development (LID) is a superb example of sustainability being led from the grassroots which has emerged from the UK and is blossoming around the world. Whilst planners and policy makers wring their hands over climate change, affordable housing and rural decline, for over a decade across the UK, LIDers have quietly got on with building a greener future from the ground up. LID is one of the few approaches offering holistic solutions to climate change, peak oil and sustainability. It has the potential to help revitalize rural and urban communities and help the world's nations feed and power themselves. Given LID's background as a grass roots movement, it should come as no surprise that its definition has not stood still, but has continued to evolve. Development under the LID umbrella is generally:

- locally adapted, diverse and unique
- made from natural, local materials
- of an appropriate scale
- visually unobtrusive
- biodiverse
- based on renewable resources
- autonomous in terms of energy, water and waste
- inclusive, working with local communities providing public access to open space
- based on sustainable travel, minimizing the need to travel, prioritizing walking, cycling, public transport and car share and thus generating little traffic
- linked to sustainable livelihoods
- coordinated by a management plan

These points have been more fully discussed elsewhere.[1] There are two crucial points to make here, however. Firstly, LID will continue to evolve in a dialogue between developers on the ground and policy makers/planners. Secondly, as independent studies consistently show, LID is a rare example of truly sustainable development capable of social, economic *and* environmental benefits to society as a whole. Thus LID draws on the skills, traditions, designs and materials best suited to each site, empowering those involved and contributing to an emerging regional uniqueness and sense of place. LID housing, for example, tends to be built to very high energy efficiency standards and *also* to use locally available natural and/or reclaimed materials so that the *embodied energy* of the building itself is also very low. Both these points are illustrated by the Welsh Assembly Government's (WAG) adaptation of LID to the Welsh context by morphing it into 'One Planet Development' (OPD).[2] This exciting move offers LID, or OPD, an exemplary role within WAG's sustainability strategy as Wales becomes a One Planet Nation by 2050! As WAG acknowledges, LID demonstrates that One Planet solutions exist right now and can act as catalysts for wider change. This policy also shows what can happen when Governments are small enough to really respond to the ideas and energy of people on the ground, rather than following big business' agenda.

Why do we Need Low Impact Development?

Ecological crisis is not a future possibility but a current reality. Current rates of species extinction are at their highest since that of the dinosaurs. Ninety percent of the large fish in our seas have gone. Anthropogenic climate change is happening one hundred times faster than our best models have predicted. This change threatens not only the extinction of individual species but the collapse or death of entire ecosystems. We are faced with the question of whether it is too late for us to take any effective action. The fact that this crisis is already happening means that the question is not to do with whether it can be averted, but what we can do to stop exacerbating it and to cope with its effects. It is not too late; anyone, anywhere can take locally effective actions today.

The Case for Urgent Preparation for Energy Descent

Human development has for the last two hundred years been powered by the use of fossil fuels. We have converted from locally self-reliant agrarian focused societies to a globalised society powered predominantly by fossil fuels. This global society is mediated by interdependent international financial markets. With the exception of minor blips such as the Great Depression and recent bank collapse these markets and their various currencies have seen continuous growth simultaneous with the growth in supplies of fossil fuels. This financial growth has also been exponential due to the growth of speculation. Our financial transactions of trade are now dwarfed by the transactions of pure finances, the speculation on that trade and the speculation on that speculation. All of this financial activity is based on assumptions and predictions of continued growth. On this our global society is dependent. We cannot have infinite economic growth on a finite planet, yet we are only recently beginning to see mainstream media and politicians brave enough acknowledge this and that we can replace economic growth with sustainable, steady state economies in which everyone has healthy, fulfilling work for which they are truly valued.

Climate change gives a clear imperative to curtail our fossil fuel use. In addition, world supplies of fossil fuels are currently passing their peak of production. There is growing agreement that oil has passed its peak, gas will very shortly, and coal will peak in the next couple of decades. Uranium may not be a fossil fuel but that too will reach and pass its peak of production within the next few decades. Without these we either have to invent a new power source, make a transition to renewables, or reduce our power/fuel consumption.

Our society is operating under the assumption that economic liberalism and the free market will provide technological solutions for our future energy needs, the effects of climate change and any other problems that we might encounter. As several chapters in this book make clear, the imperative of economic growth (business as usual) has led to much of the support for carbon offsetting and its associated techno-fixes. It is true that the free market and technological progress have extended our capabilities and even solved certain problems. However, they have more often replaced one problem, with a bigger, longer-term problem whilst undermining local skills, knowledge and sustainable solutions in the process! Furthermore, all of this has been the product of increasing consumption of fossil fuels. The technological advance that would give us a replacement source of power to continue our growth is utterly unprecedented. Never before have we done what our society relies on us achieving now, by the essentially passive continuation of an unchanging method.

The belief that future technological fixes will enable continued growth is crucial for the functioning of our speculative economies. Without this belief our markets would collapse, and unlike the slow dwindling of fuel supplies, this can happen quickly as investor confidence fails. At the moment we are staving off this occurrence with increasingly creative accounting and economic manipulation including inflated housing prices, and increased public borrowing. Already we are seeing how precarious this approach has been with the collapse of over extended banking giants and the beginning of the 'global economic downturn'.

There is a significant chance that replacement energy sources will not be realized before we lose the economic buoyancy that makes such technological progress possible. If this happens we will have no options but to make a radical transition to a non-growth paradigm and much lower energy ways of living. This will require major adaptations. The sooner we can begin to make these adaptations, the slower the transition will be and the more chance we have of positively managing the subsequent energy descent as an equitable and comfortable process. If wisely managed we still have a wealth of resources and powerful technology in our hands. With discerning use these assets could help us address our most fundamental needs for a long time to come.

Reducing our Energy Dependency

To reduce our energy dependency we will not only have to reduce our consumption, we will have to dramatically increase the productivity of our land and ecosystems. The most crucial parts of our society are our food production and distribution systems. Particularly in the 'minority world', our agricultural and food supply systems are heavily fossil fuel dependent. Most of our food is either imported, or has travelled many miles within the country. It is largely produced by industrial farming techniques which require both heavy machinery and fossil fuel derived fertilizers and pesticides. Calorifically, all of these inputs are many times greater than the outputs, meaning that we are constantly feeding energy into agriculture and most Westerners currently put more fossil fuel calories into their mouths than into heating their homes.[3] Clearly, without fossil fuels, agriculture needs to be a net donor of energy to human society. Before the use of fossil fuels in agriculture, the vast majority of the population were involved in agriculture. If we are to move beyond fossil fuels this may well have to be the case again. Cuba's transition to a low carbon, low fossil fuel economy, for example, has involved the return of 20% of the population to agriculture (from less than 1% before the 'special period).[4]

There is also a need for other important land based produce, particularly fuels and fibre. Although the details vary around the world, forestry and its derived products are often central to this. As well as timber for building, tools and the making of other objects, wood is our primary renewable fuel source.

Sustainable and Resilient Communities

In order to be sustainable, a system or community must be self-reliant in all the resources it requires. The greater the number of independent sub-systems that can provide for the functions and required resources, the greater the resilience of the system. Whilst our global society still contains many different sub-systems, they are not independent, being linked by shared fossil fuel dependency, trans-national ownership and the globalized economy. Where we can replace this with independent self-sufficiency at the smallest scales we will have sustainable and resilient local communities.

Firstly, in many places we need to heal the infertility that is the legacy of ecological degradation and intensive farming. Naturally, the restoration of soil fertility will take time, as will the establishment of gardens, orchards and complex agroforestry systems. All of these forms of land based production require supporting infrastructure and processing facilities. These also need to be localized and provided in ways suitable for a carbon-capture future. Simple, low-impact homes can be built where they are needed, with natural materials and accessible methods. These buildings can easily provide high levels of comfort and efficiency at a tiny fraction of the cost of their conventional equivalents. Effective and reliable systems for water, sewage, heating, refrigeration and even modest electricity can be simply made in a multitude of locally adapted, low-tech ways with reused and natural materials.

There will always be benefits and pleasures of community co-operation and facilities. Essential supporting facilities which also need re-localization include mills, forges, tanneries, lime-kilns and carpenters workshops. These communities may also have their own independent councils, markets, and local events. Local trading systems or currencies add to community resilience by strengthening the local economy and protecting against global financial instability. Alongside the required infrastructure comes the need for many sets of skills. A lot of these are traditional skills to be revived, some will be derivative of the contemporary world, and others will be a synthesis of the two. All take time to learn, develop and share.

Permaculture

Permaculture is a set of design principles for human scale, sustainable systems. It is based on the three ethics of 'people care, earth care and fair shares'. It provides an approach that is most frequently applied to small scale agriculture, but can equally be applied to buildings, domestic systems and community interactions.

Permaculture has played a key role in Cuba's 'special period' after the collapse of the Soviet Union in 1990. Oil imports were cut in half, and food by 80%. The island underwent a transition from an industrial system to one of urban gardens using organic methods. We can choose this kind of transition to meaningfully curtail our contributions to climate change and prepare for 'peak everything'[5] and its associated economic instability. The Cubans' response, largely based on permaculture and community agriculture, was highly successful. Vegetables were planted on rooftops and abandoned car parks. Havana now produces 60% of its food from urban land within the city itself.[6]

A Force for Change

There is significant and rapidly growing energy at the grass roots for permaculture type solutions and the intentional move towards re-localization and energy descent. Organizations such as the Transition Towns Network are part of the gathering momentum in this direction. The call to energy descent is never going to come from the corporate or political arenas, as it challenges the growth paradigm. It is coming now from the grass roots, with rapidly increasing numbers of people unwilling to remain on the sinking ship of consumption and growth, waiting blindly for the techno-fix lifeboat.

The scale and power of this enthusiasm became clear to us during our various experiences of building simple low-impact homes in Wales.[7] Part of our motivation was to show others that this kind of living was possible. I (Simon) put a few photographs of our home on a simple web page to show half a dozen friends who had helped us with the construction. Within a few weeks, it had been passed on and started to appear on a few blogs. Since then the website has been receiving up to 50,000 unique visits a day and has been looked at by 2 million people. We have both also been involved since 2005 in the setting up the Lammas cooperative (see below) and have had thousands of emails from excited and inspired people. Some with tears, some with plans, some with their own stories and every single one with enthusiasm and encouragement.

The combination of this feedback cycle with the enthusiasm and innate appeal of this route makes this a powerful movement, and one that is capable of making effective change at every small step. The major obstacles holding it back are availability of land and people's time. These again are economic issues. The sort of work required to begin to make the transition to an energy descent is inherently uneconomic and shall remain so until the point at which there are no longer any other options. It is both crucial and appealing that before this time comes we do whatever we can to build local resilience. Whilst large numbers of people are pursuing this kind of work in their leisure time, it is impossible for most to follow it as a full time vocation at the same time as paying for housing and the land they are working.

Opportunities

As in many countries, the UK planning system does make allowances for farmers and seasonal forestry workers to live on their land; this is commonly subject to strict tests of their functional need to be there and proof that the enterprise is a viable business. In France, for example, this is generally a far simpler process, whilst the OPD scheme in Wales promises to open up these opportunities for all. After much lobbying even the UK Englishframework does make limited allowance for subsistence smallholders and this is slow progress in the planning system.

Small numbers of individuals and communities have been taking a direct action approach and simply moving on to agricultural land and getting on with their projects without advance planning permission. Many, though less than half of these projects, end up coming to the attention of the planners to whom they make retrospective applications, usually under the agricultural guidelines described above. Almost without exception, those who can afford the lost sleep and considerable expenses of a planning appeal get awarded permission, although it is often temporary permissions of 3-5 years.

Understandably, the majority of people are currently put off this route by its uncertainty. If the planning system gives concessions to those wishing to live on and work small pieces of agricultural land in this way, the situation would be a very different one. Once the route is established it will be appealing to sufficient numbers of people to make significant preparations for the transition to energy descent. If a workable route can be made within the planning system to grant access to land, and the right to live on it, to those wishing to make these changes, we can allow a rising tide of people to make real progress towards a sustainable society. If a workable route is not found, we will be reliant upon the increasing numbers of people ready to take to the land without planning permission. The former is preferable on three counts: firstly, it reduces the stress and energy needed to be in a constant state of flight or fight with the planners; secondly, it allows planning systems the world over to play their part in addressing climate change and peak oil. They canuse their wealth of knowledge and experience to prevent poorly thought out developments dependent on cars, for example. Finally, as the planning system embraces LID it allows it to infiltrate further into the mainstream and play its full part in fostering sustainability transitions.

Bringing Low Impact Development into the Mainstream

We stand at a cross roads. If we act now there is still time to create a world of abundance in which everyone can fulfil their highest potential, a world in which humans take their rightful place within a thriving natural world. If we continue on our current path for another ten years this choice will be lost and we will have committed ourselves, our species and our planet to catastrophic climate change. Indeed, the latest research is increasingly clear: stopping all fossil fuel use is not enough. We need to actively draw CO2 back out of the atmosphere and lock it up again.[8] The good news is that we *can* begin this process right now! LID can contribute to this new development paradigm. In order to do this on the scale required LID will need to shift from the fringes of modern development to the mainstream. LID has much to offer mainstream development if it is to become truly sustainable.

Harnessing People Power

LID has sprung *literally* from the ground up, rooted in practical projects which have drawn upon permaculture, traditional knowledge, appropriate technology and the creativity of LIDers themselves. As we transform society to address the twin challenges of climate change and peak oil it is essential that we harness the ingenuity, energy and local knowledge of people everywhere.

LID can help turn current social trends into powerful, practical forces for change. Since 2005, for example, the UK has seen an exponential rise in interest in local, seasonal, organic and home grown food, green lifestyles and global ethics, with seed sales going through the roof and allotment waiting lists bulging.

Programmes such as Channel Four's 'Grand Designs' show the growing interest in self-build eco-housing, with LIDs consistently winning best design awards.[9] Mainstreaming LID will allow us to unlock this potential. This is particularly important as the world faces the impending peak everything-fuelled economic recession. House building is slowing down throughout the Western world as mainstream builders struggle to make a profit. LID offers one way of ensuring a continuing supply of sustainable, affordable housing as it is not motivated by the need to make a profit, but people's desire to create beautiful, efficient homes for themselves. Whilst LID need not necessarily be self-built, it harnesses the skill and energy people are willing to invest into their own lives. LID simultaneously offers empowerment, employment and re-skilling.

LID can be remarkably adaptable and flexible, from a bender produced for nothing from entirely reclaimed materials through a £500 simple straw bale house such as the one built at Coed Hills,[10] to 'luxury' designer homes costing up to £40,000 such as that built by Ben Law and featured on Channel Four's 'Grand Designs'.

LID has been a seed bed for experimentation and discovery over the last fifteen years. Indeed, in a recent UK appeal decision, planning inspector Woolnough found 'there to be considerable ecological, educational and cultural benefits in further exploring permaculture' due to 'the development of and experimentation with sustainable technologies and agricultural practices which that way of life facilitates'. Ideas and approaches which are starting to appear in the mainstream such as compost toilets, reed beds, solar water panels, turf roofs and passive solar heat gain have all been tried and tested in LIDs. As these ideas begin to inform the mainstream anyway, the time is ripe for LID *as a whole* to be embraced by the mainstream, rather than just individual components cherry picked out of context.

The UK Government's target for all new housing to be carbon neutral by 2016 is ambitious given the unsustainable performance of most contemporary housing. Indeed, even its own official reports suggest Britain is already falling behind any chance of meeting this target[11] These targets, far from being too stringent, as large building companies claim, do not go far enough for two reasons. First, they ignore the embodied energy required to quarry, process, transport and dispose of current mainstream building materials. Second, they fail to respond to the latest science which shows that rather than aiming for carbon neutral development, future development needs to actively lock up carbon. LID has always been sensitive to the embodied energy of the materials it uses and through using natural materials such as wood, straw and hemp, building up soil fertility, and planting trees and perennials, LID can contribute to the process of drawing carbon out of the atmosphere again. LID can meet the real challenge of carbon *capture* development immediately.

Lessons from Lammas

Lammas Low Impact Initiatives Ltd is one example of how LID is beginning to move into the mainstream, raising its profile and making it more accessible to a wider range of people. Formed in 2005, Lammas is a cooperative limited by shares which aims to promote LID throughout the UK. A key way Lammas achieves this is through developing its flagship eco-hamlet in Pembrokeshire. This was granted planning permission on appeal in August 2009 and includes an eco-terrace of four units as well as five more traditional smallholdings. This eco-terrace follows the co-housing model, providing both private space and shared common space. It offers an attractive route into LID for single people, couples and families, including those without capital and those who are less able to build their own home.

One of Lammas' innovative features is its commitment to working *with* the planning system. It is the first time a LID hamlet has been granted planning permission prospectively. This is invaluable in moving LID towards the mainstream as more people will consider LID if they can do so with the certainty and security that planning permission provides. However, this approach entails its own set of tensions which offer insight into some of the challenges posed by mainstreaming LID. Bureaucratic hurdles are perhaps the biggest obstacles Lammas has faced. Not only has it negotiated the delays and quirks of an overburdened and archaic local planning system, it is also breaking new ground within both the LID movement and the planning system. Lammas' approach, however, has brought international support, its own free view video series and allowed it to pioneer the first mortgages for LIDs on this scale. Mortgages allow those without cash to buy the 999 year leases Lammas sells to residents as well as all the materials to create their dream LID homes and livelihoods.

The Ecological Land Co-op (ELC)

This initiative works on a slightly different model to Lammas, but again acts as a broker through the planning and land purchasing quagmires to make it easier for people to set up LIDs. By attracting investors the co-op has the funds to buy land and gain planning permission before then leasing LID smallholdings on affordable 999year leases. The ELC was also founded in 2005 and purchased its first site in 2009. It is always on the lookout for ethical investors.[12]

Educational LID Projects

Another way in which the LID movement has already begun to reach out to the mainstream is through a range of educational initiatives. Most established LID projects include an educational component, from tours, open days and courses on site to stalls, talks and web sites off site. Additionally, a new generation of LID educational projects are beginning to emerge. These have education as their central focus and whilst they do not necessarily involve people living on site and developing more traditional low impact livelihoods, their contribution to mainstreaming LID is considerable.

Such projects include Down to Earth Menter Felin Uchaf, Cae Mabon and LILI.[13] Other educational initiatives in which LID ideas and techniques feature extend well into the mainstream from the Centre for Alternative Technology to the Eden Centre, with the largest rammed earth installation in the UK. Around the world there are thousands of such projects featuring local, natural approaches to building and land management. Each time LID in any form appears in the mainstream it is an opportunity for the movement to reach out further. It is crucial, however, that these opportunities are used to convey LIDs' true message and power – that sustainable solutions are holistic and achievable by everyone, rather than discrete bits being cherry picked out of context or allowed to become the preserve of 'experts'.

Conclusion

LID has huge potential to deliver truly sustainable development immediately, helping the world feed, fuel and house itself. In addition to offering One Planet, carbon capture, rather than carbon neutral development, LID can help both rural and urban regeneration. However, if LID is to be brought into the mainstream it is vital that LIDers themselves continue to set the agenda in terms of defining and expanding what LID is. This sets a significant challenge to the planning system tasked with working in participatory ways with people, with a minimum of bureaucracy. It also presents challenges to LIDers themselves to form new and innovative partnerships, working with more mainstream organizations such as Housing Associations, Local Authorities, charities, NGOs, researchers, schools, educators and enlightened building companies. This is a challenge in which everyone, including you, have roles to play – will you join us?

*   This article has been adopted from,Pickerill, J. and L. Maxey (eds) (2009) *Low Impact Development: The Future in Our Hands,* Leeds: Footprints Workers Co-operative. http://lowimpactdevelopment.files.wordpress.com/2008/11/low-impact-development-book2.pdf.

1   See, for example, Pickerill, J. and L. Maxey (eds.) (2009) *Low Impact Development: The future in our hands,* http://lowimpactdevelopment.wordpress.com; Fairlie, S. (2009). *Low Impact Development: Planning and People in a Sustainable Countryside* 2nd enlarged edition, UK: Jon Carpenter, Charlbury, and Maxey, L. (in press) *One Planet Development: Simple Solutions for every Situation,* East Meon, Hampshire, UK: Permanent Publications.

2   And then activists and advocates have in turn run with this policy intervention, see, for example, Maxey, L. (in press) *One Planet Development: Simple Solutions for every Situation.*

3   Free Range Network (2007) 'Food, Agriculture, and Energy: The importance, and costs, of food security', available free at www.fraw.org.uk/ebo/.

4   http://www.state.gov/r/pa/ei/bgn/2886.htm.

5   Heinburg, R. (2007) *Peak Everything: Waking Up to the Century of Declines.* Vancouver, BC, Canada: New Society Publishers, see also his video at video.google.co.uk.

6   http://www.powerofcommunity.org/cm/index.php.

7   See Pickerill, J. and L. Maxey (eds.) (2009) *Low Impact Development: The future in our hands,* http://lowimpactdevelopment.wordpress.com.

8   Hansen, J., S. Mki. P. Kharecha, D. Beerling, R. Berner, V. Masson-Delmotte, M. Pagani, M. Raymo, D.L. Royer, and J.C. Zachos (2008) 'Target atmospheric CO_2: Where should humanity aim?', *Open Atmos. Sci. J.,* 2: 217-231.

9   See Ben Law's timber frame and Rachel Shiamh's 3 storey load baring straw bale houses, for example http://www.ben-law.co.uk/ and http://www.quietearth.org.uk/.

10 www.coedhills.co.uk.

11 NHF (National Housing Federation) (2007) Government 'in danger' of missing target for all new homes to be zero carbon by 2016, http://www.housing.org.uk/default.aspx?tabid=212&mid=828&ctl=Details&ArticleID=654.

12 See www.ecologicalland.coop.

13 See Pickerill, J. and L. Maxey (eds.) (2009) *Low Impact Development: The future in our hands,* http://lowimpactdevelopment.wordpress.com

27

Planning for Permaculture? Land-Use Planning, Sustainable Development, and 'Ecosystem People'*

Mike Hannis

One day, historians may view carbon offsetting as a last-ditch attempt to keep our collective head buried in the sand, ignoring the critical metabolic relationships between human 'civilization' and the rest of the world. As a large scale response to climate change, the tortuous doublethink of carbon trading makes sense only from within a mindset unshakeably convinced that globalised consumer capitalism is the inevitable destiny of humanity.

This mindset also routinely resists and frustrates smaller scale grassroots attempts to devise more sustainable lifestyles, which seek to actually reduce their ecological impact at source rather than paying to offset it elsewhere. One site of such resistance, as Maxey and Dale note elsewhere in this book, has been the British land-use planning system's response to 'low impact development', particularly where this involves residential use associated with permacultural land management.

Applicants trying to get residential permaculture projects legitimized by local authority planning departments often end up extremely frustrated. This is not just because of the bewildering bureaucracy involved, but because the two sides often seem to be speaking different languages which somehow contain the same words. The whole process can seem like a surreal comedy, though it's rarely funny for those involved. Being refused planning permission to live in your carefully crafted oasis of ecological sanity, on the grounds that you would be 'harming the environment', is not exactly calculated to lower the blood pressure.

In Britain as elsewhere, modern planning policy claims to be all about 'sustainable development'. So why is it often so hard to get permission for these harmless yet (sometimes) cutting-edge experiments in sustainable living? Prejudice and local vested interests can play a part in the disproportionate number of refusals, but there are more fundamental issues at stake too. Understanding these requires a closer look at the nearest thing there is to an official definition. In the canonical words of the Brundtland report:

> Sustainable development is development that meets the needs of the present without compromising the ability of future generations to meet their own needs.

317

This definition, of course, emerged as a clever diplomatic compromise between new environmental concerns and the demand for continued economic growth. It achieved this heroic balancing act largely by dodging the crucial question of what constitutes 'needs', as opposed to desires or expectations. Rather than looking at what kind or level of human activity might actually be ecologically sustainable, it has always been focussed on sustaining development. So what is really meant by 'development' here?

Development: Separating People from Land?

The strange but powerful idea that human societies naturally evolve or *develop* along a defined path leading from 'primitive' hunter-gatherer cultures to modern civilization, complete with oil-driven industrialization and profit-driven economic growth, has murky origins in nineteenth century social Darwinism. Most of the world, with its prodigious variety of alternative paths, had been defined by the mid-twentieth century as 'underdeveloped': and a major indicator of supposed underdevelopment, even today, is a prevalence of land-based lifestyles and subsistence agriculture. Development has always involved separating people from land, to the point where most human beings on the planet now live in urban areas.

Conservation, now considered a key aspect of sustainable development, has also contributed to this rapid and largely involuntary separation. Beginning in the UK and US in the late nineteenth century, the enforced move away from land-based livelihoods brought with it the new idea of nature conservation. Early environmentalists began arguing that 'nature' should be preserved by enclosing it, to protect it from the destructive impacts of growing and industrializing human populations. When the first national park was established in the US, native Americans were removed from Yosemite to create the 'wilderness' that city-dwellers wanted to believe in. Indigenous and other local people continue to be evicted from parks and reserves today, particularly in Africa and Asia.[1] Ironically, these are often the only people who understand how to live sustainably in such places, which is why they are sometimes referred to as 'ecosystem people'.[2]

After 1947 the new land-use planning system in Britain, and particularly in England, inherited and enforced the idea that people should be concentrated in built-up areas designed for the purpose, and the precious rural landscape conserved as 'natural heritage'.[3] Only farmers, who really *needed* to be there to produce food for everyone else, should be allowed to live outside defined towns and villages. But as farming became industrialized and food markets became global, the number of agricultural workers fell dramatically, leaving the depopulated countryside we see today.

New Ecosystem People?

Permaculture is one part of a massive and diverse global search for alternatives to the development mindset which has largely led to the current ecological crisis. It seeks to integrate human foods and dwellings into living ecosystems. In this sense, permaculture practitioners worldwide are trying to find ways of being ecosystem people in a modern context.

This is the underlying reason why permaculture projects in Britain (even more than, say, organic smallholdings) remain hard for unadventurous planners to support. Once mature, permaculture holdings can produce abundant yields, but very rarely the profitable surpluses needed to qualify as a viable business and thus justify 'new residential development in the open countryside'. Subsistence (or 'subsistence-plus') agriculture and low impact dwellings *can* help build a sustainable future: but this, of course, would be a future with many more people living on the land. This is, to put it mildly, not an easy thing for the current planning system to embrace.

The clash of mindsets is particularly dramatic when it involves designated areas of 'unspoilt' countryside. For example, in early 2009 a planning inspector refused permission for the picturesque Karuna project in Shropshire, saying that residential use of the land 'harmed the character and appearance of the surrounding Area of Outstanding Natural Beauty'.[4]

One unexpected consolation in this case was that the AONB management were more receptive than the planners to the idea of integrating sensitive human habitation within the beautiful landscape, and have subsequently revised their management plan for the area to reflect this. Such small changes reflect the slow but significant progress which has taken place since the bizarre 1995 judgement which stated

> As to the issue of sustainable development, the view is taken that the provision of groups of tents or similar residential accommodation in the open countryside merely to provide a subsistence living for the occupants, is not a practical pattern of long term land-use.

John Gummer, the Secretary of State in whose name this decision was issued, somehow failed to notice that such patterns of land use have in fact already been sustained for thousands of years. He was refusing a planning appeal by the residents of the low impact settlement at Tinkers Bubble in Somerset, who happily did get their permission in the end. Over the last fifteen years the pattern of land use there has proved to be very practical indeed. The residents have successfully fed and sheltered themselves, and inspired many others, as well as producing a considerable surplus of timber and wholesome food for sale,[5] while maintaining some of the smallest ecological footprints in the UK. Early cases such as Tinkers Bubble, and some determined lobbying, resulted in the government's landmark acceptance in 2004 that

> Some enterprises which aim to operate broadly on a subsistence basis, but which nonetheless provide wider benefits (e.g. in managing attractive landscapes or wildlife habitats), can be sustained on relatively low financial returns.[6]

Building on this new guidance, in 2007 an unusually forward-looking planning inspector gave a temporary permission for the Landmatters permaculture community in Devon, finding that

Profitability and self-sufficiency, whilst important, are not the only measures of success of a project of this kind. [...] I find there to be considerable ecological, educational and cultural benefits in further exploring permaculture.

Some other recent appeal decisions, particularly the grants of permission to the Lammas project and Steward Wood, do reflect this new willingness on the part of certain inspectors to consider the broader merits of projects seeking to explore and demonstrate more sustainable lifestyles and land management practices. However, there are also worrying signs that in England (though not, apparently, in Wales) the next round of rural planning guidance may in fact downgrade the importance of sustainability itself, in pursuit of an old-fashioned post-credit-crunch economic stimulus. Inexplicably there is now, apparently, 'no such thing as a separate rural economy'.[7]

Sustainability: Bottom-up or Top-down?

Permaculture, like the Transition Towns movement and other grassroots initiatives, tends to be associated with a vision of sustainability from the bottom up. After all, it is tempting to think that if all land were autonomously managed by local communities according to permaculture principles, many environmental problems would disappear. For many practitioners, this view often goes hand in hand with an entirely understandable perception that government in general, and planners in particular, are the problem rather than the solution. But in practice, facilitating sustainability from the top down is also important. Paradoxically, even if a radically decentralized low-carbon society is to be the long term aim, plenty of coordinated large-scale action will be needed to get there.

Land-use planning is likely to remain a key element of this co-ordination for the foreseeable future, helping to ensure that activities like housebuilding, power generation and agriculture are sensibly located and serve the common good rather than short-term profit. In a densely populated country like Britain, there will always be a need for commercial agriculture, especially if imports are to be reduced. Planning is (or should be) an essential part of managing the inevitable transition away from unsustainable oil-based agriculture. No-one knows how this transition to sustainability will happen, but we can be sure that it will not be easy, and that it will require encouraging rather than prohibiting experimental bottom-up solutions, including those based on residential permaculture. We can be equally sure that it will never be achieved by keeping people off the land.

Notes

* An earlier version of this article was first published in *Permaculture Magazine – Solutions for Sustainable Living* (Issue 61, Autumn 2009), www.permaculture.co.uk.

1 See Sian Sullivan's chapter in this book. For more detail, see Chatty, D., and M. Colchester (eds) (2002) *Conservation and Mobile Indigenous Peoples: displacement, forced settlement, and sustainable development*, Berghahn Books; Igoe, J., and S. Sullivan (2009) 'Problematizing neoliberal biodiversity conservation: displaced and disobedient knowledge' *report for IIED*, www.iied.org/pubs/display.php?o=G02526&n =2&l=3&a=J%20Igoe.

2   See e.g. Barbier, E., J. Burgess, and C. Folke (1994) *Paradise lost? The Ecological Economics of Biodiversity*. Earthscan, p. 85. They describe ecosystem people as those who 'subsist [...] largely on resources produced or gathered from their immediate vicinity [...] [and] often develop a strong social culture that reflects their close interaction and interdependence on the environment, both within and across generations'.

3   See Fairlie, S. (2009) *Low Impact Development: Planning and People in a Sustainable Countryside (2nd edition)*. Jon Carpenter.

4   Planning Inspectorate reference numbers for decision letters cited are as follows: Karuna (2009) APP/B3220/C/07/2060815;Tinkers Bubble (1995) APP/R3325/C/94/635596-607; Landmatters (2007) APP/K1128/C/06/2032148; Lammas (2009) APP/N6845/A/09/2096728; Steward Wood (2009) APP/J9497/C/08/2083419-28.

5   As Gummer also failed to notice, Tinkers Bubble was always about much more than subsistence.

6   Department for Environment, Food and Rural Affairs (DEFRA) (2004) *Planning Policy Statement 7: Sustainable Development in Rural Areas (annex A)*.

7   Department for Communities and Local Government (DCLG) (2009) *Planning Policy Statement 4: Planning for Sustainable Economic Development (introduction)*. For discussion, see Fairlie, S. (2009) 'Rural Policy Rift Along Offa's Dyke', *The Land*, 7.

28

Cycles of Sustainability:
From Automobility to Bicycology

Chris Land

Introduction

It is often claimed that 'there is no alternative' to capitalism and market managerialism; that 'free' markets and representative democracy are the most efficient and egalitarian ways through which to organize the economy and society; and that to suggest otherwise is ideologically suspect, naïve, conceals a hidden agenda, or is just plain ignorant. This same argument is often raised in relation to the environment and carbon management: 'Yes, there is a problem, but we have to deal with it through a combination of state regulation and markets.' State regulation works through governmental representation in global forums like United Nations Framework Convention on Climate Change, the Kyoto protocols, and the Copenhagen climate summit. Markets function through activities like off-setting and carbon trading to try to reintroduce indirect costs that would conventionally be treated by organizations and individuals as externalities. Either way, the assumption behind this perspective on carbon management is that the solution to the problem of climate change and sustainability lies with relatively small-scale modifications to current systems of state and market governance. The fundamental assumptions on which governments' and corporations' success indicators are based – continued economic growth and accumulation – are rarely questioned.

Against this idea that climate change can be solved with more of the same have been arrayed a range of more radical critiques that suggest a need to challenge the dominant logic of government, markets and economic growth. Whether these critiques are 'realistic' or not, or whether such radical critiques only have a material effect through a kind of radical flank effect, whereby moderate reforms are rendered more acceptable to the mainstream by the existence of a more radical critique (see also Paterson in this volume), is not the concern of this chapter. Rather, this chapter explores some of the ways in which more radical critiques have moved beyond challenging the ability of states and markets to solve the problems of climate change, and have begun constituting 'alternatives' to these systems, and to the climate solutions they propose, that build sustainability into everyday life. At the heart of these alternatives is a combination of material artefacts and technologies, distinctive forms of social and economic organization, and a belief that genuine sustainability requires a change to our quotidian practices of economic, cultural and social reproduction. To explore this, the chapter works through one specific example of a material technology that has become emblematic of sustainability – the bicycle – and its reappropriation by groups and individuals seeking an alternative to corporate and State solutions to climate change.

Against Automobility

> … all you habitual motorists are suckers. You've all been hoodwinked. Your automobility is expensive, annoying, and anti-social. My bicycle is cheap, fun and at times a travelling party.[1]

'The car – and particularly the 4x4 SUV – has become emblematic for many of the worst excesses of Western consumerism'. With a high price tag, poor fuel efficiency, few safety features for those outside of the vehicle, and macho, aggressive styling, the SUV combines an immediate threat to the environment with a powerful symbol of Western capitalism.[2] In the UK, road transport in 2002 accounted for over 18% of all greenhouse gas emissions, approximately half of which was from private transport, significantly more than was contributed by air travel.[3] This figure continues to increase and, in the current context of economic recession, increasing sales of cars and defending the automotive industries has been a lynch-pin of the government's recovery strategy, with financial assistance being offered to individuals buying new cars and to automotive manufacturers. This should perhaps not be a surprise though. In the course of the 20th century the automobile was often associated with publicly funded economic growth. The building of the Autobahns in Germany and the Interstates in the USA are good examples of governments investing heavily in the infrastructure necessary to ensure that private car ownership was a realistic and desirable option for the masses. In both cases the direct creation of employment in the building projects helped to pull the economies into growth and created the conditions for a booming industry that would ensure such growth continued after the initial investment was made.

Despite a long historical association with big government, the automobile has become synonymous with an ideology of individualism and independence. As Rajan has argued, the idea of autonomy associated with the automobile reflects quite specific, Western, neo-liberal conceptions of what a human, political subject is.[4] The ideal of being able to decide on a path and to direct one's own movement, even as that movement is constrained by consideration for others and for the law, gives the automobile a particular affinity with Western liberal ideas of political subjectivity and the relations between the individual and the state. Of course, the parallel between mass, public transport and more communist forms of government are equally strong and ideologically affective.

On top of this, the automobile is the consumer status symbol par excellence. Our cars define who we are. They are an expression of our personal tastes, individuality and social-economic status. The car we drive says much about us whether it is a Porsche, a Lamborghini, a BMW, an old Citroen 2CV, a VW campervan, or a modern hybrid electrical car. In particular, driving an expensive, fuel-inefficient car proclaims economic success and wealth. In advertising we are repeatedly sold the image of an automobile as giving us independence and autonomy. The ubiquitous image of the solitary car flying along a mountain road with apparently no regard for speed limits or traffic cops sells an illusion of unfettered desire. And yet the lived reality for most drivers is of heavily regulated mobility, or even immobility. Cars must travel along specific routes circumscribed by transport planning, speed limits, traffic lights, regulations and traffic density. Drivers are controlled and governed by the transport police, traffic wardens, private clamping companies and other drivers. The illusion of independence is thus a sham.

As the above quote from *Resist* suggests, not everyone accepts the dominant image of automobility as independence, autonomy and success. For an increasing number of people concerned with the environment the car is a symbol of the worst excesses of capitalism. It is expensive and wasteful of scarce resources, inefficient and anti-social. The pollution from cars has been associated with breathing problems, particularly in young children.[5] The maintenance of an effective road system in an era of expanding car-use requires massive road building projects that destroy natural spaces of greenbelt, forests and farmland. New roads separate communities and create dangerous obstacles for people walking to school, to work, to the shops, to the park, or to visit friends. With road traffic casualties in the UK in 2007 in the order of 8 fatalities per day, there is also a significant immediate human cost associated with road transport.[6]

On top of these costs, road transport is inefficient. Average travel speeds in highly automobilized cities like London and LA are the same as, or lower than, the average speed of horse drawn transport more than a century ago. If you factor in the amount of time spent working to pay for petrol, vehicle purchase and upkeep, the average speed of a car would be below walking pace.[7]

The car, then, is more than just a source of greenhouse gases that can be fixed by off-setting and more ecologically friendly technologies. It is a social system of production and consumption that connects models of political subjectivity, social hierarchy, large scale government, individual consumption and the nuclear family into a particular image of the social. The refusal of the car as a mode of transport thus has much broader significance than just the immediate reduction of greenhouse gases that might result from moving to public transport, walking, or a bicycle. It is a challenge to a material symbol and bedrock of the dominant social order: the very order that has brought us to the point of immanent climate catastrophe, producing along the way a series of human and social catastrophes that we have come to accept as a part of everyday life.

Automobility and Progress

The combination of economic growth, the promise of ever increasing speed, its technological complexity and its affinity with progressive-liberal concepts of political subjectivity mean that the car has become emblematic of progress. For the Italian Futurists and their followers in the early 20th century, the automobile symbolized the modern promise of progress and speed.[8] For social planners in the early to mid-century, the automobile was a centrepiece of modern planning that was even encapsulated by the name of one of the most famous and successful car-manufacturers: Fordism. When seeking to understand the changes in society and economic production in the late 20th century, social theorists turned to concepts of post-Fordism or neo-Fordism, taking examples from the restructuring of the automobile industry as a paradigm for broader and deeper socio-economic changes.[9] In all cases, the car has long been associated with the very concept of 'progress.' To challenge the car is thus to question the very idea of technological and economic progress: the belief that social history develops along a pre-set trajectory from 'primitive' tools and tribal societies, to more complex machineries and associated forms of social organization. To question this complex of social, economic and technological progress is to invite accusations of 'Luddism', of being against civilization and seeking to throw humanity back to a primitive, pre-enlightenment dark-age by rejecting science and technology tout court.

Such conceptions of progress are dependent upon a naïve understanding of innovation and technological change as a politically neutral process that develops independently of social and political factors, as if nature simply revealed itself to science and from this scientific advance, inevitable technological developments could be read off.[10] As Langdon Winner has argued, however, technological artefacts have politics.[11] In the case of the automobile we can see this in two ways. First, the automobile is a condensed symbol of a particular conception of politics and subjectivity. Second, for its effective and on-going use, the automobile depends upon a strong central government and taxation system that can provide the infrastructure, administration and policing necessary to regulate the use of automobiles and their often deleterious consequences. It also requires a global system of oil production, refining and distribution that itself depends upon a very specific geo-political system to ensure the 'free' flow of automobiles. Rather than an abstract form of technological progress, therefore, the car is both the product and sedimentation of quite specific geo-political, social and economic formations.

There are two main responses to this kind of analysis. One is to reject technology and civilization altogether.[12] The problem with this approach is that it accepts the dominant idea that technological progress and 'civilization' are linear processes that must be either accepted or rejected in their entirety. A more subtle form of analysis suggests that there are many different forms of technology, each with its own specific politics, and that the desirability of these should be considered and judged on the basis of the forms of association and social organization that they presume or make possible,[13] including their material consequences for the environment. From this perspective, the idea of off-setting, or reducing, the carbon emissions produced by road transport is only dealing with one small part of a much deeper problem. It is not doing anything to challenge the underlying forms of social, political, and economic organization that gave rise to, and clearly cannot deal with, the environmental catastrophe we find ourselves living through today.

Towards a Constitutive Politics

These days we are no longer satisfied with symbolic protest – which can almost be seen as militant lobbying. Our movement is leaning toward a more constitutive politics. People are beginning to work out what they want, what they are for, not only what they are against. What is more, people are actually 'acting' for what they want: practice not just theory. People are realizing that 'we live in a world of our own making' and attempting to consciously (re)make it.[14]

Environmental protest and radical critique are often perceived, especially by the mainstream media, as negative affairs: positions of hostility to a social order without a clear idea of what might replace it. Thus ecological activists are characterized, sometimes even by themselves, as neo-luddites opposed to technology and progress. Even anarchism has become characterized in the popular imagination as a negative rejection of government, rather than an alternative logic of governance founded on mutual aid and autonomy. Whether this oppositional logic is attributed from outside these movements, by politicians and the media, or is actively embraced by those within the movements, as in some of the anarcho-primitivist literature, there is a danger that such positions slide into destructive nihilism, at worst, or become a form of 'militant lobbying,' at best. If it is the former, then protest can easily be dismissed through the stock conservative response that, whilst what we have may not be perfect, it is the best we can hope for and there is no real alternative. If it is the latter then the hierarchies of government are reinforced as those who seek a voice in protest articulate their concerns to those in power, expecting them to solve political, economic, social and environmental problems. This effectively reproduces disempowerment and political disenfranchisement so that when politicians and business-people fail to solve these problems, cynicism sets in and, again, we are left with an acceptance of the status quo.

In contrast to these positions, many protest organizations have moved in recent years toward a more 'constitutive' form of politics. This approach, as the quote above from the Leeds May Day Group suggests, moves beyond an oppositional form of critique to actively engage in the recreating the world and bringing alternative social, economic, political and organizational forms into being. There is an enormous and diverse range of constitutive political strategies adopted by organizations within the radical political milieu. The Leeds May Day Group were referring to the organization of the protests against the G8 and other institutions of global governance, and how that had spilled over into political organization and dialogue through social forums around the world, as well as in the internal organization of the protest movement and its camps. Another example, specifically focused on the environment, is the Climate Camp movement, now a global phenomenon whereby environmental activists get together to protest against particularly powerful 'climate criminals', for example Drax coal-fired power station or Heathrow airport, but in doing so create a temporary eco-village, with its own, lo-tech, sustainable systems for grey-water and sewage treatment, power generation, communication and community policing. Rather than trying to examine the full range of constitutive politics, however, the rest of this paper focuses on the role of the bicycle within such movements, and how it has come to be a tool not only of protest, but for forging alternative socialities and cultures.

Critical Mass: Constitutive Politics Reclaims the Streets

> During Critical Mass we pedal out the kinds of lifestyle and society we want, in the present.[15]

Critical Mass is a movement that started in San Francisco in 1992 and has since spread around the globe. The basic form is a regular, monthly bicycle ride, often including walkers, skateboarders and roller-bladers as well as cyclists. Participants ride en-masse around a city, usually at rush hour on a Friday evening. For a short period this mass of riders can reclaim the streets, asserting the value and significance of 'other' forms of mobility that, in the daily run of transport and road planning, are marginalized by cars and an ideology of 'efficient', point-to-point transportation which subordinates the process of travel to its destination.

The idea behind Critical Mass is to reclaim the public space of the streets and use it for a different purpose. In part, it can be seen as a protest or demonstration: a protest against the domination of urban space and transport by the automobile; a demonstration to assert the right of cyclists to the roads, to make their presence – often rendered imperceptible to drivers and transport planners alike, with fatal consequences – visible, audible and impossible to ignore. As a form of protest, Critical Mass has been used to attract public attention or prevent access to specific locations by slowing or preventing the passage of vehicles. For example, Critical Mass style rides have become a regular part of protests against London's biannual arms fair, Defence Systems and Equipment International (DSEI). Even on regular Critical Masses, fliers for a range of political causes, protests and events are handed out to both participants and passersby, suggesting that the ride is a forum for contentious politics or a protest tool of the new social movements.

As the oft-used catchphrase – 'We aren't blocking traffic, we are traffic!' – suggests, at least part of the function of Critical Mass is to raise the visibility of cycling as a viable form of everyday transport with a right to a place alongside, or instead of, cars. Despite this, Critical Mass is an event without a distinctive political purpose. There is no clear attempt to lobby government or promote a particular political agenda or even a specific transport policy. In this sense it departs from the strategy of 'militant lobbying'. As a collective, Critical Mass has no leaders, no spokespeople, and presents no demands to politicians and the press. Whilst this makes it difficult to pin-down as a political movement, and easily leads to accusations of being an incoherent rabble, this refusal to articulate demands to a political body runs hand in hand with a refusal to wait. The demand for safer streets for cycling and a more people friendly city is put into practice by reclaiming the streets for people and bicycles right here, right now, not in some distant utopia.

In this sense both a new idea of how the city could be, and of how non-representative, democratic social organization could be, this is performed and brought into being by the simple act of riding bikes en-masse, with no clear direction and no leaders. The ride follows whoever happens to be at the front. The steady pace of riding ensures that no one is left behind. The tactic of riding as a bunched mass, taking up the whole street, clears the road ahead of cars, producing, if only for a fleeting moment, a safe space for pedestrians and cyclists, or an empty space for experiment and play, rather than the instrumental-rational conduct of transport.[16] This sense of carnival or festival, of pleasure and play over regulation and rule, is added to by the regular appearance of bikes and bike-trailers carrying powerful, battery powered sound systems, flashing lights, and fancy dress, especially on occasions like Halloween. These all contribute to the general feel of a mobile party, positioning the pleasure of cycling as a social activity against the isolated alienation of the commuting car-driver.

Critical Mass thus sets the immediate sociality of cycling against the illusory autonomy of driving; pleasure, play and creativity against instrumentality and the subordination of means to ends; non-hierarchical, spontaneous and immediate democracy against subordination to external rule, leadership and authority; the deferral of utopia to a point 'after the revolution', or once the politicians have acted, to the autonomous creation of the desired reality right here, right now. At best, however, this is an example of what Hakim Bey has called a 'temporary autonomous zone'.[17] It creates a space in which experimentation and utopia can flourish, but only for a short while. It enables the active creation of alternatives to the dominant order of automobility but is not in and of itself a sustainable order, in part because it remains tied to a logic of protest and demonstration, but also because, as a festival, it is dependent upon the order that it inverts. Despite this though, the existence of Critical Mass has enabled new social relations and networks to emerge that exist beyond the regular monthly bike rides and generate more stable and long-lived forms of association that have taken the bicycle as an emblem of an alternative, more socially and environmentally just, reality. One such organization is Bicycology.

From Automobility to Bicycology

Governments obviously have a key role in both causing and aiding solutions to Climate Change. Just as clearly, companies that are particularly damaging must change their ways and help to reduce the threat. However, we cannot rely on these institutions to do this out of goodwill: we must take action ourselves, both by pressurizing governments and companies, and by changing our own lifestyles.[18]

In the summer of 2005 a group of between 60 and 80 cyclists rode from London, England, to the site of the G8 summit in Gleneagles, Scotland. Climate change was high on the agenda for the G8 that year and the riders were joining with thousands of other people in Scotland to lobby, protest and demonstrate. Their concerns were diverse. Some wanted to lobby the leaders of the G8 to take the environment more seriously and adopt contraction and convergence policies to mitigate what they saw as an imminent climate catastrophe, others saw the G8 itself as part of the problem and incapable of offering effective solutions to this or any other problem of late capitalism. As the G8 consists of the leaders of the most polluting, and advanced capitalist, nations, these protestors saw little hope that they would be able to do anything to solve the problems that were a product of the very system they oversaw and which gave them their authority. Instead, they saw a need for a more radical change in which people took direct responsibility for the problems of climate change and sought to create a more egalitarian world in which the rapacious economic growth of the affluent capitalist nations was challenged both through protest and through a strategy of selective disengagement: a process of creating alternative ways of organizing, and developing alternative technologies, in everyday life.

Following the G8 summit, many of those involved in the protests were asking themselves what effect their demonstrations had. Much of the press coverage was negative and focused exclusively on the relatively few examples of property destruction that had taken place. Attempts to influence the politicians involved in the talks were minimal as celebrities like Bob Geldof assumed the role of unelected spokesmen for the protest movement and ended up lending legitimacy to Blair and the other politicians involved in the talks. Any direct articulation or grievance was prevented by the rows of riot police and fencing surrounding the Gleneagles venue. Attempts to disrupt the proceedings met with police blockades and violence. They had limited impact on the talks as the main delegates were flown in by helicopter, straight over the heads of the massed protestors. At a more general level concerns were raised within the movement that mobilizing around events like Davos, the G8, WTO and IMF gatherings was unwittingly ceding the initiative to the politicians and business leaders. By protesting against these organizations, and the systems they govern, the agenda that they set was itself left unchallenged. Against this many groups sought a more positive way of organizing; a way of bringing into existence the kind of world they wanted to see, in which concern for social and environmental justice was a part of everyday life.

One such initiative was Bicycology. Growing out of the G8 cycle caravan, Bicycology was a conscious effort to create an on-going organization that would take the forms of association found in summit protests – for example non-hierarchical affinity groups, participative justice and principles of consensus decision making – and to ground those principles in a more permanent organization dedicated to the promotion of sustainable alternatives to automobility, understood in the broad sense outlined above.

In the most immediate sense, Bicycology set bicycles against cars. Whilst cars pollute, kill and literally 'cost the earth', bicycles are relatively cheap and low in their environmental impact. Whilst cars foster a sense of separation, whereby the world appears on a screen (only a windscreen rather than television screen), bicycles put riders into immediate contact with their environment, with other road users and with their bodies. Where cars are elitist, insofar as not everyone can afford one, and if everyone did drive then no one would ever get anywhere because of the congestion, bikes are relatively democratic.

Crucially, from an environmental perspective, the bicycle is embraced in Bicycology as part of a broader environmental strategy of treading lightly on the planet, of reducing consumption, re-using what is already available, and recycling what must be wasted. In this Bicycology, and similar groups, have a distinct take on the bicycle. There is nothing inherent in a bicycle that can make it challenge global capitalism. Indeed, it is all too easy to find bikes that cost thousands of pounds and are made from carbon-fibre, titanium or other costly materials with a significant environmental impact in their production. A visit to a specialist bike store will quickly reveal a massive range of performance-enhancing dietary supplements, lightweight bicycle components, and sport-specific clothing, mostly made from petrochemical-derived synthetic fibres,. In such a context cycling can easily become part of a consumerist fantasy, with Dollars, Euros and Pounds being exchanged for a few grams off the weight of a bike, or a slightly more streamlined helmet. Indeed, mountain-bikes regularly appear on the roof-racks of SUVs in car showrooms, with BMW even producing a range of own-brand mountain-bikes. There is thus no essential tension between capitalism, consumerism, or even automobility, and the bicycle. But here it is necessary to distinguish between different uses of the bicycle. Whilst commercial and sports oriented cycling sees the bicycle as another commodity, subject to strategies of built-in obsolescence and disposability,[19] what Chris Carlsson refers to as the 'outlaw' bicycling scene, 'is distinctly anti-consumerist. It is a tinkering culture that spontaneously re-uses and recycles in ways environmental advocates of recycling can only dream about'.[20]

This environmental concern, and its anti-capitalist, anti-consumerist stance, is thus connected to a wider, anarchistic, anti-authoritarian, DIY culture.[21] Bicycology, like other 'outlaw' bicycling groups, distances itself from the high-tech, high-cost end of bicycle technology for two reasons. One, as already discussed, is its rejection of consumerism. The other is more concerned with autonomy and independence. In its basic form, the bicycle is an easy machine to operate, maintain and repair. Without costly hydraulic disc brakes, sealed bearing units, or complex hub-gears, all of which require specialized tools, a basic bicycle can be repaired and maintained with only a small amount of training and a handful of everyday tools. This makes it possible to recode the bicycle as a simple tool that can liberate personal transport from dependency upon the, predominantly male, mechanics found in car garages and most professional bike workshops. By taking bicycle maintenance and repair skills into the community, and providing people with access to tools and equipment, 'outlaw' bicycle groups can foster individual and community independence, rather than dependence upon commercial products, and individual helplessness. Bicycology continue this tradition by organizing events and road shows that include workshops on bicycle maintenance and repair, as well as sessions on how to build and pedal powered electricity generators and other low-carbon technologies. Their pedal powered and bike-mounted sound systems and cinema projectors, as well as their portable workshop, have also become a part of the infrastructure that enables more conventional protest events like Climate Camps.

Cycles of Conviviality

Groups like Bicycology, or Carlsson's 'outlaw' bicyclists, show that the bicycle can be what Ivan Illich called a 'tool for conviviality'. Illich defined conviviality as a term

> to designate the opposite of industrial productivity... to mean anutonomous and creative intercourse among persons, and the intercourse of persons with their environment.[22]

Illich contrasts the industrial system of production, which produces homogeneity and allows personal expression *only* through consumption, with 'conviviality', which combines human interdependence and individual freedom. For Illich, industrial machinery is much more than just a 'tool'. Precisely because it measures the value of a tool in quantitative terms – how fast, how many, how cheaply – industrial capitalism can only recognize a tool as a means to the efficient achievement of ends. Illich's 'convivial' approach to evaluating tools, in contrast, emphasizes the kinds of relations they make possible and the kinds of human beings and social relations that they presuppose. Mass industrial society necessitates disciplined, docile subjects and has developed a complex system of compulsory schooling in order to produce these subjects[23], who, in turn, are able to fit unproblematically into an efficient tool of production: the manufactory, bureaucracy or business corporation. In order to maximize utility, individual, or collective autonomous, expression and desire must be subordinated to the broader goals of a pre-determined system.

As automobility is the exemplar of industrial production, requiring disciplined producers to man the factories and disciplined consumers to purchase and drive cars, so bicycology is opposed to this industrial system, inculcating individual independence from impersonal systems of production and transport by actively producing interpersonal forms of free association, with direct intercourse with the environment. As part of a broader network of social and technical relations, the bicycle can become a tool for conviviality but this potential is not essential to the technology itself. It depends upon a much broader set of social relations to actualize it.

Conclusion: From Technical Fix to Conviviality

What this chapter has hopefully demonstrated is that technologies are never straightforwardly technical. Attempts to solve the problem of climate change with innovations like carbon capture and storage, off-setting, hybrid and hydrogen-powered automobiles, or even bicycles, will not change the fundamentally social dynamics that have created the current climate crisis. This crisis is the product of a broader technical and social system in which the environment is an 'externality', human beings are 'consumers', justice can only be conceived of in terms of the distribution of goods, and economic growth is the unquestionable measure of success. To counter such a system we do not need new markets, new products, and more government. What is required, and is already happening in communities and social networks around the world, is a new form of evaluation: A way to measure the desirability of a technological artefact, organizational form, or economic system, in terms of its consequences for the environment and quality of human association. As Illich put it, what is needed is:

> A methodology, by which to recognize the public perversion of tools into purposes [but which will inevitably encounter] resistance on the part of people who are used to measuring what is good in terms of dollars. Plato knew that the bad statesman is he who believes that the art of measurement is universal, and who jumbles together what is greater or smaller and what is more fit to the purpose. Our present attitudes towards production have been formed over the centuries. Increasingly, institutions have not only shaped our demands but also in the most literal sense our *logic*, or sense of proportion. Having come to demand what institutions can produce, we soon believe that we cannot do without them.[24]

To challenge climate change requires precisely this new kind of *logic*. It means challenging the social, economic systems that have produced climate change. This can be done through a focus on technologies and mechanisms but only so long as we recognize that technologies and mechanisms are never just technical or mechanical. It is not, therefore, simply a case of replacing petrol cars with electrical cars, or diesel engines with hydrogen fuel cells. It is not even a case of replacing cars with bicycles. But this does not mean that technologies and mechanisms do not matter. Markets reproduce social isolation and quantitive calculation rather than ethical recognition and community. Automobiles produce dependence on impersonal systems of government, policing and expertise, as well as a global system of production, warfare and the expropriation of resources[25] whilst bicycles can foster self-reliance, community and a convivial austerity which may make us all richer, as well as weigh less heavily upon the environment. But this will only happen if we can move beyond the search for technical solutions without questioning the broader social logics of which they are a part. It is in this broad sense of an alternative logic to industrial, capitalist automobility, that outlaw bicycle groups offer us bicycle-logic, or bicycology.

Notes

1   Resist #42, cited in C. Carlsson (2008) *Nowtopia: How Pirate Programmers, Outlaw Bicyclists, and Vacant-Lot Gardeners Are Inventing the Future Today!* Oakland: AK Press. p. 139.

2   Dery, M. (2006) '"Always crashing in the same car": a head-on collision with the technosphere', in S. Böhm et al (eds) *Against Automobility*. Oxford: Blackwell.

3   http://www.statistics.gov.uk/downloads/theme_environment/transport_report.pdf.

4   Rajan, S.C. (2006) 'Automobility and the liberal disposition', in S. Böhm et al (eds) *Against Automobility*. Oxford: Blackwell.

5   Zmirou, D. et al (2004) 'Traffic related air pollution and incidence of childhood asthma: results of the Vesta case-control study', *Journal of Epidemiology and Community Health*, 58: 18-23.

6   http://www.dft.gov.uk/pgr/statistics/datatablespublications/accidents/casualtiesgbar/suppletablesfactsheets/flagbfactsheet.pdf.

7   Gorz, A. (1973) 'The social ideology of the motorcar', *Le Sauvage*, reprinted at http://rts.gn.apc.org/socid.htm.

8   Thacker, A. (2000) 'E.M. Forster and the motor car', *Literature & History*, 9(2): 37-52.

9   Kumar, K. (1995) *From Post-Industrial to Post-Modern Society: New Theories of the Contemporary World*. Oxford: Blackwell.

10 For an analysis and critique of this 'technological determinism' see, for example, D. MacKenzie and J. Wajcman (eds.)(1999) *The Social Shaping of Technology, Second Edition*. Buckingham: Open University Press.

11 Winner, L. (1980) 'Do artifacts have politics?', *Daedalus*, 109(1): 121-136.

12 See, for example, anarcho-primitivist critiques of civilization such as those found in Zerzan, J. (ed.) (1999) *Against Civilization: Readings and Reflections*. Uncivilized Books.

13 For example, Illich I. (1973) *Tools for Conviviality*. Glasgow: Fontana.

14 Leeds May Day Group (2004) 'Moments of excess', at http://www.nadir.org.uk/excess.html.

15 Horton, D. (2002) 'Lancaster Critical Mass: Does it still exist?', in C. Calrlsson (ed.) *Critical Mass: Bicycling's Defiant Celebration*. Edinburgh: AK Press, p. 65.

16 On this rationality of transport see Bonham, J. (2005) in *Against Automobility*. Oxford: Blackwell.

17 Bey, H. (2003) *The Temporary Autonomous Zone: Ontological Anarchy, Poetic Terrorism*, Second Edition. Brooklyn: Autonomedia.

18 From 'Bicycology now... or the apocalypse soon', in Bicycology Guide from 2007, http://www.bicycology.org.uk/.

19 Rosen, P. (2002) *Framing Production: Technology, Culture, and Change in the British Bicycle Industry*. Cambridge, MA: The MIT Press.

20 Carlsson, C. (2008) *Nowtopia: How Pirate Programmers, Outlaw Bicyclists, and Vacant-Lot Gardeners are Inventing the Future Today!* Oakland: AK Press. p. 121.

21 McKay, G. (ed.) (1998) *DIY Culture: Party & Protest in Nineties Britain*. London: Verso.

22 Illich, I. (1973) *Tools for Conviviality*. Glasgow: Fontana, p. 24.

23 See also Foucault, M. (1977) *Discipline and Punish*. London: Penguin.

24 Illich, I. (1973) *Tools for Conviviality*. Glasgow: Fontana. p. 31, emphasis added.

25 Retort (2005) *Afflicted Powers: Capital and Spectacle in a New Age of War*. London: Verso.

29

Towards the Sustainable School: Social Accounts and Local Solutions

John Fenwick, Jane Gibbon and Ann Marie Sidhu

A public life develops only when a society realizes that reciprocity and mutual aid are worthy of cultivation both as good in themselves and as providing the basis of the individual self.[1]

Introduction

Concerns about climate change and sustainability raise questions around what can and must be done by society and the individual to ameliorate the damaging effects of overconsumption of the earth's resources. It is evident that radical reductions in carbon emissions are required if we are to avoid dangerous climate change.[2] Efforts are being made at global and international level to formulate policy and implement mechanisms to reduce carbon emissions. Unfortunately, the discourse on climate change and sustainability has privileged business and economics and our ecosystem has become a subset of the market economy. Valid perspectives on the environmental crisis such as social equity, ecological sustainability, ethics and moral belief[3] have unfortunately been largely ignored.

The formation of the emissions trading schemes (ETS) has, by placing a price on carbon, enabled business to capture the climate change agenda.[4] Research illuminates the extensive issues with mechanisms such as the European emissions trading scheme (EU ETS) and the carbon off setting schemes. Climate change is a global problem but not all countries subscribe to the Kyoto Protocol and developing countries are reluctant to restrain economic growth. The main purpose of the emissions trading scheme is to cost emissions into production output and motivate emitters to reduce carbon emissions as they would any other costs. However, there are many flaws with this solution, and many have already been covered within other chapters of this book.

Companies can meet their emissions commitments without making major investments in clean technology and a fundamental shift from reliance on fossil fuels is hampered. At country level, emission targets may be met by using either or both of the two carbon offsetting mechanisms under the Kyoto Protocol, i.e. Joint Implementation (JI) and the Clean Development Mechanism (CDM). Under the CDM Northern countries can earn Certified Emission Reducions (CER's) from emission reducing projects implemented in developing countries (Southern countries). Renewable energy projects account for approximately 5% of the carbon market and empirical evidence highlights that 75% of all carbon credits certified are projects for capturing landfill methane and hydrofluorocarbons (banned in OECD countries from 1987).[5]

These projects do not contribute to sustainable development or the local communities within which they are located. In fact environmental injustices occur as local communities are not consulted, campaigns against such projects are ignored[6] and local low-carbon community projects are undermined in favour of CDM projects.[7] The voluntary purchasing of carbon offsets encourages a false assurance that company and individual activities have negligible impacts on the environment due to offsetting and reduces the impetus for any real change in behaviour. It is clear that carbon trading will not solve the current environmental crisis. A collective, comprehensive, holistic and multi-faceted approach is required where the environment is viewed as 'a community issue which exhibits value not only for those participating in the market, but also the present and future generations and sentient life forms'.[8] Community level approaches can advance the climate change agenda[9] and it is within this context that we consider the potential contribution of the sustainable school to the current position. The sustainable school is a community of stakeholders (students, teachers, parents, suppliers, local business, local residents) with a commitment to shared values.

The chapter discusses the sustainable school and the importance of local solutions in building broader responses to the climate change and environmental agenda in general. We argue that development of the sustainable school must be based in local initiatives at school level, drawing from mutual values and a co-operative ethos. Within these values we can identify patterns of accountability based on voluntary obligation in the public interest and a shared approach to the common good. In particular, social accounting provides a valuable way of collecting information about the relationship between school and community, and about environmental impact. The sustainable school has been defined by the UK Government as follows:

> A sustainable school puts the principle and practice of 'care' at the heart of everything it does and aspires to do. This includes:
> - care for oneself
> - care for each other (distant and near, as well as for future generations)
> - care for the environment (from the school grounds to the planet)[10]

The Sustainable Schools initiative was launched in 2006 and 'places the child at the centre of its concerns for a healthy, just and sustainable society. It paints a picture of the kind of place and the kind of school culture where each learner can be healthy, stay safe, enjoy and achieve, make a positive contribution, and achieve economic well-being – all within the earth's environmental limits'.[11] The charity Sustainability and Environmental Education (SEEd) has sought to enable more of the education sector to engage with education for sustainable development and environmental education. The work of SEEd suggests that schools that have embedded sustainability report a range of positive outcomes. Schools that have successfully engaged with sustainability take a whole school approach and are outward looking, with their students actively involved in decision making and practice in the context of a broader understanding of sustainability.[12]

The barriers to sustainable schools, however, include time and money, the lack of priority given to sustainability from central government or the inspection regime, a knowledge gap, a lack of training, too many overlapping initiatives, school buildings and a lack of evidence of impact.[13] The enablers to the development of a more sustainable school include the time to create a shared vision as a whole school community, a joined up approach, distributed leadership, embedded sustainability in policy, curriculum, budgets and staff, local authority support, external partnerships (local business, community, NGOs and international), student participation and leadership, and training for active citizenship.[14] All these factors point to the value of a wider participatory approach in developing the sustainable school.

The difficulty of embedding sustainability requires schools to focus on more than one area and to develop an understanding of where their greatest impacts lie, for example through procurement. A focus can be achieved through a school policy on sustainability embedded in the schools' strategic plans along with an understanding of the eight doorways to sustainability.[15] Local clusters of schools can share knowledge, working together with the local authority, NGO representatives and the community in order to share ideas and plans toward establishing best practice.[16]

The collective nature of sharing knowledge internally – by working across the whole school – and working externally – in local clusters – is consistent with a communitarian approach underpinned by cooperative values. These values are readily stated. They include commitments to democracy, self-help, self-responsibility, equality, equity and solidarity alongside a strong ethical sensibility. The overall co-operative principles are those of voluntary and open membership, democratic control by members, members' economic participation, autonomy and independence, the provision of education and training for members, co-operation with other co-operators, and a concern for the broader community.[17] These seven widely-agreed principles are the basis of the co-operative identity, defined by the International Co-operative Alliance as follows:

> A co-operative is an autonomous association of persons united voluntarily to meet their common economic, social, and cultural needs and aspirations through a jointly-owned and democratically-controlled enterprise.[18]

While the broad values of co-operativism are well established and largely agreed, the application of these values and principles to schools is in its early stages. The Principal of the Co-operative College has talked in the following terms:

> Look back at our co-operative principles and what it says about education. It talks about educating the wider public, and young people in particular about co-operation. What way of doing that than by giving young people the opportunity to be directly engaged in running a co-operative at their school and reinforcing that by a whole series of experiences of co-operation throughout the curriculum...just think how our Movement can be reinvigorated if, in years to

come, thousands of people are leaving school each year having spent the whole of their school lives in co-op schools, putting values into practice, and wanting to work for organizations that share those values.[19]

Sustainability and Carbon Reduction in the School

National bodies in the UK have a role to play. The Sustainable Development Commission reports that England's schools system is responsible for 9.4 million tons of carbon emissions per year and recommends the UK government works with schools to halve emissions by 2020. The government is seen as a catalyst for carbon reduction whilst empowering and enabling local authorities and individual schools. The case study of the Government Office for Yorkshire and the Humber demonstrates how regional Government Offices are charged by the Department for Children Schools and Families (DCSF) with promoting sustainable schools at a regional level through networking good practice, as well as locally through improved support and helpdesk functions for schools. Government Office for Yorkshire and the Humber (GOYH) has established a Sustainable Schools Coordination Group to help fulfil this objective also. The Group facilitates sharing of good practice within and across DCSF Sustainable Schools doorway themes, helps connect schools in the region with organizations engaged in sustainable development, advises and challenges schools and other organizations as they develop their Sustainable Schools programmes, and signposts funding and support opportunities for Sustainable Schools activities.[20]

At local level in the UK, a number of initiatives have sought to address the specific carbon reduction agenda. In the North East of England, for instance, the Government Office for the region and other regional organizations have produced a Climate Change Adaptation Study and associated Action Plan, and are now working on a study of how to achieve economic growth within carbon targets.[21] Other organizations, such as the North-East Strategic Partnership for Sustainable Schools,[22] have sought to work toward achievement of government targets for sustainable development at regional level.

Alongside this, the Carbon Trust has a message firmly based upon the hard-headed business benefits of low carbon products. It describes its 'mission' as being 'to accelerate the move to a low carbon economy, by working with organizations to reduce carbon emissions now and develop commercial low carbon technologies for the future.' It places emphasis on the cost savings to be achieved by taking carbon reduction seriously, for instance through using fuel-efficient equipment and low-energy lighting. Specifically, it has produced an action plan for energy efficiency in the school[23] together with guidance on a 'whole school approach' to reduction of the school's carbon footprint.[24] The whole school approach places emphasis upon involvement of different stakeholders within the (internal) school community, including students and staff, to address sustainable development and carbon reduction in particular. This includes impact within the curriculum itself.

Recent political debate has by no means been confined to the Left and those with a critical political stance. From the Right, a prospective parliamentary candidate for the Conservative Party argues that schools can do five things 'if

337

the current crop of children is to emerge as a generation that cherishes the environment', including provision of 'good food', skills in 'cooking and growing', tackling the 'school run', saving energy and dealing with waste.[25]

We would argue that specific suggestions for what schools can do must be set within a context that brings together sustainability, accountability, governance and the local community. We see the key to success as deriving from local initiatives, based firmly upon:

- Social reporting and accounting as a means of gathering information and, more importantly, of establishing real patterns of accountability within the school's internal community (staff, and students) and (a neglected aspect) its local external community
- A framework of governance which gives priority to values of co-operation, mutuality and the common good rather than models of governance based on private sector 'partnership' and the values of the marketplace
- A focus on sustainability in practice through all aspects of the school life: the curriculum, the school ethos and the school as producer and consumer of resources.

Within this approach, we see the school as being a local solution to global issues of climate change, environment and sustainability. For such local solutions to succeed, they cannot be seen as isolated target-based activities: their basis in shared values needs to be addressed, made explicit and agreed.

Social Reporting and Accounting

The process of social accounting can inform the school and its stakeholders of the community and environmental impact of the school. The main advantage of this is to generate information on the relationship between school and community. This largely qualitative information provides a meaningful alternative to the restricted performance data demanded by government. Value-driven social accounting can provide the empirical basis that is an essential foundation for developing sustainability.

Social accounting provides a way for co-operatives to implement their principles and also to be accountable. The Social Audit Network (SAN) defined social accounting as a framework which uses existing information and patterns of reporting to account for 'social performance', to report upon this performance, to plan for future actions, to understand impacts on the community and to be accountable to 'key stakeholders'.[26] This is not a mechanical process: values are central. Social accounting can provide an approach to understanding and appreciating co-operative associational qualities using a flexible framework to develop the relations between members.

Social accounting is flexible. As well as producing specific outcomes, it reinforces the value-base of the school community by contributing to learning and self-awareness: participation in the process cements the recognition of what is important to the school. In the UK, Launceston College in Cornwall has demonstrated the use of social accounts in secondary education. Launceston is a comprehensive school with approximately 1400 pupils aged 11-18. On the basis of their social account, the school formulated local policies with a central objective being 'the environmental impact of the college' for any future cycles of accounting.[27] School values – 'a community college that values achievement, provides opportunity and promotes responsibility'[28] – are expressly referred to within the accounts.

The school's 2006 social accounts included:

Scope: investigation of impacts based on the key factor that '...pupils make a positive contribution to the community'[29]

Stakeholders: a concern with internal *and external* stakeholders including students, parents, staff, local residents, other local schools, retailers, and employers[30]

Environment: including projects including recycling, care of the environment through Citizenship within the curriculum, and project work on sustainability with the younger students

Sustainability: projects included designing a bicycle store and recycling bins, tree planting and a 'Citizenship Day'

Strengths and Weaknesses: stated strengths included renewal of the focus on social objectives and building on the self-assessment process, and development of the relationships between a broad range of stakeholders, but social accounting also involves challenges: it takes time, and it needs support. We would add that it may also provide surprising and perhaps unwelcome results: its findings cannot be assumed at the outset. There is also value in seeing social accounting as a longitudinal commitment rather than a one-off exercise, and Launceston repeated the process in 2008.

Consistent with this approach, the World Wildlife Fund (WWF) provides a further way forward in its development framework for good practice and 'learning for sustainability'.[31] The WWF emphasizes whole-school working, expansion of capacity amongst adult stakeholders within the school community in promoting sustainability, and creating the conditions where education for sustainable development may grow. The approach is explicitly 'non-prescriptive', matching its methodology to specific needs within each school.

Whatever specific approaches might be advanced in particular circumstances, we argue that the social account provides a local solution: a qualitative perspective based upon values, not just another quantitative measure imposed by government. Additionally, the strength of this ethos may be reinforced by governance arrangements that take seriously the values of mutuality and co-operativism that provides the basis for a truly sustainable school. Recent moves toward establishing the co-operative trust as a model of school governance provide a promising way forward.

Co-operative Governance

Co-operatives the world over share the values of self help, self responsibility, democracy, equality, equity and community solidarity. In the UK all types of co-operative enterprises are increasingly keen to work with schools to ensure that young people have the skills and experience they need for the workplace and to show that an ethical approach to business works. They are also keen to show that through the adoption of these values, children and young people can gain a better understanding of their role as citizens and how they can help build a fairer society.[32]

Within patterns of governance can be located the real practical opportunity for enhancing sustainability. This is particularly so within the co-operative/mutual model. While secondary schools have always been free to make links with co-operative organizations, central government is currently encouraging something more in the shape of the co-operative trust school. Such schools are run by a foundation trust and are given a greater degree of autonomy, while remaining part of the state sector. Co-operative trusts stand in contrast to existing trust schools which normally have formal partnerships with business and private sector companies. The first co-operative trust school was announced in 2008: Reddish Vale Technology College, an 11-16 comprehensive school in Stockport. Its trust is a 'membership based organization which shares the international co-operative movement's values and principles'. Partners are the local authority, local college, the Co-operative Group and the Co-operative College.[33]

A second co-operative trust school was established in 2009 when Campsmount Technology College in Doncaster became a Foundation Trust School, with partners from education and the co-operative movement. During 2009 a small number of other schools have established co-operative foundation trusts, including Bebington High Sports College in the Wirral. 'The co-operative or mutual model is based on open membership, equal democratic participation (one member, one vote) and the clear accountability of those in charge to those for whom services are provided'.[34] Again, partners are drawn from educational and charitable organizations. 'Equal', 'democratic', 'open' and 'accountable' are words that have not been commonplace within the recent vocabulary of educational policy. These draw from current government thinking where there is an intention to create 100 co-operative trust schools 'owned and controlled by the local community'.[35]

The co-operative trust is a charitable foundation. Partners are likely to be mutual, community, charitable or educational organizations: hence, the dominant values are those of co-operativism. A 'council' or forum represents community members, staff, students and other stakeholders. This council appoints trustees to the trust, and the trust appoints some governors.[36] The Head Teacher of Reddish Vale school said: 'We are very pleased to be drawing upon the values and principles of the international co-operative movement to deliver a real mutual dividend...This is more than one schools' development – it is about empowering the whole community towards self regeneration'.[37]

Governance, accountability and sustainability are closely entwined in all these processes. We would suggest that strong relationships of accountability reside within governance processes, and that within effective governance and accountability can be identified the bases for sustainability. Values are at the heart of sustainability and this has several dimensions: the curriculum, the physical operation of the school (eg, as a building consuming resources), its links to the community, and its prevailing pattern of governance. The mutual trust is a powerful alternative to the private-sector model in secondary education.[38] It is also a strong platform for the development of local solutions to the problems of the environment.

Conclusions: Carbon markets, Sustainability, Values and Social Accounts

Carbon markets are advanced as the key solution to the climate change dilemma. This is evidenced by government policies and the willingness of big business to participate in a pseudo market (and its derivative activities) for what is in reality 'the commons.' Carbon markets operate at the international, governmental and institutional level keeping the climate change discourse mainly within the purview of the political and economic elite. Carbon trading is unlikely to deliver the required emissions reductions due to the various weaknesses outlined in this book. Approaches which create awareness, enable action, cooperative behaviour and community initiatives to lead low carbon lifestyles are likely to be more effective. The co-operative school is such an approach as it emphasizes values of equity, mutuality and democracy. It is precisely these values that are a necessary precondition for the development of the sustainable school, equipped to deal with climate change and the environmental agenda in general.

> We recognised the potential to lock a values driven ethos into schools in the long term. One of the key aspects of the trust is that it not only holds the land and assets on behalf of the community, but also its ethos. We see a national network of schools sharing Co-op values driven and global perspective as a critical contribution to bringing about greater diversity in education provision.[39]

We argue that – potentially – the mutual model is significant: it depends upon the local community, the balance of local political forces, and upon who is setting the agenda. Genuine co-operativism can be enacted through active representation of local people in governance arrangements and the use of processes such as those of social accounting. This links to broadly communitarian approaches to participation and the public good. This can be an effective basis, a precondition, for promoting awareness of carbon footprints and advancing sustainability, based on the crucial link to values. The received tradition of managerialism can be subverted by promoting alternative values. These values need to engage with the lived experience of internal and external members of the school community, within a shared mutual culture that is the prerequisite of sustainability.

Based firmly on local initiatives, a broader accountability can create better relationships between individuals and the state, using a critical accounting framework in civil society 'through practical reasoning to regain a glimpse of the goals that can be pursued by communities. These are the very values which have been submerged by instrumental political structures which contend that a globalizing world market will solve our social and environmental dilemmas'.[40] The school community stakeholders will be empowered to participate in the sustainability agenda at local level and have a platform to take specific actions to reduce carbon emissions and promote sustainability.

We have suggested that sustainability and accountability within the secondary school are strengthened by reporting based on social accounts and by governance based on co-operative values. Social accounts and the mutual co-operative trusts are practical vehicles for enacting the values of sustainability. This is an alternative to the dominance of neo-liberalism and the drift toward private-sector values throughout education. Accountability, the common good and mutual values are the basis for practical action on sustainability in education. A concern with sustainable ways of living places education for young people at the centre of enquiry, and local solutions as the way forward.

The concern to reform school governance in order to link more closely to citizens and communities is not confined to the United Kingdom. Lessons can be learned from grassroots projects in the South grappling with sustainability. In contrasting political conditions, for instance, Nicaragua has sought to 'democratize' school governance[41] and no doubt different countries will move toward different solutions in the light of local conditions. In the UK, the government says it wants every school to be a 'sustainable school' by 2020.[42] With the global economic and environmental crisis leading to a rediscovery of mutualism in many spheres, the relevance of these debates to sustainable education is clear. Clearly, sustainable schools can advance and even revolutionize the way we treat our environment and connect with our community.

Notes

1   Sullivan (1982) in Thayer Scott, J. (1995) 'Some Thoughts on Theory Development in the Voluntary and Nonprofit Sector', *Nonprofit and Voluntary Sector Quarterly*, 24(1): 31-41.

2   Intergovernmental Panel on Climate Change (2007) Fourth Assessment Report, http://www.ipcc.ch/publications_and_data/publications_and_data_reports.htm; Stern, N. (2007). 'The economics of climate change', The Stern review Cambridge: Cambridge University Press, http://www.hm-treasury.gov.uk/sternreview_index.htm.

3   Milne, M. J. (1996) 'Capitalizing and appropriating society's rights to clean air: A comment on Wamberganss and Sanford's accounting proposal', *Critical Perspectives on Accounting*, 7(6): 681-695.

4   Bachram, H. (2004) 'Climate Fraud and Carbon Colonialism: The New Trade in Greenhouse Gases', *Capitalism Nature Socialism*, 15(4): 1-16.

5   Erion, G. (2005) 'What's wrong with carbon trading?', In Bond, P. and Dada. R. (eds.) Trouble in the Air: Global Warming and the Privatised Atmosphere Centre for Civil Society and Transnational Institute, http://www.thecornerhouse.org.uk/ item.shtml?x=397683.

6   Erion, G. (2005) 'What's wrong with carbon trading?'.

7   Lohmann, L. (2008) 'Carbon trading, Climate Justice and the Production of Ignorance: ten examples', *Development* 51(3): 359-365.

8   Lehman, G. (1996) 'Environmental Accounting: Pollution Permits or Selling the Environment', *Critical Perspectives on Accounting*, 7: 667-676.

9   Lehman, G. (1996) 'Environmental Accounting: Pollution Permits or Selling the Environment'.

10 DfES (2007) Strategic, challenging and accountable: a governor's guide to Sustainable Schools www.teachernet.gov.uk/publications p2.

11 DfES (2007) Strategic, challenging and accountable: a governor's guide to Sustainable Schools.

12 Symons, G. (2008) 'Practice, barriers and enablers in ESD and EE: a review of the research', Sustainability and Environmental Education (SEEd) Report, http://www.se-ed.org.uk.

13 Symons, G. (2008) 'Practice, barriers and enablers in ESD and EE: a review of the research'.

14 Symons, G. (2008) 'Practice, barriers and enablers in ESD and EE: a review of the research'.

15 DCSF (2008) 'S3: Sustainable School Self-Evaluation: Driving School Improvement Through Sustainable Development', version 2 (corrected) London: HMSO.

16 SEEd (2008) 'Communicating Sustainability: How to reach mainstream Schools?' NGO Sustainable Schools Forum Workshop, http://www.se-ed.org.uk/.

17 Co-operative College (2008) 'Co-operative Values and Principles' http://staff.co-op.ac.uk/valuesandprinciples.htm; Co-operative Party (2008) NM: Consumers with Attitude – Can We Design the Society in Which We Live? Spring, London: Co-operative Party.

18 International Co-operative Alliance (2007) 'Statement on the Co-operative Identity' http://www.ica.coop/coop/principles.html.

19 A Lesson on the Future of Co-op Schools (2008) – Co-operative News, October 16th, Manchester; http://www.thenews.coop/features/Wider%20Co-op%20Movement/1471.

20 Sustainable Development Commission (2009). 'Towards a schools carbon management plan' http://www.sd-commission.org.uk/, p.52.

21 Government Office 'Focus' (2009) 'Government Office for the North East (GONE)', Newcastle upon Tyne, UK, 21 July.

22 http://www.sustainableschools-ne.org.uk/home.htm.

23 Carbon Trust (2007) 'Sector Overview: Schools: Learning to Improve Energy Efficiency', *Report CTV019*.

24 Carbon Trust (2008) 'A Whole School Approach: Involving the School Community in Reducing its Carbon Footprint', *Report CTV037*.

25 Goldsmith, Z. (2009) 'How to be a Green School', *The Guardian*, 28 July.

26 Pearce, J. (2001) *Social Audit and Accounting Manual*, Community Business Scotland (CBS) Network Ltd., Edinburgh, p.9.

27 Gibbon, J., J. Fenwick, and J. McMillan (2008) 'Governance and Accountability: A Role for Social Accounts in the Sustainable School', *Public Money and Management*, 28: 353-360.

28 Launceston College (2006) 'Social Audit Statement: Launceston College', Launceston College: Cornwall, p.1.

29 Launceston College (2006) 'Social Audit Statement: Launceston College', p.4.

30 Launceston College (2006) 'Social Audit Statement: Launceston College', p.5.

31 World Wildlife Fund (2004) 'Pathways: A Development Framework for School Sustainability', available at http://assets.wwf.org.uk/downloads/pathways.pdf; World Wildlife Fund (2005). 'Pathways to change: Lessons learned from the WWF Schools Support Programme', http://assets.wwf.org.uk/downloads/pathways_to_change_ report.pdf.

32 DCSF (2009) 'Co-operative Schools – Making a Difference', available at http://www.beecoop.co.uk/cms/sites/trusts.beecoop.co.uk/files/4050_Co_op_leaflet_WEB .PDF, p.4.

33 http://trust.reddish.stockport.sch.uk/index.php.

34 http://www.bebingtonhigh.com/trust.html.

35 Wintour, P. (2008) 'Balls to Set out Vision of 100 Schools Becoming Co-operative Trusts', The *Guardian*, 11 September, http://www.guardian.co.uk/politics/2008/ sep/11/education.newschools.

36 Wintour, P. (2008) 'Balls to Set out Vision of 100 Schools Becoming Co-operative Trusts', p.6-7.

37 Wintour, P. (2008) 'Balls to Set out Vision of 100 Schools Becoming Co-operative Trusts', p.7.

38 Wilson, M. and C. Mills (2007) 'Co-operative Values Make a Difference in the Curriculum and the Governance of Schools', Co-operative College/Mutuo.

39 Wilson, M. and C. Mills (2007) 'Co-operative Values Make a Difference in the Curriculum and the Governance of Schools'.

40 Lehman, G. (2002) 'Global accountability and sustainability: research prospects', *Accounting Forum*, 26(3): 219-23.

41 Gvirtz, S. and L. Minvielle (2009) 'The Impact of Institutional Design on the Democratization of School Governance: The Case of Nicaragua's Autonomous School Program', *Educational Management Administration and Leadership*, 37(4): 544-565.

42 Office for Standards in Education (Ofsted) (2008) 'Schools and Sustainability: a Climate for Change?'

30

Inspiring Examples: Sustainable Living[*]

Sally Andrew

Introduction

Political organizations, actions and technology are all crucial in ensuring climate change mitigation. But organizations and technology will not take us far, unless we know where we are going. We need to be clear not only about what we want to destroy (e.g. fossil-fuel dependence) but also about what we want to create. There are people who have already begun developing and practicing sustainable ways of living. They are creating real examples of what is possible. They are building tomorrow, today. On an economic level, I believe it is the practice of gender-sensitive and environmentally-conscious democratic socialism that can best meet the needs of the Earth and the people on it. Unfortunately, I do not think that we will be able to achieve this goal in time to save the Earth, so we also need to see what is possible within the current system. We need to find ways of reducing the fossil-chomping nature of the wealthy; and we need to look at ways of improving the quality of life of the poor majority on this planet – without spewing out more greenhouse gases. In this chapter I present specific sustainability projects that set examples for us. However, there is also a lot to learn from the practices of the millions of working class and rural poor whose destruction to the Earth is very small relative to their numbers. It is also worth studying indigenous hunter-gatherer societies, who developed tools and practices (social, technological and spiritual) to live productively, in harmony with the Earth. Some of this knowledge is still alive and practiced today. Many of the governments of the developing countries argue that they need fossil fuels to 'develop'. Some of the examples below illustrate that renewable energy can be a far better option for development – for both the poor and the planet. It is up to governments to regulate and enforce emission reductions, but these examples of sustainable living practice give us some idea of how life *can* be lived in a friendlier way to the Earth. Here are some of the stories of the fire dogs that are not just barking, but doing...

Findhorn eco-village

The Findhorn Community is one of the oldest, wisest and freshest examples of an eco-village. At Findhorn, people address sustainability not only as an environmental issue, but also in social, economic and spiritual terms. Their values are manifest in their beautiful buildings and gardens, wastewater

345

treatment, organic food production, consensus decision-making, wind turbines and solar PV panels.

As well as sustaining a 500-strong resident community, Findhorn is a humming hub of international spiritual and environmental conferences, networking and education projects.

According to a 2006 study, Findhorn has the lowest ecological footprint for any settlement ever measured in the industrialized world – at about 50% of the UK national average. In specific areas it is even lower: the 'home and energy' footprint is 21% (feed-in renewable energy systems – it sells electricity to the grid); the food footprint is 37% (largely home-grown, organic, vegetarian and seasonal diet); and car mileage is 6% of the national average (car-pooling and high employment level within the community).

To check out some of the marvellous happenings at Findhorn, go to Findhorn's website and to Jonathan Dawson's weekly blog.[1]

Findhorn is one of many eco-villages across the world. Have a look at the Global Eco-village Network to read more about the numerous 'centres of innovation and inspiration, introducing new technologies and social systems that spread out into the wider society'.[2]

Transition Towns Network

> You never change things by fighting the existing reality. To change something, build a new model that makes the existing model obsolete.' (Buckminster Fuller, cited on Transition Town Totnes website)

The Transition Towns network provides a model of change for towns responding to the challenges of peak oil and climate change. They suggest mechanisms by which a community can work together to 'unleash the collective genius of their own people' to drastically reduce their carbon emissions. The founder of the UK-based movement, Rob Hopkins, outlines their approach in *The Transition Handbook: From oil dependency to local resilience.* The subject of their website is serious, but their style is lots of fun.

Already there are over 60 communities around the world that have been inspired to become an official Transition Town, City, Village or area, with 700 others mulling it over. Community representatives in my own coastal suburb of Muizenberg, Cape Town are amongst the 'mullers'.

Totnes became the first Transition Town in 2006. Totnes have a number of groups looking at everything from 'buildings', 'energy' and 'local government', to 'education' and 'heart and soul'. Their 'Energy Descent Action Plan' involves finding ways to reduce the current nine barrels of oil per person per annum (current UK average) down to one barrel (or less) per person by 2030.

They are implementing a range of projects including: their own local currency, composting toilets, low energy street-lighting, promoting local produce, effective garden use, renewable-energy electricity, nut tree planting, cycling paths, and 'story-telling the future to educate and inspire'.

'Transition Town Totnes believes that only by involving all of us – residents, businesses, public bodies, community organizations and schools – will we come up with the most innovative, effective and practical ideas, and have the energy and skills to carry them out. Our future has the potential to be more rewarding, abundant and enjoyable than today, and by working together we can unleash the collective enthusiasm and genius of our community (that means you!) to make this transition'.[3]

Urban Carbon Management

In addition to Transition Towns, there are numerous climate change initiatives in urban areas around the world. For case studies of strategies and programmes from Mexico City to London to Shanghai have a look at the urban and regional carbon management website.[4]

Feed-in Tariffs give you a Check Instead of a Bill

A strategy that has been implemented in many European countries is 'feed-in tariffs', that allow households and RE companies to sell their renewable energy back into the central electricity grid. At the end of the month, households and companies receive electricity checks rather than bills. Tariffs can be used to subsidize and encourage renewable energy use and production.

One of the countries to recently implement this practice is Switzerland. In 2008, Swiss federal government launched a full system of feed-in tariffs differentiated by 108 technology, size, and application. There are tariffs, or payments per kilowatt-hour (kWh), for solar photovoltaics, wind, hydro, geothermal, and biomass. The Swiss system, like those in Germany, France, and Spain, pays a renewable energy generator for every kWh of electricity generated.[5]

Institutions Sharing Ideas and Training

There are a number of inspiring individuals, institutions and networks that are sharing ideas and practices about how to live well, and in accord with the Earth. Some of them are educational institutions. I list a few of them below:

Gaia Education and the GEESE

Gaia Education develops courses on sustainable community design and development. The team of eco-village-based educators are known as the GEESE: Global Eco-village Educators for a Sustainable Earth. They draw on the experience and expertise of some of the most successful eco-villages and community projects across the Earth.[6]

Wiser Earth

Wiser Earth is a community directory and networking forum that maps and connects NGOs and individuals addressing the central issues of our day: climate change, poverty, the environment, peace, water, hunger, social justice, conservation, human rights and more. Their website features over 100,000 organizations, groups and individuals involved in aspects of sustainable living.[7]

Alternatives

Bioneers

Bioneers is a forum for connecting the environment, health, social justice and spirit with a broad progressive framework. They are committed to finding practical solutions for people and planet.[8]

CIFAL Findhorn

CIFAL Findhorn – the only UN-affiliated training centre in Northern Europe – is based at the Findhorn Eco-village. It operates as a hub for training, capacity-building and knowledge sharing between local and regional authorities, international organizations, the private sector and civil society on all aspects of integrated sustainable development, and other global goals of the United Nations.[9]

Ocean Arks

The mission of ocean arks is to disseminate the ideas and practices of ecological sustainability throughout the world. Their motto is 'To Restore the Lands, Protect the Seas and Inform the Earth's Stewards'. See their list of new publications on ecological design and Dr. John Todd's Comprehensive Design for a Carbon Neutral World.[10]

Worldchanging

Worldchanging was founded on the idea that 'real solutions already exist for building the future we want. It's just a matter of grabbing hold and getting moving'.[11]

Centre for Alternative Technology

CAT aims to offer practical solutions to 'some of the most serious challenges facing our planet and the human race, such as climate change, pollution and the waste of precious resources.' The key areas they work in are renewable energy, environmental building, energy efficiency, organic growing and alternative sewerage systems. They aim to show that 'living more sustainably is not only easy to attain but can provide a better quality of life'.[12]

Schumacher College: Transformative Learning for Sustainable Living

Many of the inspiring projects around the world have the participation of people who have trained at the Schumacher College, in the UK. 'The College is renowned for the excellent teachers that lead its courses. People come to Schumacher College, in the heart of the Devon countryside, to discuss sustainability. What and how they learn stays with them for a lifetime'.[13]

New Economics Foundation: Living well need not Cost the Earth

New Economics Foundation (NEF), an 'independent think-and-do tank', was founded in 1986 by the leaders of The Other Economic Summit (TOES), which forced issues such as international debt onto the agenda of the G7 and G8 summits. They are creating an economics in which 'people and planet matter'.

They aim to improve quality of life by promoting innovative solutions that challenge mainstream thinking on economic, environment and social issues.

NEF combines rigorous analysis and policy debate with practical solutions on the ground, often run and designed with the help of local people. They believe that 'living well need not cost the Earth'.[14]

Ideas Worth Spreading
'TED, ideas worth spreading' collects inspired talks by the world's greatest thinkers and doers.[15]

Diet for a Small Planet
The books and website of Frances Moore Lappe, US author of Diet for a Small Planet have inspired many people.[16]

Hope Building
Hopebuilding Wiki was created 'to share stories of achievement by ordinary people who are doing extraordinary things to make their world a better place to live in, but whose stories are not as widely known as they should be'.[17]

Award-winning Climate Change Projects
Ashden awards celebrates and rewards 'visionary champions who are finding solutions to climate change that are also bringing real social and economic benefits to their local communities. Across the UK and developing world, our award winners are inspirational examples of simple, practical ways to cut CO2 emissions while also improving quality of life. Whether harnessing technology, energy efficiency or renewable sources such as solar, wind or biomass they're all beacons that we use to encourage others to take the sustainable energy path'.

Examples of these include a 'Fruits of the Nile' solar fruit-drying project in Uganda, a Technology Informatics Design Endeavor (TIDE) project making wood-saving stoves in South India, and a community wind project in Scotland.[18]

Poor to Sell (Biogas) Electricity to the Rich
Energy Forum is using biogas to generate cheap off-grid electricity for villages in the Dry Zone in Sri Lanka. They hope to model this technology for wider replication.

In Bangladesh, the University of New South Wales is using a new finance model for RE technologies. 'An implementation agency will assist rural poor villagers in the business to sell electricity to wealthier members of the village. Poor people will get ownership of the technology after the payback period of the technology. In this project, small biogas plants connected to latrines will produce methane to generate the electricity for the rural costumers.'

The poor selling to the rich? This makes a refreshing change from the usual patterns of the fossil fuel industry.[19]

Water Wheels in the Amazon to Provide Electricity
An organization in the Amazon is installing Low Head Micro Hydropower in two villages in the Tapajos region. 'The principle of this innovative technology is to apply broad water wheels with a small diameter to the low water levels of creeks on the river.'

This could provide electricity 24 hours a day, and will replace the expensive and polluting diesel generators that currently provide energy three hours a day.[20]

Handbook on Participatory Development of Micro-Hydropower
Many others are using micro-hydropower to provide hydro-electricity. This is much friendlier to people and the environment than large dams. For example ADEID (Action pour un Dévelopement Équitable, Intégré et Durable) has had 15 years of experience in the participatory development of micro hydropower plants in rural areas of Cameroon and other African countries. They are producing a handbook to share the lessons they have learned.[21]

Non-profit Coop to run Wind Energy Project on Pacific Island
In Vanuatu, an island country in the South Pacific Vanuatu Renewable Energy and Power Association (VANREPA) has initiated a range of renewable energy projects, including wind power, solar power, solar desalination and micro-hydro technologies.

Vanrepa has launched its project 'The Answer is Blowing in the Wind,' on the islands, Aneityum and Futuna. It will begin with providing electricity for schools and other institutions, with a longer-term goal of 100% renewable energy on these islands. Vanuatu is classified by the UN as a 'least developed' country, and most of its 200,000 population live in remote rural areas, and are engaged in subsistence agriculture.

VANREPA aims to establish a Renewable Energy Service Cooperative that will provide the necessary technical and management support. The coop as a non-profit will sell renewable energy to end-users. This organization is seen as essential to ensure the sustainability of the project, as it is 'by strength of its management and support, rather than by strength of its technology' that a project succeeds.[22]

Practical Action to Reduce Poverty
Practical Action (an initiative of The Schumacher Centre for Technology & Development) uses sustainable technology to reduce poverty in developing countries. They are currently implementing over 100 projects worldwide. Combined with their consultancy and educational work they outreached to about 664,000 people in 2006/2007. In 2008 they won a UNEP prize for a project in the Eastern Andes, Peru, in which they set up 47 micro-hydro schemes bringing clean power to about 30,000 people.[23]

Barefoot Rural Woman make Great Solar Engineers
The Barefoot College in Rajasthan, India trains rural women and youth in a range of practical technical and ecological skills. One of their projects is the training of women as 'barefoot solar engineers' to fabricate, install, maintain and repair solar PV systems.

So far, they have solar-electrified at least 300 adult education centres, 870 schools and 350 villages (12,000 households). Their own college (which spreads over 80,000 square feet) was built by barefoot architects and is completely solar-electrified.

The Barefoot approach to solar electrification 'identifies indigenous knowledge and vastly under-utilized practical wisdom of the poor, upgrades their basic skills, builds up their confidence (they already have the capacity), and applies it for their own development. It builds the confidence of villagers from the very beginning, in a non-hierarchical learning environment based on learning-by-doing.'

Women are preferred to men because they are generally more stable, more likely to stay in the villages, and they usually teach other women what they have learned.

Barefoot College trains women not only from India, but also from Africa. These women have shown that language, illiteracy, and culture are no barriers to practical mastery of solar PV systems. 'As the College ramps up its solar electrification projects across India and Africa, these women are leading by example – showing how the skills of the rural poor can drive their own development'.[24]

World Bank Projects

World Bank and International Monetary Fund policies have wreaked social, environmental and economic destruction for decades, so I found myself balking at looking at the World Bank's list of 'inspiring and replicable' examples of sustainable energy projects. The World Bank is notoriously full of contradictions. Nevertheless, they may at times fund people and projects that get up to good things, so go to their website, and see for yourself. Have a look at the book (mentioned on this site) by Paul Osborn: *Sustainable Energy: Less Poverty, More Profits*.

One promising project, which the World Bank initiated (together with US Aid, the US Department of Energy, the National Renewable Energy Laboratory, Winrock and other private companies) is the Global Village Energy Partnership, which is currently hosted by Practical Action (UK). GVEP helps developing countries to set up energy action plans.[25]

BEN and the Bicycle

> Adding highway lanes to deal with traffic congestion is like loosening your belt to deal with obesity. (Louis Mumford, city planner, cited in BEN report, 2007)

Bicycles are healthy for your legs, heart, pocket and planet. Inexpensive and easy to repair, you can ride them to work and ride them to play. There are many organizations around the world that promote the use of bicycles. One of them is the Bicycle Empowerment Network (BEN) in South Africa. The mission of BEN is poverty alleviation through the use of bicycles. Together with local and international partners, BEN gets (often second-hand) bicycles from Europe, the Americas and Asia to Southern Africa; establishes community-based bicycle workshops; and sets up bicycle paths.[26]

No Till Farming Reduces CO2 Emissions

Agriculture is a significant contributor to CO2 emissions. One of the reasons for this is the CO2 released into the air by land disturbance.

Repeated tillage also destroys the soil resource base, causing adverse environmental impacts. Tillage degrades the fertility of soils, causes air and water pollution, intensifies drought stress, destroys wildlife habitat, wastes fuel energy, and contributes to global warming. The no till (or zero till) 'conservation method' is increasingly being used by big farmers because it improves soil quality, producing better crops.

South African Dirk Lesch, Swartland Canola Farmer of the year, produced more than double the yield of the average farmer in the same district. '*Ja*,' he says, 'the plough died in 1989, when Pierre Matthee… beat my yield on my own land with a no-tillage experiment block of wheat.'

The carbon content in Dirk Lesch's land is more than three times the level in neighbouring farms that practice ploughing.[27]

In a properly designed no-till system, pest (weeds, disease, and insect) control is accomplished primarily with the following cultural practices: rotation, sanitation, and competition.[28]

Organic, Free Range and Permaculture Farming

There are many other agricultural techniques, technologies and methods that are productive, friendly to the Earth and reduce CO2 emissions. Many of these have been developed in the practice of free-range, organic, biodynamic and permaculture farming. These practices work together with nature, and do not use (petroleum-based) fertilizers and toxic herbicides and pesticides. They minimize waste, GHG emissions and energy use. They also challenge, and provide alternatives to, the existing methods of (feed-lot) meat farming. Earth-friendly technologies have been used for centuries by farmers around the world, and have been developed and documented in modern literature.

For a great overview of agricultural practices that can reduce GHG emissions, see the chapter 'You the Farmer', by Linda Scott and Leonie Joubert, in *Bending the Curve*, edited by Zipplies.[29]

In her book *Animal, Vegetable, Miracle*, Barbara Kingsolver documents her own family's experience of growing and buying local, seasonal organic foods. Her book and website have a wealth of farming and food related links (mostly USA based, but they will have links to organizations across the world).[30]

Farmer Managed Natural Regeneration

Tony Rinaudo, of World Vision, Australia, sent me an interesting story of an African reforestation programme. I will go into this in some detail, as it provides important lessons for sustainable development practice, and reforestation.

Rinaudo's article on Farmer Managed Natural Regeneration states: 'Conventional methods of reforestation in Africa have often failed. Even community-based projects with individual or community nurseries struggle to keep up the momentum once project funding ends. The obstacles working against reforestation are enormous. But a new method of reforestation called Farmer Managed Natural Regeneration (FMNR) could change this situation. It has already done so in the Republic of Niger, one of the world's poorest nations, where more than three million hectares have been re-vegetated using this method.'

Similar programmes are being initiated in other African countries. The Niger was an area hit hard by desertification, as people chopped down trees and vegetation for firewood, and because they believed that trees competed with their crops. Trees provide us with many benefits in addition to reducing CO_2. For the farmers, the ecosystem of the trees provided: natural predators – which reduced the insects that ate the crops; fertilizer from the creatures that sheltered in their shade; as well as protection from extreme heat and wind. With the trees gone, crops were devastated by drought and insects, and famine spread across the land. Programmes across Africa attempted deforestation programmes that involved growing the trees from seed and planting them in the damaged areas. However, despite investing millions of dollars and 116 thousands of hours labour, there was little overall impact. The conditions were harsh for the trees, not only because of the elements and the animals, but because people continued to cut them down.

The FMNR programme had two crucial allies in the success of their project: the one was the earth and the other was the local farmers. By observing the shoots that came out of some of the felled stumps, Rinaudo became aware of the 'underground forest' that could be harnessed. Careful pruning could support the trees to regenerate. These ancient methods (coppicing and pollarding) were taught to small-scale local farmers; and accompanied by support and education programmes, as well as laws that both protected trees and allowed farmers to harvest them in a sustainable fashion.

'The benefits of FMNR quickly became apparent and farmers themselves became the chief proponents as they talked amongst themselves. FMNR can directly alleviate poverty, rural migration, chronic hunger and even famine in a wide range of rural settings. FMNR contributes to stress reduction and nutrition of livestock, and contributes directly and indirectly to both the availability and quality of fodder... The environment in general benefits as bio-diversity increases and natural processes begin to function again'. Malatin André, a Chadian farmer practicing FMNR for just two years reported: 'Food production has doubled and many people who were laughing at us, have also adopted the techniques for soil regeneration. As a result, there is always good production, the soil is protected from erosion and heat, and women can still get firewood'.[31]

Malatin Andre, a Chadian farmer practicing FMNR for just two years reported: 'Food production has doubled and many people who were laughing at us, have also adopted the techniques for soil regeneration. As a result, there is always good production, the soil is protected from erosion and heat and women can still get firewood'.[32]

The FMNR programme is an example of how appropriate sustainable practices offer climate change mitigation as well as adaptation benefits.

It reminds us of a truth that is at the core of most successful sustainable living practices (including those discussed in this chapter): the solutions to a problem are usually right in front of us – in the Earth and in the people. We should work closely with these resources, rather than imposing programmes that go against the grain.

Notes

* This chapter has been taken from *The Fire Dogs of Climate Change: An Inspirational Call to Action*, pp. 105-116. © Sally Andrew, 2007/2009 revised, enlarged edition published by Findhorn Press, Scotland.

1 www.newstatesman.com/blogs/life-at-findhorn.

2 www.findhorn.org; www.gen.ecovillage.org.

3 www.transitiontowns.org; www.totnes.transitionnetwork.org.

4 www.gcp-urcm.org; www.gcp-urcm.org/Category/UrbanCarbonManagement.

5 www.wind-works.org.

6 www.gaiaeducation.org.

7 www.wiserearth.org.

8 www.bioneers.org.

9 www.cifalfindhorn.org.

10 www.oceanarks.org.

11 www.worldchanging.com.

12 www.cat.org.uk.

13 www.schumachercollege.org.

14 www.neweconomics.org.

15 www.ted.com.

16 www.smallplanet.org.

17 www.hopebuilding.pbwiki.com.

18 www.ashdenawards.org; For more inspiring applications of clean energy, also look at the 'world clean energy awards' and the projects funded by WISIONS: www.cleanenergyawards.com and www.wisions.net.

19 www.energyforum.slt.lk; www.unsw.edu.au; www.wisions.net; Also see the Kenya Biogas project (www.itpower.co.uk) on www.wisions.net.

20 www.wisions.net.

21 www.adeid.org; www.wisions.net.

22 www.vanrepa.org; www.wisions.net.

23 www.practicalaction.org.

24 www.barefootcollege.org; For similar projects, have a look at ENERGIA (www.energia.org), which is a network on gender and sustainable energy.

25 www.worldbank.org/astae; www.gvepinternational.org.

26 www.benbikes.org.za; Also check out international cycling/transport websites: www.carfree.com; www.cycling.nl; www.itdp.org (Institute for Transport and Development Planning) and www.velo.info.

27 Farmers Weekly, June 2008.

28 www.no-till.com.

29 Zipplies, R. (ed) (2008) *Bending the Curve: Your guide to tackling Climate Change in South Africa*, Cape Town: Africa Geographic, www.bendingthecurve.co.za.

30 www.animalvegetablemiracle.com; www.permaculture.com; Many of the websites mentioned earlier (see 'Institutions sharing ideas and training') contain information about sustainable agriculture and food production.

31 www.leisa.info.

32 Rinaudo, T. (2007) 'Farmer Managed Natural Regeneration,' *Leisa Magazine*, 23(2), http://ileia.leisa.info/index.php?url=article-details.tpl&p[_id]=113390; see http://ileia.leisa.info/index.php?url=article-details.tpl&p[_id]=113390 for more ideas and practices related to the 'Restoration of Natural Capital'; see www.rncalliance.org and www.leisa.info (Low External Input and Sustainable Agriculture).

Afterword

On the Road to Copenhagen:
Urgent Action is Required

Ida Auken

Being – as we are – just weeks away from the Copenhagen climate summit, it is almost impossible to come up with new words to stress the urgency of our present situation and the disaster awaiting humankind if we fail to meet a substantial agreement. World leaders have pledged their commitment to the enormous task which lies ahead. Yet even so, and despite the stakes being as high as they are, we find ourselves in a political deadlock: if developed countries have delivered neither their fair share of reductions as agreed in the Kyoto Protocol, nor the already-agreed money for helping the poorest developing countries, why should the less developed nations trust us this time around? Especially considering that the climate problem is created by many decades of enormous emissions of greenhouse gasses by the richer nations.

While it is hard to see what could ease the situation, this is not the first time in history that things have looked extremely gloomy on the eve of international agreement (think Rio or Kyoto). Moreover: there is no time for despair. We should instead mobilize all the pragmatism, optimism and action so badly needed at this time. In a spirit of idealistic realism we must keep the ultimate goal of global transition to a low carbon path in our sights, while we work with the tools and realities we have at our disposal.

While there is broad agreement on the science, economics will be the battle ground. Whereas the developed world needs to invest large sums in its transition to a low carbon economy, the developing world needs to do much more even than that. It must simultaneously adapt to accelerating climate change and pursue the imperative of tackling devastating poverty. To do this the developing nations require help from the developed world.

While other methods of global financial transfer will be necessary, carbon markets are currently one of the most important financing mechanisms already in some kind of existence. Large developing countries like China and India who are benefiting most from the CDM system at the moment are unlikely to accept a system where there are no project-based mechanisms for transferring money from North to South. The least developed countries, many of them African nations, have an interest in CDM too. At the Nairobi meeting a pledge was made to Africa that changes would be made to the CDM rules so that they would also benefit from CDM projects. Moreover, the UN Adaptation Fund has hitherto primarily been financed by a 2 percent share of proceeds from the Certified Emission Reductions (CERs) issued for projects of the CDM. Not a lot of money at the moment, but the share of proceeds could be increased to, for example, 5 or 15 percent. For all its faults it remains one of the only financing sources for the developing countries already agreed upon and one they are not going to want to remove or reform until another credible financing mechanism has been launched.

This, however, is probably one of the best things to be said of the CDMs. Many critical voices – including that of my party – have been raised against the existing project-based CDM system. And the harsh and well documented critiques presented in this book cannot be ignored and should not be taken lightly. If the use of CDM projects by the wealthy countries is not to harm the climate, it must be 100 percent additional. This book illustrates that many projects are patently not; they would have been carried out anyway. Tragically, the CDM may even have caused severe harm to the global climate by directing attention away from domestic CO_2 cuts in rich countries without providing additional cuts in the global South. This hinders innovation in turn making 'business as usual' more, not less, likely. As many cases in this book show, even the CDM's social and economic impact could be negative, as there is no guarantee that new jobs are created for local people. And then there are the claims of 'sustainable development'. If the CDM props up big polluters in the developing world that have all sorts of negative impacts on the environment, then is it worth trading this in for often spurious claims of 'additional' greenhouse gas cuts? As this book shows, the problems with the current CDM are manifold. The question is whether reforming it will be possible or not.

As a politician I have to be practical. So, we need to realize that, despite all the flaws of project-based CDM, it's still very much on the table in the negotiations. This means we need to make the best of a bad deal and reform rather than abolish it. Several solutions have been floated. Firstly, the CMD rules could be supplemented by a project /technology based discount factor in order to reduce or eliminate the problem of additionality.

Secondly, the CDM could also be divided into geographic sectors to give e.g. Africa a fair share of projects. This can be done in the negotiations, even if it seems very complicated. It could also be done simply by setting rules for the use of CDM in the rich countries. The EU could do that using its Emission Trading System rules – thus getting much more climate for the same money – and respecting the need for deviation from business as usual in developing countries.

Thirdly, the CDM system could be supplemented with a sector-based CDM where targets could be set for large sectors such as the steel and cement industries. Such a system has to respect the principles of common but differentiated responsibilities and capabilities and could be initiated by using the so-called no-loose targets. This implies setting a business as usual emissions baseline for plants in developing countries, rewarding those plants that go under that baseline and not punishing those that do not. At the moment we have concocted a 'carbon poverty trap' whereby a country limits its availability for getting CDM projects if it improves its emissions reduction policies. A policy giving credits to countries that improve their policies and enforce them – such as by strengthening the national building codes – could change that.

Finally, an exciting new instrument is now on the table. Its aim is not to help rich countries to avoid making sufficient reductions at home but to help finance voluntary reduction actions in developing countries. The NAMA's (National Appropriate Mitigation Action) was introduced by South Korea a couple of years ago. It is a voluntary system where developing countries can report to a register which projects they are going to carry out. These NAMA are eligible for up front support from credit-short rich countries which in turn are able to purchase credits achieved by documented resulting reductions.

Let this book bring home to the world the huge problems created by the CDM and spur the effort to reform it. We must get away from business as usual in the way the CDM works. Just as importantly, let it be remembered that CDM can only be one aspect of the architecture of managing climate change. We must also continue to work on real sustainable development involving changes in lifestyle, production forms, housing, transportation, forestry and agriculture. Instead of offsetting its responsibilities, the richer countries need to take a lead in bringing about real changes at home. This is their historical duty, given that they are mostly responsible for creating the problem of climate change in the first place. But of course there is no point for the rich world to clean up its act, if the developing world just reproduces a development model that literally will 'cost the earth'.

My hope is that the Copenhagen conference will bring the world closer together and back on a track that leads us not to chaos, but to a brighter, more equitable and sustainable future.

Afterword

Time to Breathe

Zoe Young

So one element – carbon – has been transmuted from neglected-priceless-connected to exploited-tradeable-separate.

Underlying its newly invented 'market' in imaginary units seems to be the continued lust for more growth in available ease and goodies, backs still turned to other ripples of impact resulting. A fragile alchemy now quantifies certain environmental value and defers responsibility for its transformation; reproducing to some degree bankers' shifting of responsibility for financial debt. Symptoms of such beguiling and destructive practices have long been actively countered by far-sighted social movements of intelligence and compassion. But these patterns sustain nevertheless, partly due to suppression, and partly perhaps to widespread ignorance/naivety of some environmental activists about geo-politics, enclosure and/or the nature of complex corruption; plus some preference among privileged greens for experiments in self-peasantification, gesture politics and/or protest culture that simply reproduces impotence, instead (for example) of moving strategically to cut off the most dangerous flows of large-scale fear and finance at source.

In 1972, my father Wayland Young told a meeting for the UN Conference on the Human Environment that our species 'as a whole must now slightly alter course. Industry everywhere must build pollution control measures into its planning, both political and economic'. I was three years old back then, and grew up with a kind of knowing, at some level, that something was wrong with the rules of Western economies that fetishized the claims of 'objectivity' in economics, but did not listen to many actual scientists, and pursued 'growth' always without awareness of maturity, decline and death. My parents argued that democratic governments should adopt systemic preservation of what people need and value – peace, beauty, fertile ecology, justice etc – and enforce the principle of 'polluter pays' to prevent or clean up any mess or damage caused. They also argued that governments should never pay people *not* to destroy something that people need, because that would simply reward destructive intent. But in the following decades of international 'greed is good' consensus, such rewards often became the last resort of a marginalised and co-opted conservation movement. The Yasuni proposal to pay Ecuador not to sell the oil from under rich biodiversity has excited much interest but promises nothing really new. Inter-governmental preparations for a new 'financial mechanism' for climate action are also not about finally directing all our billions of international public investment into cleaner development options, instead they are once more about how much 'additional' money can be found for environmental funding as an 'additional' extra and, by now, this has to include adaptation and mitigation – because sufficient effective action was not taken in the past four decades, when there was still lots of time to change.

Nearly 40 years on from the first UN environment conference in Stockholm, institutionalized environmental protection at the global level is all about sustaining 'growth' – production, education, media, science, even waste disposal, must work harder, better, faster, stronger – and most all of it fired still with dark materials drilled from the earth's crust. From the 70s until lately, North Atlantic elites aggressively pushed neo-liberal 'free market solutions' to resource allocation questions both new and old, including environmental protection. Maintaining growth in profits for some by disembedding and privatizing ever more, while quietly maintaining subsidy and tariff regimes that benefit their allies, they were able to counter the late 1960s disturbances and transformation to science, culture and democratic politics from peace, popular liberation, ecology, gay and women's movements. The emerging freedoms and adventurous engagement with life of the people involved in those movements, had fed a growing perception in this West at least that Earth and all those dwelling with her may in some sense be alive. My beloved father, aka Lord Kennet, who lies not long in his grave, as author of 'Eros Denied', helped spawn this resurgence of the erotic connection with life. He and my mother Liz were friendly with fellow author Bill Golding, who suggested the goddess' name Gaia for scientist James Lovelock's 'earth feedback hypothesis'. The insights from these developments in experimental science and culture seem to have been marginalised by economic ideologies adopted by North Atlantic elites for narrow political reasons. The UN Environment Programme set up in the early 1970s was exiled to Nairobi, underfunded, and given no power to make international institutions like the World Bank invest their influential funds for broader public good. The UN's office on regulating multinational corporations was shut down under corporate pressure at the end of the 1970s. Now corporations that do something 'good' for culture, science or the environment, say sponsor an art exhibition or well-known groups like Conservation International and WWF, gain kudos even as they carry on strip mining, deforesting, extracting oil and polluting somewhere else.

Scientific unease about growing environmental destruction rose again during the late 1980s, leading to important UN treaties on protecting biodiversity and preventing climate change in 1992. But still the modality of implementation was limited to a small percentage of public finance targeted to sweeten investments in environmental protection: and next to nothing for the very difficult and complex process of setting fair, effective and flexible international standards for using resources sustainably. So it was that in 1995 I found myself with a job researching the Global Environment Facility, a World Bank-UN fund intended to create 'global environmental benefits' – effectively supporting expensive islands of publicly paid-for clean-up, in a sea of continued resource extraction and pollution from unregulated sources – few of which even pay their share of tax. To be fair, some players in big business have called for clear and binding standard regulations to make their global business simpler – but professional communities interested in 'environmental' investment now understandably guard their own scarce resources too.

As we have seen, it was in this context that 'carbon markets' were created. The sale of climate-related 'indulgences' for energetic and emitting 'sinners' certainly opens up lucrative new streams of products for devotees of the kind of magical thinking embodied in financial sector-dominated economies: if the work of a Goldman Sachs employee can be valued a thousand times more highly than that of a nurse or farmer, then it is not much of a stretch to believe that the value of CO_2 sequestered in a monoculture pine, eucalyptus or oil palm plantation really is worth more than the complex old growth forest grubbed up to lay it out. This logic that defies scientific understanding can be extended to biodiversity – of course it makes sense for a company to drown or poison a valley full of endangered wildlife with a mine or dam, as long as you can claim that nobody will do the same to the next valley too on your watch. Makes total sense – as long as you are not a tree, beast or mycelium in that first valley. Or a scientist, tourist, or local resident who studies, or loves, or gardens, hunts, and gathers in that valley sustainably. In fact, only as long as you somehow do well out of mines or dams, whether from your pension fund's investments, or the minerals and energy to make the products you choose to consume, it makes sense. But otherwise, not much. Watch out if anyone tries to apply this logic to your family. 'OK, let me torture your little boy, and I'll pay for your girl's education.' This appeal to the desperate, could only come from the unethical mighty – who may well do something that starves you and your little girl later anyway... if it benefits the bottom line.

In 2009 Prince Charles and the World Bank were among those declaring 'war' on climate change – a little sibling perhaps for the existing pretend 'wars' to disguise failure in our governments' policies on poverty, cancer, drugs, 'terror' etc? So who is supposed to be the enemy this time – Gaia? Profit maximizing CEOs and the shareholders they report to? Opponents of energy efficiency regulation? An ubiquitous atmospheric gas called CO_2? Winning a real war with an engaged enemy demands troop numbers, strategy, preparation, luck, intelligence, learning from history, taking territory and indulging the confrontational machineries of violence. Real war leaves fighters and victims mutilated, shellshocked, grieving and displaced. War of any kind lies, breaks hearts and minds, and benefits few. Oh dear, not more...

Like bankers' bonuses, critical researchers' grants and the World Bank's Carbon Investment Funds, the costs of any war must be conjured out of debt and/or extracted from taxes on the surplus of others' increasing 'productivity'. Most production is still not regulated for energy efficiency, so the money to wage a climate 'war', as for salaries of consultants etc in the 'carbon market', will likely be derived from activities generating greenhouse gases. To protect the (energy) 'security' of their enduring freedom to grow their markets, ideological and armed militarists suck trillions from society; while good food, clean air and water, and spiritual, ecological and job security are neglected. So to live well on this bright gem of a planet, is it not now time to turn our backs on these priests and warriors of old gods too greedy for tribute?

Breathe.. Take a moment just to breathe,

(illness and injustice thrive on our crude reaction)

I turn my back, step away, and make time to reflect.

To look over what has passed and seek a place of graceful action..

Breathe, connect, and move...

Swaying, walking, dancing, loving life and the understanding that everything changes, passes away.

Still, taking action to reduce, reuse and recycle, to confront, stop, and reverse to causes of environmental destruction in whatever form they arise. And always, building and supporting what gives sustenance to many in the long term, and sanctuary where it's needed now. Seeking where possible to be fair and kind to self and others. Learning to forgive those in government, big business, army, police, even bankers for the years ongoing of pain and sorrow and loss and struggle of so many. It's hard.

Stepping through fears of failures and futures into fuller self-responsibility, exploring old/new ways of living and dying – whatever happens – that give as much as we take from the web of life, and love it all the more.

And this I can only do for myself... not for others. We each choose.

Turning down the volume not only of produce/consume but also of the emergency climate crisis nature destruction panic injustice activist mode. For too long, too long now, some of us who feel and study and labour with compassion for all beings have been driven by cycles of stress and sorrow and strain, so now I try to turn my back on that too.... To listen to the earth, the air, the fire, the water, to hear and experience what is needed now, to break the cycles of pollution, inside and out... moving always onwards, offering thanks, completely facing up to and grateful for what is... taking time, making space to break outdated norms and laws and reimagine values, where practicable in council circle, with scientists advocates and ancestors and brave fools to shake us up with laughter. That's a different challenge.

Drawing to a close 'the great fracture' between bull-headed children growing fast and furious and our bearlike mother earth who will not give humans any more than she can give. Only connecting with what is, as it is, before seeking to change or make demands of it so that maybe one day, three year old offspring of engaged environmentalists need no longer grow up haunted by the fear of big mama Gaia turning on her babies and refusing any more goodies, ever, at all.